Adventures in
Cosmology

Adventures in Cosmology

Editor

David Goodstein
California Institute of Technology

NEW JERSEY · LONDON · SINGAPORE · BEIJING · SHANGHAI · HONG KONG · TAIPEI · CHENNAI

Published by

World Scientific Publishing Co. Pte. Ltd.
5 Toh Tuck Link, Singapore 596224
USA office: 27 Warren Street, Suite 401-402, Hackensack, NJ 07601
UK office: 57 Shelton Street, Covent Garden, London WC2H 9HE

British Library Cataloguing-in-Publication Data
A catalogue record for this book is available from the British Library.

ADVENTURES IN COSMOLOGY

Copyright © 2012 by World Scientific Publishing Co. Pte. Ltd.

All rights reserved. This book, or parts thereof, may not be reproduced in any form or by any means, electronic or mechanical, including photocopying, recording or any information storage and retrieval system now known or to be invented, without written permission from the Publisher.

For photocopying of material in this volume, please pay a copying fee through the Copyright Clearance Center, Inc., 222 Rosewood Drive, Danvers, MA 01923, USA. In this case permission to photocopy is not required from the publisher.

ISBN-13 978-981-4313-85-8
ISBN-10 981-4313-85-8

Typeset by Stallion Press
Email: enquiries@stallionpress.com

Printed by FuIsland Offset Printing (S) Pte Ltd Singapore

CONTENTS

1. Introduction ... 1
2. Galaxy Formation: From Start to Finish ... 5

   Andrew Benson

   2.1 Historical Perspective ... 5
   2.2 The Universe Before Galaxies ... 7
   2.3 The Story So Far ... 9
       2.3.1 The end of the dark ages ... 9
       2.3.2 Population III and the first galaxies ... 13
       2.3.3 The reionization of the universe ... 15
       2.3.4 Establishing the Hubble sequence ... 17
       2.3.5 The rise of the supermassive black holes ... 25
   2.4 What the Future Holds ... 32
   References ... 35

3. The Reionization of Cosmic Hydrogen by the First Galaxies ... 41

   Abraham Loeb

   3.1 Introduction ... 42
       3.1.1 Observing our past ... 42
       3.1.2 The expanding Universe ... 46
   3.2 Galaxy Formation ... 49
       3.2.1 Growth of linear perturbations ... 49
       3.2.2 Halo properties ... 53
       3.2.3 Formation of the first stars ... 56

	3.2.4	Gamma-ray bursts: probing the first stars one star at a time	61
	3.2.5	Supermassive black holes	63
	3.2.6	The epoch of reionization	66
	3.2.7	Post-reionization suppression of low-mass galaxies	71
3.3	Probing the Diffuse Intergalactic Hydrogen	73	
	3.3.1	Lyman-alpha absorption	73
	3.3.2	21-cm absorption or emission	75
		3.3.2.1 The spin temperature of the 21-cm transition of hydrogen	75
		3.3.2.2 A handy tool for studying cosmic reionization	79
3.4	Epilog	86	
References			87

4. Clusters of Galaxies   89

Elena Pierpaoli

4.1	What are Galaxy Clusters? Why are They Interesting? . .	89
4.2	Structure Formation	91
4.3	How do We Observe Clusters?	93
4.4	Clusters in Cosmology	97
4.5	Dark Matter or Modified Gravity?	105
4.6	Gas, Galaxies and Their Evolution	107
References		109

5. Reionizing the Universe with the First Sources of Light   111

Volker Bromm

5.1	The End of The Dark Ages	113
5.2	Formation of a Population III Star	115
5.3	Feedback in the Early Universe	120
5.4	Brief History of Reionization	124
5.5	Empirical Probes for Reionization	127

	5.6	Explosions at Cosmic Dawn	129
	5.7	The First Black Holes	132
	5.8	Toward the First Galaxies	134
	References	136

6. **Mapping the Cosmic Dawn** 139

    Steven Furlanetto

6.1	A Brief History of Our Universe: From Soup to Galaxies	140
6.2	The Hidden Cosmic Dawn	143
6.3	The Solution: Flipping Spins	145
6.4	The Spin-Flip Transition as an Astronomical Tool	146
6.5	Foiled!: Early Cosmology with the Spin-Flip Transition	148
6.6	Spin-Flip Radiation Holds the Key to Observing the Cosmic Dawn	149
6.7	The Spin-Flip Background: The First Stars	153
6.8	The Spin-Flip Background: The First Black Holes	156
6.9	The Spin-Flip Background: The Epoch of Reionization	158
6.10	FM Radio Antennae as Cosmic Observatories	159
6.11	Piles and Tiles of Antennae: Mapping the Spin-Flip Background	161
6.12	Mountains to Scale: Challenges to Observing the Spin-Flip Background	164
6.13	Sound and Fury, Signifying Statistics	167
6.14	An Explosion of Telescopes	167
6.15	Dreams for the Future	169
6.16	An Unfinished Story	171

7. **Neutrino Masses from Cosmology** 173

    Ofer Lahav and Shaun Thomas

7.1	A Brief History of Cosmological Neutrinos and Hot Dark Matter	174
7.2	Insights from Particle Physics	175

7.3	Background to Cosmology	178
7.4	The Physics of Cosmological Neutrinos	181
7.5	Observational Methods	184
	7.5.1 The cosmic microwave background	184
	7.5.2 Galaxy clustering	185
	7.5.3 Weak gravitational lensing	186
	7.5.4 The Lyman-alpha forest	187
7.6	Observational Limits as of 2010	187
References		191

8. Measuring the Expansion Rate of the Universe — 193
Laura Ferrarese

8.1	Introduction	193
8.2	Twinkle Twinkle Little Star: Cepheid Variables	199
	8.2.1 A brief historical overview	201
	8.2.2 The theoretical perspective	203
	8.2.3 The Leavitt law	206
	8.2.4 An independent check	208
	8.2.5 Hubble (the telescope) observes Cepheids	209
8.3	The Aborted Explosion of Stars: The Tip of the Red Giant Branch	210
8.4	Bumpy Galaxies: The Surface Brightness Fluctuation Method	214
8.5	The Orderly Nature of Early Type Galaxies: The Fundamental Plane	217
8.6	The (Not Quite As) Orderly Nature of Spiral Galaxies: The Tully–Fisher Relation	220
8.7	Stellar Explosions: Type Ia Supernovae	224
8.8	Further Reflections and Future Directions	229
	8.8.1 Eighty years of Hubble constants	229
	8.8.2 The age of the Universe	232
	8.8.3 Room for improvement	236
References		237

9. Particles as Dark Matter 241
   Dan Hooper

   9.1 The Evidence for Dark Matter 241
   9.2 The Production of Dark Matter in the Early
       Universe 244
       9.2.1 Case example — the thermal abundance
             of a light or heavy neutrino 247
   9.3 Beyond the Standard Model Candidates
       for Dark Matter 248
       9.3.1 Supersymmetry 249
       9.3.2 Kaluza–Klein dark matter in models with
             universal extra dimensions 251
       9.3.3 A note on other possibilities for TeV-scale
             dark matter 254
   9.4 Direct Detection 254
       9.4.1 Direct detection of neutralino dark matter ... 256
       9.4.2 Direct detection of Kaluza–Klein
             dark matter 257
       9.4.3 Some model-independent comments regarding
             direct detection 258
   9.5 Indirect Detection 259
   9.6 Dark Matter at Particle Colliders 264
   9.7 Conclusions 265
   References 265

10. Detection of WIMP Dark Matter 269
    Sunil Golwala and Dan McKinsey

    10.1 Introduction 269
    10.2 Direct Detection of WIMPs via WIMP–Nucleon
         Elastic Scattering 269
    10.3 Inelastic Scattering 274
    10.4 Background Sources 274
    10.5 Backgrounds and WIMP Sensitivity 279

10.6 Direct Detection Techniques 283
    10.6.1 Detectors without nuclear-recoil discrimination . . . 283
        10.6.1.1 Germanium spectrometers 284
        10.6.1.2 Scintillation-mediated detectors 285
    10.6.2 Detectors with nuclear-recoil discrimination 289
        10.6.2.1 Metastable or threshold detectors 289
        10.6.2.2 Sub-Kelvin detectors 293
        10.6.2.3 Liquid noble detectors 302
        10.6.2.4 Two-phase detectors 308
        10.6.2.5 Single-phase argon detectors 313
    10.6.3 Direction-sensitive detectors 316
References 318

11. The Accelerating Universe   321
Dragan Huterer

11.1 Introduction and History: Evidence for the Missing Component 321
11.2 Type Ia Supernovae and Cosmology 322
11.3 Parametrizations of Dark Energy 331
11.4 Other Probes of Dark Energy 337
11.5 The Accelerating Universe: Summary 347
References 349

12. Frontiers of Dark Energy   355
Eric V. Linder

12.1 Introduction to Dark Energy 355
12.2 The Dynamics of Nothing 356
12.3 Knowing Nothing 364
12.4 The Frontiers of Nothing 370
12.5 Conclusions 375
References 376

13. The First Supermassive Black Holes in the Universe   379

   Xiaohui Fan

   13.1 Supermassive Black Holes and Galaxy Formation   380
   13.2 Observations of the Most Distant Quasars   386
   13.3 Growing the First Supermassive Black Holes
        in the Universe   395
   13.4 Future Prospects   399
   References   404

Index   407

CHAPTER 1

INTRODUCTION

Many non-scientists tend to think that cosmology became a science only in our own time, prior to which it was a vague system of beliefs that did not amount to much. But in fact the Greeks were cosmological scientists (Plato, Aristotle and many who predated those two), and the tradition continued with Ptolemy in second-century Alexandria, and especially with the likes of Copernicus, Kepler, Galileo, and Isaac Newton. Each of those thinkers had a vision of how the Universe works and tried to arrange the empirical facts to fit that vision. That is what cosmology is all about. This book presents essays on the current state of the art by their modern counterparts.

To begin with, Andrew Benson, a senior research fellow in theoretical cosmology in the TAPIR (Theoretical Astrophysics Including Relativity) group at Caltech and a member of Caltech's Moore Center for Theoretical Cosmology and Physics, tells us all about how galaxies were formed in the early Universe. Galaxy formation, he says, is a process, as opposed to being an event, and is one that is on-going. The formation occurs as the galaxy's constituent matter is drawn together by the force of gravity in spite of the accelerating expansion of the Universe. Dark matter halos (the dark matter surrounding a galaxy) and small density fluctuations all play an important role in galaxy formation.

Abraham Loeb, director of the Institute for Theory and Computation at Harvard University, describes the period, starting 400 000 years after the Big Bang, when the plasma of subatomic particles had cooled enough for them to combine into hydrogen atoms and a lesser amount of helium atoms, along with a few other light elements. It was at just this moment when small fluctuations in the density of matter started to grow into galaxies.

Then Elena Pierpaoli tells us about clusters of galaxies, including their structure formation, how we observe them and their role in cosmology.

Volker Bromm, Assistant Professor in the Department of Astronomy, University of Texas at Austin, considers the period of reionization that brought an end to the cosmic dark ages, when "reionization" brought forth the first stars, galaxies, and even spawned the black holes that are thought to exist everywhere in the Universe.

Steven Furlanetto, Associate Professor of Physics and Astronomy at UCLA, gives us a picture of the formation of the earliest galaxies taken largely from the study of spin-flip interactions (whose 21 cm wavelengths are easily detectable as radio-waves).

Laura Ferrarese, Senior Research Officer of the National Research Council of Canada, shows us how to measure the expansion of the Universe using various methods to get the distances to stars. Type Ia supernovae and cepheid variable stars both serve as "standard candles" to calculate distances; they give us one important scale. Other methods involve the Tip of the Red Giant Branch (TRGB), the Surface Brightness Fluctuation (SBF), and the Tully–Fisher relation.

Dan Hooper, Assistant Professor of Astronomy and Astrophysics at the University of Chicago, speculates on the possible nature of dark matter — exotic, nonbaryonic material that is thought to constitute 80–85% of the matter in the Universe. He argues against a modification of Newtonian dynamics and for such the favorite candidate is WIMPs (Weakly Interacting Massive Particles). He brings forward the view that the mutual annihilation of these particles and their antiparticles slowed as the Universe expanded, leaving large numbers of them present.

Sunil Golwala and Dan McKinsey, respectively Associate Professor of Physics at Caltech and at Yale University, review the methods by which we may be able to detect WIMPS — if indeed they exist.

Dragan Huterer, Assistant Professor of Physics at the University of Michigan tells us about the current acceleration of the Universe's expansion: How was it discovered? How long has this present epoch been? And how is the dark energy causing the acceleration phenomenologically described?

Eric Linder, codirector of the Institute for Nuclear and Particle Astrophysics at the Lawrence Berkeley National Laboratory, writes regarding what is yet unknown about dark energy: whether it is uniform, dynamic, disappears at early times, or whether its origin is of quantum or gravitational nature. They are valid possibilities, and carry profound implications for the frontiers of physics and the fate of the Universe. Even though our knowledge of dark matter is very limited, we are taking initial steps for such advancement.

Xiaohui Fan, Professor of Astronomy at the University of Arizona, looks into the matter of supermassive black holes that, while accreting mass, often outshine entire galaxies. Supermassive black holes are thought to power the quasistellar radio sources known as quasars — the most luminous objects in the Universe — and related objects known as AGNs (Active Galactic Nuclei).

All in all these modern cosmologists give us a remarkably detailed view of what the Universe is all about and what we do not yet know about it.

CHAPTER 2

GALAXY FORMATION: FROM START TO FINISH

ANDREW BENSON

Theoretical Astrophysics Including Relativity (TAPIR)
California Institute of Technology
Pasadena, CA 91125, USA

It was once observed[1] that galaxy formation "is a process, not an event", meaning that galaxies began forming rather quickly (by cosmological standards) after the Big Bang and are still forming today. Galaxy formation is a consequence of the remarkable ability of gravity to organize matter over cosmological scales, even in the face of cosmic expansion. In this chapter, we will explore the physical processes that control the process of galaxy formation and how they shape the properties of galaxies, both at the present day and in the early Universe.

2.1 Historical Perspective

The first detailed observations of galaxies external to our own (neglecting the Magellanic Clouds, which have undoubtedly been viewed by human eyes for many millennia) were made by the Earl of Rosse, using a 72-inch telescope which he had constructed in Birr, Ireland. Lord Rosse made sketches showing the spiral structure of galaxies such as M51, such as the one shown in Fig. 2.1. Galaxies, or simply "spiral nebulae" as they were then known, have been the subject of much debate ever since.

---

[1]The comment is usually attributed to Simon White.

Fig. 2.1 A sketch of M51, the Whirlpool Galaxy, made by Lord Rosse in 1845 as observed though his 72-inch telescope. Rosse did not know that this was an external galaxy, but could clearly see the spiral arms and even the smaller, companion galaxy NGC 5195 (to the right) which may well have triggered the formation of the strong spiral pattern as it passed through the disk of M51.

Their true nature remained uncertain for a long time. For example, at the start of the 20th century many believed that they were proto-planetary systems, with Sir Robert Ball of Cambridge University writing in his book *In The High Heavens* (all, 190):

> Probably this nebula [M51] will in some remote age gradually condense down into more solid substances. It contains, no doubt, enough material to make many globes as big as our Earth.

In fact, both Thomas Wright and Immanuel Kant had suggested that the spiral nebulae were actually "island-universes", i.e. stellar systems comparable to our own Milky Way, rather than small gaseous nebulae within the Milky Way. The debate as to whether spiral nebulae were galactic or extragalactic was famously taken up by Shapley, who favored a galactic explanation, and Curtis, who took the opposite view (Shapley and Curtis, 1921). The debate ran for nearly five years in the early 1920s, and was not fully resolved until 1924 when Edwin Hubble measured the distance to the Andromeda galaxy (M31) by observing a Cepheid variable star[2] in the outskirts of that galaxy (Hubble, 1929). The distance to M31, which is now known to be around 1 Mpc (Megaparsec; 1 Mpc $\approx 3.1 \times 10^{22}$ m), proved conclusively that it lay outside of the Milky Way, which was already known to be much smaller than this (Shapley, 1919).

2.2 The Universe Before Galaxies

Galaxy formation has been an ongoing process throughout most of the history of the Universe. If, however, we examine the history of the Universe on a logarithmic time axis, from the Planck time at 5×10^{-44} s to the present day at 4.3×10^{17} s, galaxy formation occupies just slightly over two of those 61 logarithmic decades. A lot happened in the remaining 59 decades, including inflation (which, importantly, laid down the density perturbations, from which galaxies would eventually form, in the otherwise homogeneous Universe), nucleosynthesis, recombination (during which the Universe first become sufficiently cool for electrons and protons to combine to form hydrogen) and the extended "dark ages" after recombination but prior to the first stars forming (to be discussed in Sec. 2.3.1). Those topics are discussed elsewhere in this volume.

Looking out into the Universe we are also looking back in time, due to the finite speed with which light travels. Cosmologists usually specify the distance to a galaxy (and, therefore, the time at which the light we receive from it was emitted) in terms of *redshift*. The redshift, denoted by z, measures the factor by which the wavelengths of photons emitted by a distant galaxy have been stretched due to the expansion of the Universe in the intervening time, such that the observed and emitted wavelengths, λ_o and λ_e respectively, are related by $\lambda_o = (1 + z)\lambda_e$. Referring to some point in cosmic history as "redshift z" implies a time at which the Universe was smaller by a factor of $1 + z$ in each linear dimension (such that today

---

[2]A Cepheid star exhibits periodic variations in its luminosity, with a period that correlates closely with the mean luminosity of the star. Thus, by measuring the period of luminosity variations from a Cepheid, one can infer its intrinsic luminosity. This, combined with the flux of light received from it and the usual inverse-square relation between flux, luminosity and distance, allows one to infer the distance to the star.

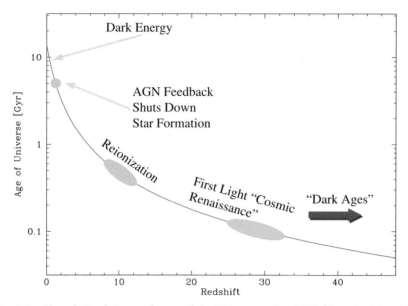

Fig. 2.2 The relation between the age of the Universe and redshift (the extent to which wavelengths of light have been shifted from their values at emission by the expansion of the Universe). The present day corresponds to $z = 0$ and an age of approximately 13.8 Gyr. Also shown are some of the key epochs in cosmic history: the "dark ages" (see Sec. 2.3.1), the epoch of first light (see Sec. 2.3.2), the epoch of reionization (see Sec. 2.3.3), the time at which feedback from active galactic nuclei (AGN) begins to shut down star formation (see Sec. 2.3.5) and the time at which dark energy begins to dominate the expansion of the Universe.

corresponds to $z = 0$ and the Big Bang[3] to $z = \infty$). Figure 2.2 shows the relation between redshift and the age of our Universe, and indicates a few key epochs in the Universe's history that will be discussed below.

Galaxy formation studies usually take as their initial conditions the Universe at redshifts of around $z \approx 50$–100, prior to the formation of any galaxy, when the material content of the Universe consists of almost uniformly distributed dark matter and gas, with small density perturbations in the dark matter component gradually growing under their own self-gravity. The statistical properties of those perturbations are predicted in

---

[3]The Big Bang occurs a finite time, approximately 13.8 Gyr, in the past. At that time, the Universe had zero size (ignoring any complications arising from quantum gravity physics) which implies an infinite redshift.

the inflationary paradigm in a Universe dominated by cold dark matter,[4] and are measured directly from the cosmic microwave background (CMB; the relic radiation left over from the Big Bang).

The goal of galaxy formation studies then is to understand how the near uniform distribution of dark matter and gas evolved over a period of around 13 billion years to form the population of galaxies with a rich phenomenology as seen today.

2.3 The Story So Far

2.3.1 *The end of the dark ages*

After protons and electrons recombined[5] at $z \approx 1100$ and the CMB radiation began its journey toward us, the Universe entered a long period of quiescence known as the Dark Ages during which no luminous sources, stars or accreting black holes, existed. This apparent quiescence neglects the fact that much activity was underway in the dark sector. The ripples in the density of dark matter seeded by inflation and which were tiny at the epoch of recombination were gradually growing under their own self-gravity during this period. Gravitational collapse is usually a runaway process, with density increasing exponentially in time. This rapid growth is mitigated cosmologically by the expansion of the Universe which reduces that exponential growth to a mere power-law in time. Nevertheless, by redshifts of $z \approx 50$, perturbations in the dark matter on mass scales of around $10^6 M_\odot$ ($M_\odot \approx 2 \times 10^{30}$ kg is the mass of our Sun) have grown enough to become nonlinear, decouple from the expansion of the Universe and collapse to high density, forming approximately spherical systems close to virial equilibrium known as dark matter halos.[6]

---

[4]Dark matter is thought to make up around 22% of the energy density of the Universe. "Cold" dark matter is thought to consist of fundamental particles that are sufficiently massive that they were non-relativistic when they decoupled from the rest of the Universe (i.e. when their timescale for interactions with other particles first exceeded the age of the Universe).
[5]"Recombined" is the standard nomenclature, even though these protons and electrons had never before been "combined".
[6]The term "halo" refers to the fact that galaxies will form in the centers of these systems, so the dark matter appears as a halo surrounding a galaxy.

The process of dark matter halo growth turns out to be hierarchical in nature such that halos are built through the merging together of earlier generations of less massive halos. While for a long time numerical calculations indicated that all trace of earlier generations of halos was erased during the merging process (Katz and White, 1993; Summers *et al.*, 1995) it was understood on analytical grounds that this was likely a numerical artifact rather than a physical result (Moore *et al.*, 1996). Beginning in the late 1990s, computer simulations which follow the gravitational interactions of large numbers of particles representing dark matter (N-body simulations) clearly demonstrated that this was indeed the case (Klypin *et al.*, 1999; Moore *et al.*, 1999; Tormen *et al.*, 1998). Unlike earlier generations of simulations, they found that halos can persist as *subhalos* within larger halos into which they merge. The current highest resolution simulations of individual halos (Kuhlen *et al.*, 2008; Springel *et al.*, 2008) show almost 300 000 subhalos[7] and even show multiple levels of subclustering (i.e. subhalos within subhalos within subhalos etc.).

The deep dark matter potential wells associated with each halo and subhalo are natural locations for gas (a primordial mixture of approximately 75% hydrogen and 25% helium at this time) to pool, and therefore for stars and galaxies to form. The densities obtained in dark matter halos are around 200 times the mean density of the Universe, still far too low to form a star or galaxy. Barring any cooling processes in the gas, it will be shock-heated as it accretes to the characteristic temperature ($k_{\rm B}T \approx GMm_{\rm H}/r$ for a halo of mass M and radius r) of the halo and settle into a hydrostatic, pressure-supported atmosphere. Further progress toward forming stars therefore relies on the ability of gas to lose energy by cooling and thereby collapse further within the dark matter potential well.

---

[7]This is a lower limit due to the limited resolution of the simulations. The earliest generations of cold dark matter halos may have masses as low as $10^{-12} M_\odot$ (depending on the particle nature of the dark matter) while state of the art simulations resolve only halos with masses greater than around $10^5 M_\odot$. The ability of even lower mass halos to survive is a subject of much debate (Angus and Zhao, 2007; Berezinsky *et al.*, 2006; Elahi *et al.*, 2009; Goerdt *et al.*, 2007; Zhao *et al.*, 2007).

At the low temperatures characteristic of these early halos ($T \sim 10^3 \, \mathrm{K}$) the gas remains mostly unionized and the primary coolant turns out to be molecular hydrogen. Cosmologically, molecular hydrogen forms via the gas phase reactions[8] (Mcdowell, 1961):

$$\mathrm{H} + \mathrm{e}^- \rightarrow \mathrm{H}^- + \gamma,$$
$$\mathrm{H}^- + \mathrm{H} \rightarrow \mathrm{H}_2 + \mathrm{e}^-, \qquad (2.1)$$

and

$$\mathrm{H}^+ + \mathrm{H} \rightarrow \mathrm{H}_2^+ + \gamma,$$
$$\mathrm{H}_2^+ + \mathrm{H} \rightarrow \mathrm{H}_2 + \mathrm{H}^+. \qquad (2.2)$$

Once sufficient molecular hydrogen has formed, cooling times in the gaseous atmosphere can become sufficiently short (shorter than the heating time due to continued accretion of mass onto the halo) that the atmosphere loses pressure support and collapses to higher density.

What happens next becomes significantly more complicated, but depends on the density, cooling rate and angular momentum of the gas. Fortunately, this problem of galaxy formation theory is almost unique in having well-posed initial conditions and so is ideal for tackling numerically. Beginning in the late 1990s simulations following the hydrodynamics and molecular chemistry of the gas became possible and showed that each of these early halos began to form a high density core containing hundreds of solar masses of gas which accretes toward the center (see, for example, Abel *et al.*, 2002). This seems to imply that such halos will form a single, perhaps very massive star at their center. Unfortunately, the simulations currently have to stop when densities become so high that the gas becomes optically thick to its own radiation, as the codes used do not include radiative transfer. This radiation becomes important as molecular hydrogen

---

[8]In galaxies at lower redshifts molecular hydrogen forms more readily on the surfaces of dust grains in the interstellar medium. However, at these early times this is not possible as dust is formed from heavy elements — carbon and silicon — which have yet to be created.

can be dissociated by photons in Lyman–Werner bands in the energy range 11.2–13.6 eV, which may limit future cooling and star formation. The formation of this first star signals the end of the Dark Ages. Understanding the formation of second and later generations of stars becomes significantly more challenging, as they may be affected by the products of earlier generations (radiation, kinetic energy and cosmic rays for example).

Assuming that these calculations are correct, Population III stars may contain up to several hundred solar masses of material. More massive stars are much more effective at fusing their supply of hydrogen into heavier elements than are lower mass stars (due to their higher central temperatures). As a result, despite starting out with more hydrogen to fuse, a massive star lives shorter than a less massive star. The extremely high masses of Population III stars therefore makes them extremely short-lived (with lifetimes of order one million years — compare to the Sun's lifetime of around ten billion years). As with all massive stars, the death of these Population III objects is likely to be spectacular. In fact, over much of the expected mass range, Population III stars are expected to end with a pair-instability supernova. If the core of a Population III star becomes hot enough, gamma rays produced by nuclear fusion will lead to the production of electron-positron pairs. This reduces the pressure in the core (which was supplied by the photons) and leads to collapse, which heats the core further thereby enhancing the rate of pair production. The result is a a catastrophic loss of pressure and complete collapse of the star followed by a supernova explosion. In the mass range 130–250$M_\odot$ (approximately) this can lead to the liberation of more energy than the stars' gravitational binding energy and, therefore, complete disruption of the star. Unlike the supernovae of lower mass stars, such an explosion would leave no remnant (white dwarf, neutron star or black hole). Experimentally confirming the existence of Population III stars may therefore have to focus on the material that returns to the interstellar medium during their life and death. In particular, the relative amounts of different heavy elements that they produce has a unique signature (they produce relatively little of elements with odd nuclear charge and almost no elements heavier than zinc; Heger and Woosley, 2002) which may be detectable if these elements were incorporated into later, longer-lived stars.

2.3.2 *Population III and the first galaxies*

Stars forming from primordial gas are known as Population III stars. These, like all stars, fuel themselves by nuclear fusion in their core, thereby turning hydrogen into heavier elements, some of which they return to their surroundings as a result of stellar winds blown from their surfaces during their lifetimes or as a result of supernova explosions at the ends of their lives. Once the surrounding gas becomes sufficiently enriched in heavy elements, or "metals",[9] any stars which form from it are known as Population II stars. Since the presence of metals in gas provides significant new cooling channels, the process of star formation in metal-enriched gas is thought to proceed quite differently from that in metal-free gas. Theoretical reasoning suggests that the stars which form from this metal-enriched gas will have lower masses than their Population III counterparts — comparable to the masses of stars which we see in our own galactic neighborhood.

The first galaxies to form would have been very different from those that we see today. Due to the hierarchical nature of dark matter halo growth, the dark matter halos in which they formed were much smaller and lower mass than those present in today's Universe. Consequently, the first galaxies are expected to be small, containing a few hundred thousand solar masses of gas and stars, or much less if they lose significant mass in outflows driven by supernovae explosions (as will be discussed below). Additionally, since the Universe was much denser in the past (at $z = 30$ the Universe was almost 30 000 times denser than today) these galaxies should have been much denser too. This, coupled with the fact that successive generations of hierarchically growing halos were much more closely spaced in time in the early Universe than today meant that early galaxies experienced a rapid rate of growth and encounters with other galaxies. As we will discuss in Sec. 2.3.4, this would make it very difficult for these galaxies to grow disks as seen in present day counterparts. Instead, they were likely very messy, highly turbulent systems, never able to reach an equilibrium state.

---

[9]In astrophysics, the term "metals" refers to all elements heavier then helium.

Since galaxies are collections of stars, an inevitable consequence of being a galaxy is experiencing supernova explosions when massive constituent stars exhaust their nuclear fuel and collapse. It has long been realized (Dekel and Silk, 1986) that the energy and momentum deposited into the interstellar medium (the gas between stars) of a galaxy by supernovae is sufficient to unbind material from the galaxy, at least in lower mass galaxies. This forms a negative feedback loop in which stars form from the gaseous content of a galaxy, the resulting supernovae explosions expel some of that gas from the galaxy and, as a result, inhibit the rate of future star formation. The lower the mass of a galaxy, the shallower its gravitational potential well, and the easier it will be for supernovae explosions to eject material from the galaxy. This leads to a differential effect, with galaxy formation being suppressed more efficiently in lower mass galaxies and, therefore, lower mass dark matter halos.

This feedback turns out to be a crucial ingredient in the process of galaxy formation. As early as 1974 (Larson, 1974; see also Dekel and Silk, 1986; White and Rees, 1978) it was realized that star formation could not proceed with 100% efficiency in all dark matter halos. Evidence for this comes from multiple observed facts, but the two crucial ones are:

(1) The total mass density[10] in stars is $\Omega_\star = (2.3 \pm 0.34) \times 10^{-3}$ (Cole et al., 2001), much less than the total baryonic mass density of the Universe $\Omega_b = 0.0462 \pm 0.0015$ (Dunkley et al., 2009). Therefore, only a small fraction of all baryons have been able to turn into stars.

(2) The distribution of galaxy luminosities is very different from the distribution of dark matter halo masses. In particular there are many fewer faint galaxies relative to bright galaxies than there are low mass to high mass dark matter halos. If each halo contained baryons at the universal mix and turned all of them into stars we would expect these two ratios to be equal (Benson et al., 2003).

---

[10]In cosmology, densities are often expressed relative to the critical density of the Universe — the energy density required to eventually halt the expansion of the Universe and cause a recollapse. At present, the critical density corresponds to the mass of just over five hydrogen atoms per cubic meter. Densities expressed in this way are denoted by Ω.

Needless to say then, studying the formation of the first galaxies is complicated, although no more complicated than modeling the formation of later generations. In the next decade new facilities such as JWST and 30-meter class telescopes will begin to open a window on the earliest generations of galaxies and, as such, there is a need for theoretical understanding and predictions for this epoch of cosmic history. In particular, this represents a real opportunity to make testable predictions from galaxy formation theory which can be confronted with data in the near future.

2.3.3 *The reionization of the universe*

As mentioned above, the process of structure formation in the dark matter component continues in a hierarchical manner, with more and larger halos forming through the conglomeration of earlier generations of halos. As a result, there are more and more potential sites of galaxy formation. Young, massive stars in galaxies emit copious quantities of ultraviolet photons. While many of these may be absorbed in the dense gas of the galaxy itself some will escape into the intergalactic medium (the low density gas that suffuses the Universe between galaxies), where they may ionize a hydrogen atom. As time goes on, more and more galaxies form, producing more and more UV photons. Simultaneously, the expansion of the Universe is reducing the density of intergalactic hydrogen and so its recombination timescale grows longer. Eventually, photoionization overtakes recombination and the Universe becomes reionized and simultaneously photo-heated to a temperature of order 10^4 K. Evidence[11] from the Wilkinson Microwave Anisotropy Probe (WMAP) suggests that this *reionization* happened at $z \approx 10$.

The process of reionization is discussed in more detail in Chapter 3 of this volume. Reionization is of interest here because it has consequences for later generations of forming galaxies. The increased temperature of the

---

[11]Cosmic microwave background photons traveling through the post-reionization Universe see a non-negligible optical depth due to Thomson scattering from free electrons. This optical depth can be measured from the shape of the CMB power spectrum and used to infer the column density of free electrons and so the epoch of reionization.

intergalactic medium raises the Jeans mass[12] in the gas preventing it from being pulled into the potential wells of smaller dark matter halos. This prevents halos with characteristic velocities of less than around 30 km/s (temperatures of a few tens of thousands of Kelvin) from forming galaxies, and thereby suppresses the formation of low mass, low luminosity galaxies.

This may be particularly relevant for our own cosmological backyard, the Local Group. The Local Group consists of the Milky Way, M31 (the Andromeda galaxy) together with a host of satellite galaxies gravitationally bound to one of these larger systems and ranging in luminosity from about one billion to just a few hundred solar luminosities. Around forty of these satellites are known, most detected in the Sloan Digital Sky Survey, but estimates (Tollerud *et al.*, 2008) suggest that many more are yet to be found, such that the total population may consist of several hundred satellites. However, calculations suggest that the halo of the Milky Way should contain tens of thousands of smaller "subhalos" — remnants of the process of hierarchical merging from which it formed. Naively one might expect each of these subhalos to have been able to form a small galaxy. This would be in conflict with the observational data and so was put forward as a possible failing of the cold dark matter cosmological paradigm. The suppression of low mass galaxy formation as a consequence of reionization had been known for some time, and was quickly demonstrated to be an effective cure for this proposed ailment of the cold dark matter paradigm (Benson *et al.*, 2002; Bullock *et al.*, 2000; Busha *et al.*, 2009; Macció *et al.*, 2009; Somerville, 2002), with some help from supernovae feedback. More recent work has suggested that other physics, such as cosmic ray pressure (Wadepuhl and Springel, 2010), may also play an important role, but the effects of reionization are clearly an important part of our picture of low mass galaxy formation.

These low mass galaxies turn out to be interesting for another reason. Analysis of their internal dynamics shows that their stars move much too fast to remain bound without the presence of copious amounts of dark

---

[12] In an expanding Universe with time-varying temperature the Jeans mass is technically not the correct quantity to consider, but the principles are the same nevertheless.

matter. While almost all galaxies show similar evidence for dark matter, the Local Group satellite galaxies contain orders of magnitude more dark matter relative to their stellar content, to such extent that the gravitational field of their stars is entirely negligible. As such, these may be ideal systems in which to test theories of dark matter. For example, scenarios in which the dark matter particle has a mass of just a few eV predict that the distribution of dark matter should be smoother on small scales than in the standard cold dark matter model. This would lead to a reduction in the mass of dark matter that could collect in the centers of small galaxies. Recent measurements (Strigari *et al.*, 2008) have been able to measure the dark matter content Local Group satellite galaxies and show that they all have around $10^7 M_\odot$ of dark matter within a radius of 300 pc. This is consistent with the expectation from cold dark matter theory, providing the galaxies formed in dark matter halos spanning a relatively narrow range of masses, but future measurements have the potential to place strong constraints on the particle properties of dark matter.

2.3.4 *Establishing the Hubble sequence*

The variety of shapes (or, in astronomical parlance, "morphologies") of galaxies is, perhaps, the most obvious observed characteristic of galaxies. Traditionally, morphology has been measured "by eye" by a trained observer who classifies each galaxy into a different morphological class based upon (amongst other things) the prominence of any central bulge, how concentrated the light distribution is and the presence or otherwise of lanes of dusty material that obscure starlight. Hubble (Hubble, 1936; see also de Vaucouleurs, 1959) placed galaxies into a morphological classification scheme — his so-called "tuning-fork diagram" as shown in Fig. 2.3 — using such an approach, and this basic morphological classification has persisted to the present (de Vaucouleurs *et al.*, 1991). Applying this type of morphological classification to today's large datasets is difficult, but has been achieved by utilizing "crowdsourcing" techniques (Lintott *et al.*, 2008) in which citizen-scientists learn how to classify galaxies and are able to perform such classifications on datasets provided via the internet. A key

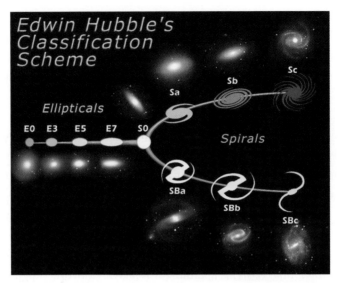

Fig. 2.3 The classic Hubble "tuning fork" diagram of galaxy morphological types. To the left (the handle of the tuning fork) are the elliptical galaxies, arranged in the order of ellipticity. To the right are two branches of spiral galaxies: barred (lower tine of the fork) and unbarred (upper tine of the fork), arranged in the order of spiral arm strength. Where these sequences meet the so-called S0 galaxies are located. Image courtesy of NASA and ESA.

observational goal of galaxy formation studies is to assess how the Hubble sequence evolves over time, as this should provide insight into how the physics of galaxy formation was able to assemble such a variety of galaxy morphologies. This is observationally challenging, but progress is being made (e.g. Kriek *et al.*, 2009).

A fundamental difficulty in assessing the ability of any given theoretical model to explain the morphological properties of galaxies is that the definition of morphology itself is very complicated, and somewhat nebulously defined. This problem is beginning to be circumvented, both by the ability of numerical simulations of galaxy formation to produce realistic "mock images" of galaxies (Jonsson, 2006) which can, in principle, be classified by eye just as a real galaxy, and by the use of more quantitative morphological measures such as directly measuring the contribution of disk and spheroid components of a galaxy to its total luminosity (Benson *et al.*, 2007; Ratnatunga *et al.*, 1999; Schade *et al.*, 1996; Simard *et al.*, 2002), or by defining empirical parameters that measure properties such as the concentration, asymmetry and smoothness of a galaxy (Abraham *et al.*, 1996; Bell *et al.*, 2003; Watanabe *et al.*, 1985).

Understanding the build-up and evolution of the Hubble sequence of morphologies is, nevertheless, an important task for galaxy formation theory. The morphological structure of a galaxy clearly tells us something about its formation history and therefore captures information which its stellar population does not. For example, two galaxies could contain the exact same stars, but one may have formed those stars "in place" (i.e. from gas already settled into the galaxy), while the other may have formed by the merging together of a large number of smaller galaxies each containing their own population of already formed stars. Those two galaxies would be indistinguishable in their luminosities and spectra, but, as we will discuss below, we would expect the first to most likely be disk-dominated, while the second would be expected to be an elliptical galaxy.

It is unclear whether Hubble intended his tuning-fork diagram to be suggestive of a sequence of physical transformation, or merely a useful tool for categorization. However, it has often been viewed as a sequence of snapshots of a transformation process (with time running in different directions depending on one's preferred theory!). Current thinking suggests that, at least for reasonably massive galaxies, the process of galaxy formation may begin with disk-dominated galaxies, which are later transformed into elliptical galaxies. Galaxy disks are formed from collections of stars which all orbit around in a very orderly manner — most of the stars moving in the same direction around the center of the galaxy and on approximately circular orbits. While there is some dispersion in the orbits of disk stars it is typically quite small. Ellipticals, on the other hand, consist of stars which move around the galactic center on a wide variety of different orbits, many highly elliptical and all oriented in different directions. Any process that transforms a disk into an elliptical must therefore rearrange the orbits of the constituent stars significantly. As can be imagined from the above description, stellar orbits in an elliptical galaxy are much more "thermal" than those in a disk, in the sense that they look like the motions of atoms in a gas, all moving around in random directions rather than with any coherent motion. Indeed, elliptical galaxies are often referred to as being "pressure supported" by virtue of this analogy. As a result, we might guess that an elliptical galaxy is a natural high entropy state, while a disk has a much lower entropy, and that any process able to "stir up" a disk might naturally lead to an elliptical galaxy being formed. While the concept of entropy in gravitating systems becomes somewhat slippery (Lynden-Bell and Wood, 1968), this line of reasoning essentially holds, and we are left with the need to find a physical process sufficient to disrupt a disk galaxy.

The hierarchal nature of dark matter halo growth leads to the possibility of galaxies in different subhalos colliding with each other and possibly merging. To cause gravitationally bound interactions between subhalos and their galaxies typically requires a dissipative process to reduce their orbital energies. The usual suspect for such a process is dynamical friction,[13] in which an orbiting galaxy gravitationally deflects particles of dark matter that pass by it focusing them into a wake behind it. The gravitational pull of this wake on the orbiting galaxy decelerates the galaxy, causing it to lose orbital energy. As a result dynamical friction tends to drag subhalos down toward the center of their host halo, where they may merge with any other galaxy which finds itself there. The classic derivation of dynamical friction acceleration from Chandrasekhar (1943) has been used extensively to estimate dynamical friction timescales within dark matter halos. For example, Lacey and Cole (1993) applied this formula to estimate merging timescales for subhalos in simple models of dark matter halos, finding:

$$T_{\rm df} \approx \frac{T_{\rm Universe}}{20} \left(\frac{M_{\rm v}}{m_{\rm v}}\right), \qquad (2.3)$$

where $T_{\rm Universe}$ is the age of the Universe, and $M_{\rm v}$ and $m_{\rm V}$ are the total (i.e. including dark matter) masses of the host and satellite systems respectively. This implies, roughly speaking, that satellites that are less than 1/20th of the mass of their host are unlikely to merge within the age of the Universe.

---

[13]Dynamical friction can be shown to be related to the spectrum of density fluctuations in a system and, as such, can be seen as a consequence of the fluctuation–dissipation theorem (Nelson and Tremaine, 1999).

When a merger occurs, the process of "violent relaxation" (Lynden-Bell, 1967; see also Tremaine *et al.*, 1986), in which the energy of orbits undergoes order unity changes due to significant time-variable fluctuations in the gravitational potential, leads to a randomization of the orbits leading to a Maxwellian distribution of energies but with temperature proportional to stellar mass. This can turn the ordered motions of disks into the random motions seen in ellipticals. This process is, however, rather poorly understood — it seeks an equilibrium state which maximizes the entropy of the system, but the usual entropy is unbounded in gravitating systems implying that no equilibrium state exists. Arad and Lynden-Bell (2005) demonstrate this problem by showing hysteresis effects in violently relaxed systems (i.e. the final state depends on how the system goes from the initial to final states). Nevertheless, numerical calculations show that the end result of this process looks rather like an elliptical galaxy.

The remnants of major mergers of purely stellar disk systems, while spheroidal, do not look like precisely elliptical galaxies. As shown by Hernquist *et al.* (1993) their phase-space densities are too low in the central regions compared to observed ellipticals. On the scale of galaxies, stars act as a collisionless (i.e. zero pressure) fluid — the cross-section for collisions between stars is tiny. This suggests that mergers between reasonably gas-rich galaxies might be required — the presence of gas allows for dissipation and the formation of higher phase-space density cores. Furthermore, elliptical galaxies as a population show a remarkable degree of similarity — when examined in a three-dimensional space made up of their size, mass and internal velocity dispersion they are found to lie along a very well defined plane, known as the Fundamental Plane. Robertson *et al.* (2006) find that similar gas fractions and the subsequent dissipation are required to produce the observed orientation of the fundamental plane of elliptical galaxies — mergers of purely stellar systems instead predict a slightly different orientation that is inconsistent with that which is observed.

Mergers are often separated into *major* (mergers between galaxies of comparable mass) and *minor* (mergers in which one galaxy is significantly less massive than the other). Numerical simulations (e.g. Bournaud *et al.*, 2005) show that mergers with a mass ratio $M_2/M_1 < 0.25$ are able to destroy any disks in the ingoing galaxies and leave an elliptical remnant, while mergers with lower mass ratio tend to leave disks in place although somewhat thickened since some of the energy from the satellite galaxy's orbit will be transferred to the motions of stars in the disk. Mergers are expected to trigger a, possibly very large, enhancement in the star formation rate in the merging system by generating torques that drain gas in the interstellar medium of its angular momentum, driving it to the center of the galaxy and higher densities.

It has recently become clear that not all major mergers lead to the formation of an elliptical, however. Under certain conditions, major mergers of very gas rich systems can lead to the reformation of a disk after the merger is over (Barnes, 2002; Robertson *et al.*, 2006; Springel and Hernquist, 2005). This requires a high gas fraction just prior to the final coalescence of the merging galaxies and therefore may preferentially occur under conditions which prevent the rapid depletion of gas after the first passage of the galaxies.

Major mergers are not the only way to form an elliptical or spheroidal galaxy. Internal, secular processes[14] in galaxies can also disrupt the cold and relatively fragile disks (Kormendy and Kennicutt, 2004). In particular, many disks possess a central bar (a radially oriented density enhancement from the ends of which spiral arms emerge) that can efficiently redistribute mass and angular momentum and lead to the build up of dense central mass concentrations, reminiscent in many ways of bulges formed through mergers. These processes involve some very interesting gravitational dynamics as we will discuss below. To distinguish such secularly formed bulges from their merger-formed (or "classical") counterparts, they are referred to as "pseudobulges" (Kormendy and Kennicutt, 2004).

Such secular processes are the result of quite generic dynamical considerations[15] and so most likely operate in all galaxies. Whether or not

---

[14]Generically, any internal dynamical process operating on a timescale significantly longer than the dynamical time.

[15]As pointed out by Kormendy and Kennicutt (2004) disks are fundamentally prone to spreading in the presence of any dissipative process, where mass is transported inwards and angular momentum outwards, because this lowers the energy of the system while conserving angular momentum (Lynden-Bell and Kalnajs, 1972). This result can be

they are important depends upon their timescale. For example, relaxation due to star-star encounters in a galaxy operates on a timescale many orders of magnitude longer than the age of the Universe and so can be safely neglected. Instead, most relevant secular processes involve the interaction of stars (or gas elements) with collective phenomena such as bars or spiral density waves.

Spiral arms and other non-axisymmetric features such as bars in galactic disks are a visually impressive reminder that these systems possess interesting dynamics. These spiral features are simply density waves in the disk[16] as was first proposed by Lin and Shu (1964; see also Binney and Tremaine, 2008; Marochnik and Suchkov, 1996). From our current standpoint, the question in which we are interested in is whether these perturbations to an otherwise smooth disk are stable or unstable and, if unstable, how they affect the evolution of the galaxy as a whole.

Toomre (1964; see also Goldreich and Lynden-Bell, 1965) derived an expression for the local stability of thin disks to axisymmetric modes which turns out to be extremely useful (and often approximately correct even in regimes where its assumptions fail). These results showed that, generically, disks become unstable when the gravitational field of their own stars becomes strong (allowing for the possibility of gravitational growth of any small perturbations) and are stabilized by having significant random motions of their constituent stars (since such random motions tend to smooth away any perturbations).

Toomre's criterion applies to short-wavelength perturbations — those on scales much smaller than the size of the disk itself. Perturbations on the scale of the disk can occur also. Study of these global instabilities is less amenable to analytic treatment (since one can no longer ignore

---

traced back to the negative specific heat of gravitating systems, and is analogous to the process of core collapse in three-dimensional systems (Binney and Tremaine, 2008; Lynden-Bell and Kalnajs, 1972).

[16]The usual analogy that is made is to traffic on a freeway during rush hour. Slow downs in the traffic lead to a bunching up of cars, making the density of cars along one section of road higher. Such density enhancements can persist for a long time, even though individual cars enter the slow down, crawl through it and accelerate away at the other side. This is the same in spiral arms — individual stars continuously enter and then leave the spiral arm, but the density enhancement itself persists.

contributions from distant parts of the disk). Significant work on this subject began in the 1970s. The classic result from that time is due to Ostriker and Peebles (1973) who found that rapidly rotating, self-gravitating stellar systems could become violently unstable to nonaxisymmetric perturbations if the ratio of the kinetic energy associated with their rotation to their gravitational binding energy becomes too large.[17] Such systems can be unstable to global perturbations even if they satisfy Toomre's stability criterion and so are stable against short-wavelength perturbations. Numerical studies of these processes are difficult (since one must be sure that numerical artifacts are not the cause of any instability found), but the results described above have now been approximately confirmed by numerous numerical studies.

Global instabilities most likely lead to the formation of a very strong bar which effectively disrupts the disk leaving, after a few dynamical times, an elliptical bulge (Athanassoula, 2008). This may therefore be a possible formation scenario for pseudobulges (Kormendy and Kennicutt, 2004), i.e. bulges formed through secular processes in the disk rather than as the result of a merger event. A general picture of how secular evolution leads to the formation of pseudobulges has emerged. As a bar spontaneously begins to form[18] it transfers angular momentum to the outer disk and increases in amplitude. The response of gas to this bar is crucial — gas accelerates as it enters the bar potential and decelerates as it leaves. This leads to shocks forming in the gas which lie approximately along the ridge-line of the bar.

---

[17]Dark matter halos therefore help to stabilize galaxy disks, by increasing their gravitational binding energy.

[18]The bar instability in galactic disks involves some fascinating dynamics. Binney and Tremaine (2008) give a clear explanation of the physics involved. Briefly, the bar instability involves the joint actions of the swing amplifier and a feedback mechanism. A randomly occurring leading spiral density wave in a disk will unwind and, as it does so, will rotate faster. As it swings from leading to trailing it reaches a maximum rotation speed which is close to the average orbital speed of stars in the disk. This leads to a resonance condition, in which the wave can strongly perturb the orbits of those stars, the self-gravity of which enhances the bar further, leading to an amplification of the wave. If there is some mechanism to convert the amplified trailing wave that results into a leading wave once more (e.g. if the wave can pass through the center of the galaxy and emerge as a leading wave, or if nonlinear couplings of waves can generate leading waves) the whole process can repeat and the wave will grow stronger and stronger.

These shocks lead to dissipation of orbital energy and, consequently, inflow of the gas. The enhancement in the gas density as it is concentrated toward the galactic center inevitably leads to star formation and the build up of a pseudobulge. Bars eventually destroy themselves in this way — the increase in the central mass of the galaxy effectively prevents the bar instability from working.

Several studies have explored how the Hubble sequence grows and changes in our present galaxy formation theory (Baugh *et al.*, 1996a,b; Benson and Devereux, 2009; Firmani and Avila-Reese, 2003; Governato *et al.*, 1999, 2007; Kauffmann, 1996; Lucia and Blaizot, 2007; Lucia *et al.*, 2006; Parry *et al.*, 2009) with the general conclusion that hierarchical cosmologies can plausibly give rise to the observed mix of morphological types, although with significant uncertainties remaining in both the modeling of morphological transformation and in the comparison with observed morphologies. The consensus opinion is that massive elliptical galaxies form most of their stars in smaller progenitor galaxies and only assemble them into a single galaxy significantly later, while spiral galaxies are dominated by *in situ* star formation.

2.3.5   *The rise of the supermassive black holes*

White and Rees (1978) first proposed that galaxies would form as gas cools inside of dark matter halos and demonstrated that this provided a reasonable estimate of the typical mass scale of galaxies. This simple picture has a long standing problem however. The mass function of dark matter halos (i.e. the number of halos per unit volume in some small range of mass) rises steeply (approximately as $M^{-1.9}$; Reed *et al.*, 2007) at low masses. Since cooling is very efficient in these low mass halos we might expect the galaxy mass and/or luminosity function to show a similar slope at the low mass/luminosity end. In fact, measured slopes are much shallower (typically around -1; e.g. Cole *et al.*, 2001). Rectifying this discrepancy is usually achieved by postulating some form of feedback, typically from supernovae, which can inhibit star formation in these low mass systems (e.g. by driving a wind out of the galaxy). However, as shown by Benson *et al.* (2003)

this causes another problem — too much gas is now left over to accrete into massive halos at late times wherein it cools and forms over-massive galaxies (much more massive than any galaxy observed).

This "overcooling" problem is not easy to solve, for the simple reason that the energy scales involved are much larger than for lower mass systems (the characteristic potential well depth of a dark matter halo scales as $M^{2/3}$ for halos of the same mean density). Several possible solutions have been proposed however, ranging from thermal conduction (which draws energy from the hot outer regions of the halo into the cooling inner regions and thereby offsets that cooling; Benson et al., 2003; Dolag et al., 2004; Pope et al., 2005), massive outflows (Benson et al., 2003), multi-phase cooling (in which gas cools but is locked up into clouds which are inefficiently transferred to the galaxy; Kaufmann et al., 2009; Maller and Bullock, 2004) and energy input from active galactic nuclei (accreting supermassive black holes in the centers of galaxies which are highly efficient at converting the rest mass energy of material that flows into them into radiation and mechanical outflow).

Over the past ten years it has become possible to measure the masses of supermassive black holes residing at the centers of galaxies for relatively large samples. The existence of strong correlations between the masses of these black holes and the properties of their host galaxy — such as spheroid mass (Magorrian et al., 1998), velocity dispersion (Ferrarese and Merritt, 2000; Gebhardt et al., 2000a,b), number of globular clusters (Burkert and Tremaine, 2010) or even host dark matter halo (Ferrarese, 2002) — is suggestive of some interaction between forming galaxies and supermassive black holes. Of course, correlation does not imply causation and Jahnke and Macciò (2010) show that relation between black hole and host galaxy mass can arise from uncorrelated initial conditions via simple merging, but the theoretical need for large amounts of energy to inhibit galaxy formation in massive halos naturally leads to the idea that supermassive black holes and galaxy formation are connected (Benson et al., 2003). In addition, understanding the formation of these most massive of black holes is interesting in its own right and has important observational consequences for both studies of active galactic nuclei and gravitational wave detection experiments such as the *Laser Interferometer Space Antenna.*[19]

In light of these reasons, several studies have attempted to follow the process of supermassive black hole formation within forming galaxies (Begelman and Nath, 2005; Kauffmann and Haehnelt, 2000; King, 2003, 2005; Malbon et al., 2007; Matteo et al., 2003, 2008; Monaco et al., 2000; Sijacki et al., 2009; Silk and Rees, 1998; Volonteri et al., 2003, 2008; Volonteri and Natarajan, 2009; Wyithe and Loeb, 2003).

Feedback from AGN can be divided into two categories: quasar mode and radio mode. The quasar mode is triggered when large amounts of gas are funneled toward the black hole from much larger scales by, for example, a merger between two galaxies in which torques act to transfer angular momentum from the gas, causing it to flow inward. This is likely the dominant mechanism for black hole mass growth and results in a high accretion rate — most likely through a thin accretion disk — and significant optical luminosity. In contrast, radio mode feedback occurs when the black hole is accreting at a more modest rate from a diffuse hot atmosphere of gas and is in an optically dim but radio-loud phase. In this phase, the black hole is thought to drive powerful jets (Benson and Babul, 2009; Komissarov et al., 2007; Meier, 1999, 2001; Nemmen et al., 2007; Villiers et al., 2005) which can reach to large distances and have been seen to have a significant impact on their surroundings (Birzan et al., 2004). The mechanism via which energy from the jets is efficiently coupled to the hot atmosphere of gas remains poorly understood: a combination of observational evidence (e.g. Owen et al., 2000) and theoretical insights suggest that jets inflate bubbles or cavities in the hot atmosphere which then transfer energy to their surroundings (by exciting sound waves in the atmosphere or doing pdV work on the atmosphere as they expand) as they rise buoyantly through the atmosphere. Despite these uncertainties, simple analytic treatments which assume an efficient coupling have demonstrated that this can effectively shut down cooling in massive halos, resulting in a reduction in the masses

---

[19]http://lisa.nasa.gov/

of the largest galaxies and good agreement with luminosity functions and the bimodal distribution of galaxy colors (Bower *et al.*, 2006; Croton *et al.*, 2006; Somerville *et al.*, 2008). Observational evidence in support of AGN feedback is beginning to emerge (Schawinski *et al.*, 2007) and seems to favor a radio mode scenario (Schawinski, 2009).

Before supermassive black holes can grow via accretion or merging, there must be some pre-existing (probably not supermassive) *seed* black holes. Most plausibly, these seeds form at high redshifts as the remnants of the earliest generation of Population III stars which have reached the end of their stellar lifetimes. Details of the formation of these first stars remain incompletely understood, but hydrodynamical simulations suggest that they have masses in the range of a few hundred solar masses (Abel *et al.*, 2002; Bromm *et al.*, 2002), leaving intermediate mass black hole remnants (as discussed in Sec. 2.3.1).

To determine the rate at which gas accretes onto a black hole we must consider how the black hole affects the gas through which it is moving. The gravity of a fast moving black hole will deflect gas that passes by it, focusing it into a wake behind the black hole which will then accrete onto the black hole. This problem was first studied by Hoyle and Lyttleton (1939) and Bondi and Hoyle (1944) who showed that it leads to an accretion rate of

$$\dot{M}_{\rm BHL} = \frac{4\pi {\rm G}^2 M_\bullet^2 \rho}{(c_{\rm s}^2 + v^2)^{3/2}}, \qquad (2.4)$$

where $M_\bullet$ is the mass of the black hole, $c_{\rm s}$ is the sound speed in the gas, v the relative velocity of black hole and gas and ρ is the density of the gas. The growth of black holes will be enhanced by any process which increases the density in the central regions of the galaxy in which they reside. At early times, this may occur due to gravitationally unstable disks forming bars and driving gas toward the center (Volonteri *et al.*, 2008), while at later times galaxy-galaxy mergers can result in dissipation and gas flows to the center (Matteo *et al.*, 2005). This Bondi–Hoyle–Lyttleton accretion causes gas to flow toward the black hole. At some point, the angular momentum of the gas will become important and the accreting gas must form a disk. The final stage of accretion is then governed by this accretion disk, in which matter flows in while angular momentum is transported outward (by viscous forces). The gravitational energy liberated by this inflow can be very large. Accretion flows may be geometrically thin, and emit liberated gravitational energy locally as thermal radiation (Shakura and Sunyaev, 1973) or may be geometrically thick and inefficient radiators (also known as an "advection dominated accretion flow" or ADAF; Narayan and Yi 1994) with liberated energy being advected with the flow. Which form a given disk takes depends on the rate at which mass is accreted.

Galaxy-galaxy mergers can also lead to galaxies containing two (or potentially more) supermassive black holes, resulting in the potential for black hole mergers. The process of bringing two supermassive black holes together begins by dynamical friction against the background of dark matter (the same process which is causing the black hole host galaxies to merge). The subsequent merging process was originally outlined by Begelman *et al.* (1980) and assumes that the two black holes orbit each other as a binary system. Around this binary, the stellar distribution is expected to form a cusp with density profile $\rho_\star \propto r^{-7/4}$ (Bahcall and Wolf, 1976). Initially, the binary star system's orbit shrinks, becoming more strongly gravitationally bound, due to dynamical friction against the stellar background acting on each black hole individually. As the binary shrinks this process becomes less effective as perturbations from distant stars tend to perturb the center of mass of the binary without changing the semi-major axis, a, of the binary orbit. However, once the binary becomes sufficiently tightly bound, it can shrink further by three-body interactions in which a passing star is captured and then ejected at high velocity. If this process continues long enough, the binary eventually becomes sufficiently tightly bound that gravitational radiation dominates the evolution of the system which then coalesces on a timescale of (Peters, 1964)

$$t_{\rm GR} = \frac{5c^4 a^4}{256 G^3 M_1 M_2 (M_1 + M_2)}, \quad (2.5)$$

where M_1 and M_2 are the masses of the two black holes. The fly in the ointment of this neat picture is that the above estimates, for the rate of shrinking of the orbit by stellar encounters, assume a fixed stellar background. In reality, as stars interact with the black hole binary, they are kicked into new orbits which cause them to leave the vicinity of the black hole. This reduces the number of stars with the required positions and orbits to efficiently interact with the black holes. This inevitably slows the shrinking process. The past ten years have seen numerous studies of this process and examination of various mechanisms by which this depletion may be counteracted. For example, Yu (2002) finds that in triaxial potentials, changes in the orbits of stars can efficiently replenish those that were lost. Several other processes may also help here. While the details remain uncertain it seems that this basic process can lead to black holes merging in less than 10 Gyr.

In addition to their mass, cosmological black holes are characterized by one other parameter, their angular momentum,[20] the magnitude of which can range from zero (as in the so-called Schwarzchild black hole) to GM^2/c (an "extreme Kerr" black hole). This spin angular momentum is therefore often parameterized by j, which is just the magnitude of the angular momentum in units of GM^2/c. The spin of a black hole can have a strong influence on the radiative efficiency and jet power of accretion flows around black holes since the jet can draw energy from the rotational energy of the black hole[21] in addition to that which it derives from the accretion flow. Consequently, the cosmological evolution of this quantity is important to understand. There are fundamentally two mechanisms which change the spin of a black hole: merging with another hole and accretion of material.

The outcomes of binary black hole mergers have proven very difficult to simulate numerically. However, recent advances in numerical techniques have allowed for successful simulation of the entire merging process (e.g. Tichy and Marronetti, 2007) and, therefore, determination of the final spin

---

[20]They will not possess any significant charge, as this would be quickly neutralized by accretion of oppositely charged material.

[21]Neglecting quantum mechanical effects, energy cannot be extracted from a non-rotating black hole. As first demonstrated by Penrose (1969), energy *can* be extracted from a rotating black hole, due to the existence of orbits close to the black hole which have negative total (i.e. including rest mass) energy. Astrophysically, it is thought that energy can be extracted if the material accreting onto the black hole is threaded by magnetic fields which can be thought of as torquing the black hole, spinning it down and simultaneously transferring the energy of its rotation outwards into the accretion flow.

of the merger product. While the number of simulations carried out to date is small, Boyle et al. (2008) exploit symmetries of the problem to construct a simple model which accurately predicts the spin of the final black hole as a function of the incoming black hole masses, spins and orbital properties.

As first considered by Bardeen (1970), material accreted from an accretion disk carries with it some angular momentum (approximately equal to its angular momentum at the last stable circular orbit, before it began its plunge into the black hole) and so will cause the black hole to spin up. Unchecked, accretion from a thin disk will eventually spin up the black hole to the maximal spin of $j = 1$. Thorne (1974) found a small correction to this result due to the fact that the hole preferentially swallows negative angular momentum photons. Consequently, the black hole is limited to $j \approx 0.998$. Although this is close to $j = 1$ it can nevertheless lead to significant changes in the radiative efficiency of the accretion flow. For a thick accretion flow the result is somewhat different (since the flow is no longer supported against gravity by its rotation alone as it has significant thermal and magnetic pressure). Benson and Babul (2009) compute the spin up by ADAFs and additionally compute how the magnetic torques which allow black holes to drive jets result in a braking torque on the black hole, spinning it down and resulting in an equilibrium spin of $j \approx 0.93$ for a hole accreting from an ADAF.

The relative importance of mergers and accretion for determining the spins of cosmological black holes depends upon the rate of galaxy mergers, the supply of gas to the black hole and additional factors, such as the alignment of accretion disks and black hole spins and merging black hole spins and orbits. Many of these factors are not too well understood. However, cosmological calculations (Berti and Volonteri, 2008; see also Volonteri et al., 2005) suggest that accretion dominates over mergers in terms of determining the spins of supermassive black holes, with the consequence that most such holes are predicted to be rapidly spinning.

In summary, the AGN feedback scenario has gained considerable favor in the past few years for a variety of reasons. Firstly, observations have indicated that all galaxies seem to contain a central supermassive black hole with a mass that scales roughly in proportion to the mass of the galaxy (Ferrarese and Merritt, 2000; Gebhardt et al., 2000a,b; Magorrian et al.,

1998; Tremaine *et al.*, 2002). The formation of these black holes must have involved the liberation of large amounts of energy as material sinks deep into the potential well of the black hole in some form of accretion flow (Benson *et al.*, 2003). If this energy can be successfully utilized to counteract the overly rapid cooling of gas onto the galaxy it would provide a natural source of sufficient energy available in every galaxy. Additionally, some feedback loop connection between supermassive black hole and galaxy formation of this type is attractive as it provides a means to explain the correlation between galaxy and black hole properties.

2.4 What the Future Holds

Observational evidence and theoretical understanding suggests that the process of galaxy formation is well past its peak and is quickly declining. This is seen in the mean rate of star formation per unit volume of the Universe, which has declined by a factor of 10–20 since a redshift of 2 as shown in Fig. 2.4. At first sight, this is somewhat surprising, as galaxies have turned only a small fraction of the total gas content of the Universe into stars — there should be plenty more fuel for star formation. The problem can be traced to (at least) two factors. Firstly, current evidence suggests that the expansion of our Universe is accelerating, due to the presence of dark energy (see Fig. 2.2). This accelerated expansion severely limits the growth of dark matter structures, preventing new dark matter halos from forming and old ones from gaining more mass. As a result, galaxies are no longer receiving new gas from the intergalactic medium at the rates they once were and so there is no supply of new fuel for star formation. If dark energy behaves as we expect, this acceleration will continue. An additional cause is the presence of supermassive black holes in galaxies. These black holes are incredibly efficient at extracting energy from material that accretes into them (typically extracting 10% or more of the rest mass energy of accreted material and converting it into electromagnetic energy and radiatively driven outflowing winds). This has a negative effect on galaxies and their surroundings, expelling gas from galaxies and preventing surrounding gas from being able to cool and condense into the galaxy (as discussed in Sec. 2.3.5).

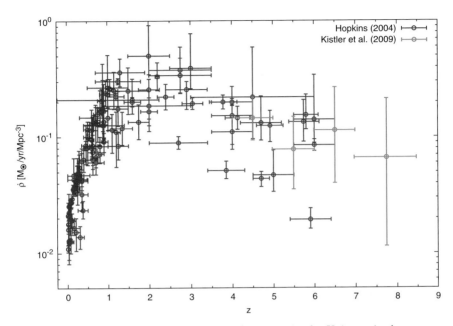

Fig. 2.4 The volume-averaged rate of star formation in the Universe is shown as a function of redshift. Volume here is defined in terms of comoving coordinates which expand along with the Universe. Expressing results in this coordinate system removes the overall increase in density at high redshifts that arises from the simple fact that the Universe was smaller in the past. The red points show a compilation of observational estimates from Hopkins (2004) while the green points show estimates based on the rate of gamma-ray bursts by Kistler et al. (2009). These results come from a variety of observational techniques, but all are uncertain due to complexities arising from how to convert an observed luminosity of a galaxy into a rate of star formation. Nevertheless, it is clear that star formation activity reaches a peak at around $z = 2$ and has declined rapidly since then.

Consequently, it seems likely that the bulk of galaxy formation is over, and what remains will proceed at a much more leisurely pace. Of course, our Universe seems set to expand forever, giving infinite time to form new stars. Estimates based on current state-of-the-art models of galaxy formation suggest that future star formation could increase the total number of stars formed by around 60% compared to the present number, but on timescales of hundreds of billions of years.

For our own Milky Way, there is still some excitement to come. The Andromeda galaxy (M31) is plunging toward us at 120 km/s. Current estimates suggest that it will collide with the Milky Way in 4–5 billion years. The result of such a collision is likely to be the formation of a massive elliptical galaxy (the fragile disks of the two galaxies being destroyed in the violence of their collision) and perhaps a short burst of star formation during which much of the remaining gas content of the two galaxies is turned into stars (Cox and Loeb, 2008). After that, the combined Milky Way/Andromeda galaxy will mostly likely just fade into old age — the lowest mass stars can live for around one trillion years before running out of fuel, so the fading will be a slow process. Beyond that, a nice review of what the vastly distant future may hold for galaxies and the Universe as a whole is given by Krauss and Starkman (1999); for a more technical discussion, see also Krauss and Starkman (2000).

Observationally, the next few decades promise to be exciting ones for studies of galaxy formation. The next generation of observational facilities will include the successor to the Hubble Space Telescope, the James Webb Space Telescope, which is due to be launched to the L2 Lagrangian point (about 1 million miles from Earth) in 2013 and which will have an 8 m diameter mirror and be optimized for observing in the infrared. These features will allow it to peer back to the earliest stages of galaxy formation, potentially revealing the first generations of galaxies to us. In addition to this new space telescope, plans are underway for the next generation of ground-based telescopes. These facilities will combine giant mirrors (typically 30 m in diameter) with adaptive optics to combat the distortion of images by the Earth's atmosphere. This will give them an unprecedented clarity of vision and permit detailed study of the most distant and faintest galaxies.

On the theoretical side, studies of galaxy formation have reached the point where a coherent and consistent model of galaxy formation exists that is able to explain a wealth of galaxy phenomenology in the low and intermediate redshift Universe. The real test of this theory, as with any theory, is to now see if it has true predictive power. The high redshift Universe is the natural regime in which to make such predictions and have them tested within a relatively short time. While we can make predictions about what these galaxies might look like with a reasonable degree of confidence the history of galaxy formation studies teaches that there are always more surprises in store, so this author expects the next decade to be an exciting one for galaxy formation theorists.

References

Abel, T., Bryan, G. L. and Norman, M. L. (2002). *Science* **295**, pp. 93–98.
Abraham, R. G. *et al.* (1996). *Monthly Notices of the Royal Astronomical Society* **279**, p. L47.
Angus, G. W. and Zhao, H. (2007). *Monthly Notices of the Royal Astronomical Society* **375**, pp. 1146–1156.
Arad, I. and Lynden-Bell, D. (2005). *Monthly Notices of the Royal Astronomical Society* **361**, pp. 385–395.
Athanassoula, E. (2008). *Monthly Notices of the Royal Astronomical Society* **390**, pp. L69–L72.
Bahcall, J. N. and Wolf, R. A. (1976). *Astrophysical Journal* **209**, pp. 214–232.
Ball, R. S. (1901). *Great Astronomers*, Isbister, London, p. 1.
Bardeen, J. M. (1970). *Nature* **226**, p. 64.
Barnes, J. E. (2002). *Monthly Notices of the Royal Astronomical Society* **333**, pp. 481–494.
Baugh, C. M., Cole, S. and Frenk, C. S. (1996a). *Monthly Notices of the Royal Astronomical Society* **282**, pp. L27–L32.
Baugh, C. M., Cole, S. and Frenk, C. S. (1996b) *Monthly Notices of the Royal Astronomical Society* **283**, pp. 1361–1378.
Begelman, M. C., Blandford, R. D. and Rees, M. J. (1980). *Nature* **287**, pp. 307–309.
Begelman, M. C. and Nath, B. B. (2005). *Monthly Notices of the Royal Astronomical Society* **361**, pp. 1387–1392.
Bell, E. F. *et al.* (2003). *Astrophysical Journal Supplement Series* **149**, pp. 289–312.
Benson, A. and Devereux, N. A. (2009). *Monthly Notices of the Royal Astronomical Society* **402**, pp. 2321–2334.
Benson, A. J. and Babul, A. (2009). Maximum spin at black holes driving jets.
Benson, A. J. *et al.* (2003). *Astrophysical Journal* **599**, pp. 38–49.
Benson, A. J. *et al.* (2007). *Monthly Notices of the Royal Astronomical Society*, **379**, pp. 841–866.
Benson, A. J. *et al.* (2002). *MNRAS* **333**, pp. 156–176.
Berezinsky, V., Dokuchaev, V. and Eroshenko, Y. (2006). *Physical Review D* **73**, p. 63504.
Berti, E. and Volonteri, M. (2008). *Astrophysical Journal* **684**, pp. 822–828.
Binney, J. and Tremaine, S. (2008). *Galactic Dynamics*, 2nd edn., Princeton University Press.
Birzan, L. *et al.* (2004). *Astrophysical Journal* **607**, pp. 800–809.
Bondi, H. and Hoyle, F. (1944). *Monthly Notices of the Royal Astronomical Society* **104**, p. 273.
Bournaud, F., Jog, C. J. and Combes, F. (2005). *Astronomy and Astrophysics* **437**, pp. 69–85.

Bower, R. G. et al. (2006). *MNRAS* **370**, pp. 645–655.
Boyle, L., Kesden, M. and Nissanke, S. (2008). *Physical Review Letters* **100**, p. 151101.
Bromm, V., Coppi, P. S. and Larson, R. B. (2002). *Astrophysical Journal* **564**, pp. 23–51.
Bullock, J. S., Kravtsov, A. V. and Weinberg, D. H. (2000). *Astrophysical Journal* **539**, pp. 517–521.
Burkert, A. and Tremaine, S. (2010). *Astrophysical Journal* **720**, pp. 516–521.
Busha, M. T. et al. (2009). *Astrophysical Journal* **710**, pp. 408–420.
Chandrasekhar, S. (1943). *Astrophysical Journal* **97**, p. 255,
Cole, S. et al. (2001). *Monthly Notices of the Royal Astronomical Society* **326**, pp. 255–273.
Cox, T. J. and Loeb, A. (2008). *Monthly Notices of the Royal Astronomical Society* **386**, pp. 461–474.
Croton, D. J. et al. (2006). *MNRAS* **365**, pp. 11–28.
de Vaucouleurs, G. (1959). *Handbuch der Physik* **53**, p. 275.
de Vaucouleurs, G. et al. (1991). *Third Reference Catalogue of Bright Galaxies*, Springer.
Dekel, A. and Silk, J. (1986). *Astrophysical Journal* **303**, pp. 39–55.
Dolag, K. et al. (2004). *Astrophysical Journal* **606**, pp. L97–L100.
Dunkley, J. et al. (2009). *Astrophysical Journal Supplement Series* **180**, pp. 306–329.
Elahi, P. J. et al. (2009). *Monthly Notices of the Royal Astronomical Society* **395**, pp. 1950–1962.
Ferrarese, L. (2002). *Astrophysical Journal* **578**, pp. 90–97.
Ferrarese, L. and Merritt, D. (2000). *Astrophysical Journal* **539**, pp. L9–L12.
Firmani, C. and Avila-Reese, V. (2003). pp. 107–120.
Gebhardt, K. et al. (2000). *Astrophysical Journal* **539**, pp. L13–L16.
Gebhardt, K. et al. (2000). *Astrophysical Journal* **543**, pp. L5–L8.
Goerdt, T. et al. (2007). *Monthly Notices of the Royal Astronomical Society* **375**, pp. 191–198.
Goldreich, P. and Lynden-Bell, D. (1965). *Monthly Notices of the Royal Astronomical Society* **130**, p. 125.
Governato, F. et al. (1999). *Astronomical Journal* **117**, pp. 1651–1656.
Governato, F. et al. (2007). *Monthly Notices of the Royal Astronomical Society* **374**, pp. 1479–1494.
Heger, A. and Woosley, S. E. (2002). *Astrophysical Journal* **567**, pp. 532–543.
Hernquist, L., Spergel, D. N. and Heyl, J. S. (1993). *Astrophysical Journal* **416**, p. 415.
Hopkins, A. M. (2004). *Astrophysical Journal* **615**, pp. 209–221.
Hoyle, F. and Lyttleton, R. A. (1939). *Proc. Cambridge Phil. Soc.*, p. 405.
Hubble, E. P. (1929). *Astrophysical Journal* **69**, pp. 103–158.
Hubble, E. P. (1936). *Realmet the Nebulae*, Yale University Press, New Haven.
Jahnke, K. and Maccio, A. (2010). Submitted to *Astrophysical Journal Letters*, arXiv: 1006.0482.

Jonsson, P. (2006). *Monthly Notices of the Royal Astronomical Society* **372**, pp. 2–20.
Katz, N. and White, S. D. M. (1993). *Astrophysical Journal* **412**, pp. 455–478.
Kauffmann, G. (1996). *MNRAS* **281**, pp. 487–492.
Kauffmann, G. and Haehnelt, M. (2000). *MNRAS* **311**, pp. 576–588.
Kaufmann, T. *et al.* (2009). *Monthly Notices of the Royal Astronomical Society* **396**, pp. 191–202.
King, A. (2003). *Astrophysical Journal* **596**, pp. L27–L29.
King, A. (2005). *Astrophysical Journal* **635**, pp. L121–L123.
Kistler, M. D. *et al.* (2009). arXiv: *0906.0590*.
Klypin, A. *et al.* (1999). *The Astrophysical Journal* **516**, pp. 530–551.
Komissarov, S. S. *et al.* (2007). *Monthly Notices of the Royal Astronomical Society* **380**, pp. 51–70.
Kormendy, J. and Kennicutt, R. C. (2004). *Annual Review of Astronomy and Astrophysics* **42**, pp. 603–683.
Krauss, L. M. and Starkman, G. D. (1999). *Scientific American* **281**, pp. 36–43.
Krauss, L. M. and Starkman, G. D. (2000). *The Astrophysical Journal* **531**, pp. 22–30.
Kriek, M. *et al.* (2009). *The Astrophysical Journal* **705**, pp. L71–L75.
Kuhlen, M. *et al.* (2008). *Journal of Physics Conference Series* **125**, p. 2008.
Lacey, C. and Cole, S. (1993). *Monthly Notices of the Royal Astronomical Society* **262**, pp. 627–649.
Larson, R. B. (1974). *Monthly Notices of the Royal Astronomical Society* **169**, pp. 229–246.
Lin, C. C. and Shu, F. H. (1964). *The Astrophysical Journal* **140**, p. 646.
Lintott, C. J. *et al.* (2008). *Monthly Notices of the Royal Astronomical Society* **389**, pp. 1179–1189.
Lucia, G. D. and Blaizot, J. (2007). *Monthly Notices of the Royal Astronomical Society* **375**, pp. 2–14.
Lucia, G. D. *et al.* (2006). *MNRAS* **366**, pp. 499–509.
Lynden-Bell, D. (1967). *Monthly Notices of the Royal Astronomical Society* **136**, p. 101.
Lynden-Bell, D. and Kalnajs, A. J. (1972). *Monthly Notices of the Royal Astronomical Society* **157**, p. 1.
Lynden-Bell, D. and Wood, R. (1968). *Monthly Notices of the Royal Astronomical Society* **138**, p. 495.
Macciò, A. V. *et al.* (2009).
Magorrian, J. *et al.* (1998). *Astronomical Journal* **115**, pp. 2285–2305.
Malbon, R. K. *et al.* (2007). *MNRAS* **382**, pp. 1394–1414.
Maller, A. H. and Bullock, J. S. (2004). *Monthly Notices of the Royal Astronomical Society* **355**, pp. 694–712.
Marochnik, L. S. and Suchkov, A. A. (1996). *The Milky Way Galaxy*, Gorden and Breach.
Matteo, T. D. *et al.* (2008). *Astrophysical Journal* **676**, pp. 33–53.

Matteo, T. D. et al. (2003). *Astrophysical Journal* **593**, pp. 56–68.
Matteo, T. D., Springel, V. and Hernquist, L. (2005). *Nature* **433**, pp. 604–607.
McDowell, M. R. C. (1961). *The Observatory* **81**, pp. 240–243.
Meier, D. L. (1999). *Astrophysical Journal* **522**, pp. 753–766.
Meier, D. L. (2001). *Astrophysical Journal* **548**, pp. L9–L12.
Monaco, P., Salucci, P. and Danese, L. (2000). *Monthly Notices of the Royal Astronomical Society* **311**, pp. 279–296.
Moore, B. et al. (1999). *Astrophysical Journal* **524**, pp. L19–L22.
Moore, B., Katz, N. and Lake, G. (1996). *Astrophysical Journal* **457**, p. 455.
Narayan, R. and Yi, I. (1994). *Astrophysical Journal* **428**, pp. L13–L16.
Nelson, R. W. and Tremaine, S. (1999). *Monthly Notices of the Royal Astronomical Society* **306**, pp. 1–21.
Nemmen, R. S. et al. (2007). *Monthly Notices of the Royal Astronomical Society* **377**, pp. 1652–1662.
Ostriker, J. P. and Peebles, P. J. E. (1973). *Astrophysical Journal* **186**, pp. 467–480.
Owen, F. N., Eilek, J. A. and Kassim, N. E. (2000). *The Astrophysical Journal* **543**, pp. 611–619.
Parry, O. H., Eke, V. R. and Frenk, C. S. (2009). *Monthly Notices of the Royal Astronomical Society* **396**, pp. 1972–1984.
Penrose, R. (1969). *Nuovo Cimento Rivista Serie* **1**, p. 252.
Peters, P. C. (1964). *Physical Review* **136**, pp. 1224–1232.
Pope, E. C. D. et al. (2005). *Monthly Notices of the Royal Astronomical Society* **364**, pp. 13–28.
Ratnatunga, K. U., Griffiths, R. E. and Ostrander, E. J. (1999). *Astronomical Journal* **118**, pp. 86–107.
Reed, D. S. et al. (2007). *MNRAS* **374**, pp. 2–15.
Robertson, B. et al. (2006). *Astrophysical Journal* **645**, pp. 986–1000.
Robertson, B. et al. (2006). *Astrophysical Journal* **641**, pp. 21–40.
Schade, D. et al. (1996). *Astrophysical Journal* **464**, p. L63.
Schawinski, K. (2009). To appear in the proceedings of The Monsters's Fiery Breath, *AIP Conf. Proc.*
Schawinski, K. et al. (2009). *Monthly Notices of the Royal Astronomical Society* **382**, pp. 1415–1431.
Shakura, N. I. and Sunyaev, R. A. (1973). *Astronomy and Astrophysics* **24**, pp. 337–355.
Shapley, H. (1919). *The Astrophysical Journal* **49**, pp. 249–265.
Shapley, H. and Curtis, H. D. (1921). Technical Report 2.
Sijacki, D., Springel, V. and Haehnelt, M. G. (2009). Accepted in *MNRAS*.
Silk, J. and Rees, M. J. (1998). *Astronomy and Astrophysics* **331**, pp. L1–L4.
Simard, L. et al. (2002). *Astrophysical Journal Supplement Series* **142**, pp. 1–33.
Somerville, R. S. (2002). *ApJ* **572**, pp. L23–L26.
Somerville, R. S. et al. (2008). *Monthly Notices of the Royal Astronomical Society* **391**, pp. 481–506.
Springel, V. and Hernquist, L. (2005). *Astrophysical Journal* **622**, pp. L9–L12.

Springel, V. et al. (2008). *Monthly Notices of the Royal Astronomical Society* **391**, pp. 1685–1711.
Strigari, L. E. et al. (2008). *Nature* **454**, pp. 1096–1097.
Summers, F. J., Davis, M. and Evrard, A. E. (1995). *Astrophysical Journal* **454**, p. 1.
Thorne, K. S. (1974). *Astrophysical Journal* **191**, pp. 507–520.
Tichy, W. and Marronetti, P. (2007). *Physical Review D* **76**(6), pp. 061502–5.
Tollerud, E. J. et al. (2008). *Astrophysical Journal* **688**, pp. 277–289.
Toomre, A. (1964). *Astrophysical Journal* **139**, pp. 1217–1238.
Tormen, G., Diaferio, A. and D. Syer, (1998). *Monthly Notices of the Royal Astronomical Society* **299**, pp. 728–742.
Tremaine, S., Henon, M. and Lynden-Bell, D. (1986). *Monthly Notices of the Royal Astronomical Society* **219**, pp. 285–297.
Tremaine, S. et al. (2002). *Astrophysical Journal* **574**, pp. 740–753.
Villiers, J. D. et al. (2005). *Astrophysical Journal* **620**, pp. 878–888.
Volonteri, M., Haardt, F. and Madau, P. (2003). *Astrophysical Journal* **582**, pp. 559–573.
Volonteri, M., Lodato, G. and Natarajan, P. (2008). *Monthly Notices of the Royal Astronomical Society* **383**, pp. 1079–1088.
Volonteri, M. et al. (2005). *Astrophysical Journal* **620**, pp. 69–77.
Volonteri, M. and Natarajan, P. (2009). Accepted in *MNRAS*.
Wadepuhl, M. and Springel, V. (2010). ArXiv: 1004.3217.
Watanabe, M., Kodaira, K. and Okamura, S. (1985). *Astrophysical Journal* **292**, pp. 72–78.
White, S. D. M. and Rees, M. J. (1978). *MNRAS* **183**, pp. 341–358.
Wyithe, J. S. B. and Loeb, A. (2003). *Astrophysical Journal* **595**, pp. 614–623.
Yu, Q. (2002). *Monthly Notices of the Royal Astronomical Society* **331**, pp. 935–958.
Zhao, H. et al. (2007). *Astrophysical Journal* **654**, pp. 697–701.

CHAPTER 3

THE REIONIZATION OF COSMIC HYDROGEN BY THE FIRST GALAXIES

ABRAHAM LOEB

Department of Astronomy, Harvard University
60 Garden St., Cambridge MA, 02138, USA

Cosmology is by now a mature experimental science. We are privileged to live at a time when the story of genesis (how the Universe started and developed) can be critically explored by direct observations. Looking deep into the Universe through powerful telescopes, we can see images of the Universe when it was younger because of the finite time it takes for light to travel to us from distant sources.

Existing data sets include an image of the Universe when it was 0.4 million years old (in the form of the cosmic microwave background), as well as images of individual galaxies when the Universe was older than a billion years. But there is a serious challenge: in between these two epochs was a period when the Universe was dark, stars had not yet formed, and the cosmic microwave background no longer traced the distribution of matter. And this is precisely the most interesting period, when the primordial soup evolved into the rich zoo of objects we now see.

The observers are moving ahead along several fronts. The first involves the construction of large infrared telescopes on the ground and in space, that will provide us with new photos of the first galaxies. Current plans include ground-based telescopes which are 24–42 m in diameter, and NASA's successor to the Hubble Space Telescope, called the James Webb Space Telescope. In addition, several observational groups around the globe are constructing radio arrays that will be capable of mapping the three-dimensional distribution of cosmic hydrogen in the infant Universe. These arrays are aiming to detect the long-wavelength (redshifted 21 cm) radio emission from hydrogen atoms. The images from these antenna arrays will reveal how the non-uniform distribution of neutral hydrogen evolved with cosmic time and eventually was extinguished by the ultraviolet radiation from the first galaxies. Theoretical research has focused in recent years on predicting the expected signals for the above instruments and motivating these ambitious observational projects.

3.1 Introduction

3.1.1 *Observing our past*

When we look at our image reflected off a mirror at a distance of 1 m, we see the way we looked 6.7 ns ago, the light-travel time to the mirror and back. If the mirror is spaced 10^{19} cm $\simeq 3$ pc away, we will see the way we looked 21 years ago. Light propagates at a finite speed, and so by observing distant regions, we are able to see what the Universe looked like in the past, a light-travel time ago (Fig. 3.1). The statistical homogeneity of the Universe on large scales guarantees that what we see far away is a fair statistical representation of the conditions that were present in in our region of the Universe a long time ago.

This fortunate situation makes cosmology an empirical science. We do not need to guess how the Universe evolved. Using telescopes we can simply see how it appeared at earlier cosmic times. In principle, this allows the entire 13.7 billion year cosmic history of our universe to be reconstructed by surveying the galaxies and other sources of light to large distances (Fig. 3.2). Since a greater distance means a fainter flux from a source of a fixed luminosity, the observation of the earliest sources of light requires the development of sensitive instruments and poses challenges to observers.

As the universe expands, photon wavelengths get stretched as well. The factor by which the observed wavelength is increased (i.e. shifted toward the red) relative to the emitted one is denoted by $(1 + z)$, where z is the cosmological redshift. Astronomers use the known emission patterns of hydrogen and other chemical elements in the spectrum of each galaxy to measure z. This then implies that the universe has expanded by a factor of $(1 + z)$ in linear dimension since the galaxy emitted the observed light,

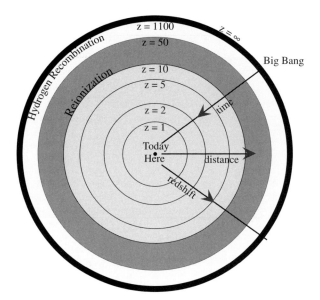

Fig. 3.1 Cosmic archaeology of the observable volume of the Universe, in comoving coordinates (which factor out the cosmic expansion). The outermost observable boundary ($z = \infty$) marks the comoving distance that light has travelled since the Big Bang. Future observatories aim to map most of the observable volume of our Universe, and improve dramatically the statistical information we have about the density fluctuations within it. Existing data on the CMB probes mainly a very thin shell at the hydrogen recombination epoch ($z \sim 10^3$, beyond which the Universe is opaque), and current large-scale galaxy surveys map only a small region near us at the center of the diagram. The formation epoch of the first galaxies that culminated with hydrogen reionization at a redshift $z \sim 10$ is shaded grey. Note that the comoving volume out to any of these redshifts scales as the distance cubed. **Figure credit:** A. Loeb, *How Did the First Stars and Galaxies Form?* (Princeton University Press, 2010).

and cosmologists can calculate the corresponding distance and cosmic age for the source galaxy. Large telescopes have allowed astronomers to observe faint galaxies that are so far away that we see them more than twelve billion years back in time. Thus, we know directly that galaxies were in existence as early as 500 million years after the Big Bang, at a redshift of $z \sim 10$ or higher.

We can in principle image the Universe only if it is transparent. Earlier than 400 000 years after the big bang, the cosmic hydrogen was broken into its constituent electrons and protons (i.e. "ionized") and the Universe was opaque to scattering by the free electrons in the dense plasma. Thus,

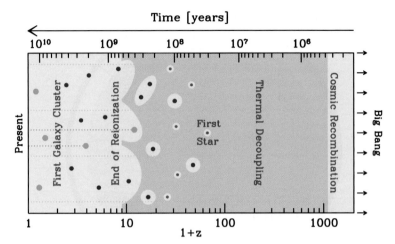

Fig. 3.2 Overview of cosmic history, with the age of the Universe shown on the top axis and the corresponding redshift on the bottom axis. Yellow represents regions where the hydrogen is ionized, and gray, neutral regions. Stars form in galaxies located within dark matter concentrations whose typical mass grows with time, starting with $\sim 10^5\ M_\odot$ (red circles) for the host of the first star, rising to 10^7–$10^9\ M_\odot$ (blue circles) for the sources of reionization, and reaching $\sim 10^{12}\ M_\odot$ (green circles) for present-day galaxies like our own Milky Way. Astronomers probe the evolution of the cosmic gas using the absorption of background light (dotted lines) by atomic hydrogen along the line of sight. The classical technique uses absorption by the Lyman-α resonance of hydrogen of the light from bright quasars located within massive galaxies, while a new type of astronomical observation will use the 21-cm line of hydrogen with the cosmic microwave background as the background source. **Figure credit:** R. Barkana and A. Loeb, *Rep. Prog. Phys.* **70**, 627 (2007).

telescopes cannot be used to electromagnetically image the infant Universe at earlier times (or redshifts $> 10^3$). The earliest possible image of the Universe was recorded by the COBE and WMAP satellites, which measured the temperature distribution of the cosmic microwave background (CMB) on the sky.

The CMB, the relic radiation from the hot, dense beginning of the universe, is indeed another major probe of observational cosmology. The universe cools as it expands, so it was initially far denser and hotter than it is today. For hundreds of thousands of years the cosmic gas consisted of a plasma of free protons and electrons, and a slight mix of light nuclei, sustained by the intense thermal motion of these particles. Just like the plasma in our own Sun, the ancient cosmic plasma emitted and scattered a strong field of visible and ultraviolet photons. As mentioned above, about

400 000 years after the Big Bang the temperature of the universe dipped for the first time below a few thousand degrees Kelvin. The protons and electrons were now moving slowly enough that they could attract each other and form hydrogen atoms, in a process known as cosmic recombination. With the scattering of the energetic photons now much reduced, the photons continued traveling in straight lines, mostly undisturbed except that cosmic expansion has redshifted their wavelength into the microwave regime today. The emission temperature of the observed spectrum of these CMB photons is the same in all directions to one part in 100 000, which reveals that conditions were nearly uniform in the early universe.

It was just before the moment of cosmic recombination (when matter started to dominate in energy density over radiation) that gravity started to amplify the tiny fluctuations in temperature and density observed in the CMB data. Regions that started out slightly denser than average began to develop a greater density contrast with time because the gravitational forces were also slightly stronger than average in these regions. Eventually, after hundreds of millions of years, the overdense regions stopped expanding, turned around, and eventually collapsed to make bound objects such as galaxies. The gas within these collapsed objects cooled and fragmented into stars. This process, however, would have taken too long to explain the abundance of galaxies today, if it involved only the observed cosmic gas. Instead, gravity is strongly enhanced by the presence of dark matter — an unknown substance that makes up the vast majority (83%) of the cosmic density of matter. The motion of stars and gas around the centers of nearby galaxies indicates that each is surrounded by an extended mass of dark matter, and so dynamically-relaxed dark matter concentrations are generally referred to as "halos".

According to the standard cosmological model, the dark matter is cold (abbreviated as CDM), i.e. it behaves as a collection of collisionless particles that started out at matter domination with negligible thermal velocities and have evolved exclusively under gravitational forces. The model explains how both individual galaxies and the large-scale patterns in their distribution originated from the small initial density fluctuations. On the largest scales, observations of the present galaxy distribution have indeed found the same statistical patterns as seen in the CMB, enhanced as expected by billions of years of gravitational evolution. On smaller scales, the model describes how regions that were denser than average collapsed due to their enhanced gravity and eventually formed gravitationally-bound halos, first on small spatial scales and later on larger ones. In this hierarchical model of galaxy formation, the small galaxies formed first and then merged or accreted gas to form larger galaxies. At each snapshot of this cosmic evolution, the abundance of collapsed halos, whose masses are dominated by dark matter, can be computed from the initial conditions using numerical simulations. The common understanding of galaxy formation is based on the notion that stars formed out of the gas that cooled and subsequently condensed to high densities in the cores of some of these halos.

Gravity thus explains how some gas is pulled into the deep potential wells within dark matter halos and forms the galaxies. One might naively expect that the gas outside halos would remain mostly undisturbed. However, observations show that it has not remained neutral (i.e. in atomic form) but was largely ionized by the UV radiation emitted by the galaxies. The diffuse gas pervading the space outside and between galaxies is referred to as the intergalactic medium (IGM). For the first hundreds of millions of years after cosmological recombination, the so-called cosmic "dark ages", the universe was filled with diffuse atomic hydrogen. As soon as galaxies formed, they started to ionize diffuse hydrogen in their vicinity. Within less than a billion years, most of the IGM was reionized. We have not yet imaged the cosmic dark ages before the first galaxies had formed. One of the frontiers in current cosmological studies aims to study the cosmic epoch of reionization and the first generation of galaxies that triggered it.

3.1.2 *The expanding Universe*

The modern physical description of the Universe as a whole can be traced back to Einstein, who assumed for simplicity the so-called "cosmological principle": that the distribution of matter and energy is homogeneous and isotropic on the largest scales. Today isotropy is well established for the distribution of faint radio sources, optically-selected galaxies, the

X-ray background, and most importantly the cosmic microwave background (hereafter, CMB). The constraints on homogeneity are less strict, but a cosmological model in which the Universe is isotropic but significantly inhomogeneous in spherical shells around our special location, is also excluded.

In General Relativity, the metric for a space-time which is spatially homogeneous and isotropic is the Friedmann–Robertson–Walker metric, which can be written in the form

$$ds^2 = c^2 dt^2 - a^2(t)\left[\frac{dr^2}{1-k\,r^2} + r^2(d\theta^2 + \sin^2\theta\, d\phi^2)\right], \qquad (3.1)$$

where c is the speed of light, $a(t)$ is the cosmic scale factor which describes expansion in time t, and (r, θ, ϕ) are spherical comoving coordinates. The constant k determines the geometry of space; it is positive in a closed Universe, zero in a flat Universe (Euclidean space), and negative in an open Universe. Observers at rest remain at rest, at fixed (r, θ, ϕ), with their physical separation increasing with time in proportion to $a(t)$. A given observer sees a nearby observer at physical distance D receding at the Hubble velocity $H(t)D$, where the Hubble constant at time t is $H(t) = d\,a(t)/dt$. Light emitted by a source at time t is observed at $t = 0$ with a redshift $z = 1/a(t) - 1$, where we set $a(t = 0) \equiv 1$ for convenience.

The Einstein field equations of General Relativity yield the Friedmann equation

$$H^2(t) = \frac{8\pi G}{3}\rho - \frac{k}{a^2}, \qquad (3.2)$$

which relates the expansion of the Universe to its matter-energy content. For each component of the energy density ρ, with an equation of state $p = p(\rho)$, the density ρ varies with $a(t)$ according to the thermodynamic relation

$$d(\rho c^2 r^3) = -p\,d(r^3). \qquad (3.3)$$

With the critical density

$$\rho_c(t) \equiv \frac{3H^2(t)}{8\pi G} \qquad (3.4)$$

defined as the density needed for $k = 0$, we define the ratio of the total density to the critical density as

$$\Omega \equiv \frac{\rho}{\rho_c}. \tag{3.5}$$

With Ω_m, Ω_Λ, and Ω_r denoting the present contributions to Ω from matter [including cold dark matter as well as a contribution Ω_b from ordinary matter (baryons) made of protons and neutrons], vacuum energy (cosmological constant), and radiation, respectively, the Friedmann equation becomes

$$\frac{H(t)}{H_0} = \left[\frac{\Omega_m}{a^3} + \Omega_\Lambda + \frac{\Omega_r}{a^4} + \frac{\Omega_k}{a^2}\right], \tag{3.6}$$

where we define H_0 and $\Omega_0 = \Omega_m + \Omega_\Lambda + \Omega_r$ to be the present values of H and Ω, respectively, and we let

$$\Omega_k \equiv -\frac{k}{H_0^2} = 1 - \Omega_m. \tag{3.7}$$

In the particularly simple Einstein–de Sitter model ($\Omega_m = 1$, $\Omega_\Lambda = \Omega_r = \Omega_k = 0$), the scale factor varies as $a(t) \propto t^{2/3}$. Even models with non-zero Ω_Λ or Ω_k approach the Einstein–de Sitter scaling-law at high redshift, i.e. when $(1+z) \gg |\Omega_m^{-1} - 1|$ (as long as Ω_r can be neglected). In this high-z regime the age of the Universe is

$$t \approx \frac{2}{3H_0\sqrt{\Omega_m}}(1+z)^{-3/2} \approx 10^9 \text{yr}\left(\frac{1+z}{7}\right)^{-3/2}. \tag{3.8}$$

Recent observations confine the standard set of cosmological parameters to a relatively narrow range. In particular, we seem to live in a universe dominated by a cosmological constant (Λ) and cold dark matter, or in short a ΛCDM cosmology (with Ω_k so small that it is usually assumed to equal zero) with an approximately scale-invariant primordial power spectrum of density fluctuations, i.e. $n \approx 1$ where the initial power spectrum is $P(k) = |\delta_\mathbf{k}|^2 \propto k^n$ in terms of the wavenumber k of the Fourier modes $\delta_\mathbf{k}$ (see Sec. 3.2.1 below). Also, the Hubble constant today is written as

$$H_0 = 100\, h\, \text{km}\,\text{s}^{-1}\,\text{Mpc}^{-1}, \tag{3.9}$$

in terms of h, and the overall normalization of the power spectrum is specified in terms of σ_8, the root-mean-square amplitude of mass fluctuations in spheres of radius $8\,h^{-1}$ Mpc. For example, the best-fit cosmological parameters matching the WMAP data together with large-scale surveys of galaxies and supernovae are $\sigma_8 = 0.81$, $n = 0.96$, $h = 0.72$, $\Omega_m = 0.28$, $\Omega_\Lambda = 0.72$ and $\Omega_b = 0.046$.

3.2 Galaxy Formation

3.2.1 *Growth of linear perturbations*

As noted in Sec. 1, observations of the CMB show that the Universe at cosmic recombination (redshift $z \sim 10^3$) was remarkably uniform apart from spatial fluctuations in the energy density and in the gravitational potential of roughly one part in $\sim 10^5$. The primordial inhomogeneities in the density distribution grew over time and eventually led to the formation of galaxies as well as galaxy clusters and large-scale structure. In the early stages of this growth, as long as the density fluctuations on the relevant scales were much smaller than unity, their evolution can be understood with a linear perturbation analysis.

As before, we distinguish between fixed and comoving coordinates. Using vector notation, the fixed coordinate $\mathbf{r}$ corresponds to a comoving position $\mathbf{x} = \mathbf{r}/a$. In a homogeneous Universe with density ρ, we describe the cosmological expansion in terms of an ideal pressureless fluid of particles, each of which is at fixed $\mathbf{x}$, expanding with the Hubble flow $\mathbf{v} = H(t)\mathbf{r}$ where $\mathbf{v} = d\mathbf{r}/dt$. Onto this uniform expansion we impose small perturbations, given by a relative density perturbation

$$\delta(\mathbf{x}) = \frac{\rho(\mathbf{r})}{\bar{\rho}} - 1, \tag{3.10}$$

where the mean fluid density is $\bar{\rho}$, with a corresponding peculiar velocity $\mathbf{u} \equiv \mathbf{v} - H\mathbf{r}$. Then the fluid is described by the continuity and Euler equations in comoving coordinates:

$$\frac{\partial \delta}{\partial t} + \frac{1}{a} \nabla \cdot [(1 + \delta)\mathbf{u}] = 0, \tag{3.11}$$

$$\frac{\partial \mathbf{u}}{\partial t} + H\mathbf{u} + \frac{1}{a}(\mathbf{u} \cdot \nabla)\mathbf{u} = -\frac{1}{a}\nabla \phi. \tag{3.12}$$

The potential ϕ is given by the Poisson equation, in terms of the density perturbation:

$$\nabla^2 \phi = 4\pi G \bar{\rho} a^2 \delta. \tag{3.13}$$

This fluid description is valid for describing the evolution of collisionless cold dark matter particles until different particle-streams cross. This "shell-crossing" typically occurs only after perturbations have grown to become nonlinear, and at that point the individual particle trajectories must in general be followed. Similarly, baryons can be described as a pressureless fluid as long as their temperature is negligibly small, but nonlinear collapse leads to the formation of shocks in the gas.

For small perturbations $\delta \ll 1$, the fluid equations can be linearized and combined to yield

$$\frac{\partial^2 \delta}{\partial t^2} + 2H \frac{\partial \delta}{\partial t} = 4\pi G \bar{\rho} \delta. \tag{3.14}$$

This linear equation has in general two independent solutions, only one of which grows with time. Starting with random initial conditions, this growing mode comes to dominate the density evolution. Thus, until it becomes nonlinear, the density perturbation maintains its shape in comoving coordinates and grows in proportion to a growth factor $D(t)$. The growth factor in the matter-dominated era is given by

$$D(t) \propto \frac{(\Omega_\Lambda a^3 + \Omega_k a + \Omega_m)^{1/2}}{a^{3/2}} \int_0^a \frac{a'^{3/2}\, da'}{(\Omega_\Lambda a'^3 + \Omega_k a' + \Omega_m)^{3/2}}, \tag{3.15}$$

where we neglect Ω_r when considering halos forming in the matter-dominated regime at $z \ll 10^4$. In the Einstein–de Sitter model (or, at high redshift, in other models as well) the growth factor is simply proportional to $a(t)$.

The spatial form of the initial density fluctuations can be described in Fourier space, in terms of Fourier components

$$\delta_{\mathbf{k}} = \int d^3 x\, \delta(x) e^{-i\mathbf{k}\cdot\mathbf{x}}. \tag{3.16}$$

Here we use the comoving wave-vector $\mathbf{k}$, whose magnitude k is the comoving wavenumber which is equal to 2π divided by the wavelength.

The Fourier description is particularly simple for fluctuations generated by inflation. Inflation generates perturbations given by a Gaussian random field, in which different **k**-modes are statistically independent, each with a random phase. The statistical properties of the fluctuations are determined by the variance of the different **k**-modes, and the variance is described in terms of the power spectrum $P(k)$ as follows:

$$\langle \delta_{\mathbf{k}} \delta_{\mathbf{k}'}^* \rangle = (2\pi)^3 P(k) \, \delta^{(3)}(\mathbf{k} - \mathbf{k}'), \tag{3.17}$$

where $\delta^{(3)}$ is the three-dimensional Dirac-delta function. The gravitational potential fluctuations are sourced by the density fluctuations through Poisson's equation.

In standard models, inflation produces a primordial power-law spectrum $P(k) \propto k^n$ with $n \sim 1$. Perturbation growth in the radiation-dominated and then matter-dominated Universe results in a modified final power spectrum, characterized by a turnover at a scale of order the horizon cH^{-1} at matter-radiation equality, and a small-scale asymptotic shape of $P(k) \propto k^{n-4}$. The overall amplitude of the power spectrum is not specified by current models of inflation, and it is usually set by comparing to the observed CMB temperature fluctuations or to local measures of large-scale structure.

Since density fluctuations may exist on all scales, in order to determine the formation of objects of a given size or mass it is useful to consider the statistical distribution of the smoothed density field. Using a window function $W(\mathbf{r})$ normalized so that $\int d^3r\, W(\mathbf{r}) = 1$, the smoothed density perturbation field, $\int d^3 r \delta(\mathbf{x}) W(\mathbf{r})$, follows a Gaussian distribution with zero mean. For the particular choice of a spherical top-hat, in which $W = 1$ in a sphere of radius R and is zero outside, the smoothed perturbation field measures the fluctuations in the mass in spheres of radius R. The normalization of the present power spectrum is often specified by the value of $\sigma_8 \equiv \sigma(R = 8\,h^{-1}\,\text{Mpc})$. For the top-hat, the smoothed perturbation field is denoted δ_R or δ_M, where the mass M is related to the comoving radius R by $M = 4\pi \rho_m R^3/3$, in terms of the current mean density of matter ρ_m. The variance $\langle \delta_M \rangle^2$ is

$$\sigma^2(M) = \sigma^2(R) = \int_0^\infty \frac{dk}{2\pi^2} k^2 P(k) \left[\frac{3 j_1(kR)}{kR} \right]^2, \tag{3.18}$$

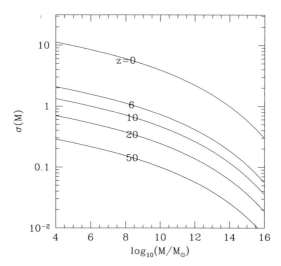

Fig. 3.3 The root-mean-square amplitude of linearly-extrapolated density fluctuations σ as a function of mass M (in solar masses $M_\odot$, within a spherical top-hat filter) at different redshifts z. Halos form in regions that exceed the background density by a factor of order unity. This threshold is only surpassed by rare (many-σ) peaks for high masses at high redshifts. **Figure credit:** A. Loeb, *How Did the First Stars and Galaxies Form?* (Princeton University Press, 2010).

where $j_1(x) = (\sin x - x \cos x)/x^2$. The function $\sigma(M)$, plotted in Fig. 3.3, plays a crucial role in estimates of the abundance of collapsed objects.

Different physical processes contributed to the perturbation growth. In the absence of other influences, gravitational forces due to density perturbations imprinted by inflation would have driven parallel perturbation growth in the dark matter baryons and photons. However, since the photon sound speed is of order the speed of light, the radiation pressure produces sound waves on a scale of order the cosmic horizon and suppressed sub-horizon perturbations in the photon density. The baryonic pressure similarly suppressed perturbations in the gas below the (much smaller) so-called baryonic *Jeans* scale. Since the formation of hydrogen at recombination had decoupled the cosmic gas from its mechanical drag on the CMB, the baryons subsequently began to fall into the pre-existing gravitational potential wells of the dark matter.

Spatial fluctuations developed in the gas temperature as well as in the gas density. Both the baryons and the dark matter were affected on small scales by the temperature fluctuations through the gas pressure. Compton heating due to scattering of the residual free electrons (constituting a fraction $\sim 10^{-4}$) with the CMB photons remained effective, keeping the gas temperature fluctuations tied to the photon temperature fluctuations, even for a time after recombination. The growth of linear perturbations can be calculated with the standard CMBFAST code (http://www.cmbfast.org), after a modification to account for the fact that the speed of sound of the gas also fluctuates spatially.

After recombination, two main drivers affect the baryon density and temperature fluctuations, namely, the thermalization with the CMB and the gravitational force that attracts the baryons to the dark matter potential wells. The density perturbations in all species grow together on scales where gravity is unopposed, outside the horizon (i.e. at $k < 0.01\,\mathrm{Mpc}^{-1}$ at $z \sim 1000$). At $z = 1200$ the perturbations in the baryon-photon fluid oscillate as acoustic waves on scales of order the sound horizon ($k \sim 0.01\,\mathrm{Mpc}^{-1}$), while smaller-scale perturbations in both the photons and baryons are damped by photon diffusion and the drag of the diffusing photons on the baryons. On sufficiently small scales, the power spectra of baryon density and temperature roughly assume the shape of the dark matter fluctuations (except for the gas-pressure cutoff at the very smallest scales), due to the effect of gravitational attraction on the baryon density and of the resulting adiabatic expansion on the gas temperature. After the mechanical coupling of the baryons to the photons ends at $z \sim 1000$, the baryon density perturbations gradually grow toward the dark matter perturbations because of gravity. Similarly, after the thermal coupling ends at $z \sim 200$, the baryon temperature fluctuations are driven by adiabatic expansion toward a value of 2/3 of the density fluctuations. By $z = 200$ the baryon in-fall into the dark matter potentials is well advanced and adiabatic expansion is becoming increasingly important in setting the baryon temperature.

3.2.2 Halo properties

The small density fluctuations evidenced in the CMB grow over time as described in the previous subsection, until the perturbation δ becomes of order unity, and the full nonlinear gravitational problem must be considered. A dark matter halo forms in a region where matter relaxes dynamically to a make a stable object that is much denser than the mean cosmic density and is held together by its own self-gravity. The dynamical collapse of a dark matter halo can be solved analytically only in cases of particular symmetry. If we consider a region which is much smaller than the horizon cH^{-1}, then the formation of a halo can be formulated as a problem in Newtonian gravity, in some cases with minor corrections coming from General Relativity. The simplest case is that of spherical symmetry, with an initial ($t = t_i \ll t_0$) top-hat of uniform overdensity δ_i inside a sphere of radius R. Although this model is restricted in its direct applicability, the results of spherical collapse have turned out to be surprisingly useful in understanding the properties and distribution of halos in models based on cold dark matter.

The collapse of a spherical top-hat perturbation is described by the Newtonian equation (with a correction for the cosmological constant)

$$\frac{d^2 r}{dt^2} = H_0^2 \Omega_\Lambda \, r - \frac{GM}{r^2}, \tag{3.19}$$

where r is the radius in a fixed (not comoving) coordinate frame, H_0 is the present-day Hubble constant, M is the total mass enclosed within radius r, and the initial velocity field is given by the Hubble flow $dr/dt = H(t)r$. The enclosed δ grows initially as $\delta_L = \delta_i D(t)/D(t_i)$, in accordance with linear theory, but eventually δ grows above δ_L. If the mass shell at radius r is bound (i.e. if its total Newtonian energy is negative) then it reaches a radius of maximum expansion and subsequently collapses. As demonstrated in the previous section, at the moment when the top-hat collapses to a point, the overdensity predicted by linear theory is $\delta_L = 1.686$ in the Einstein–de Sitter model, with only a weak dependence on Ω_m and Ω_Λ. Thus a top-hat collapses at redshift z if its linear overdensity extrapolated to the present day (also termed the critical density of collapse) is

$$\delta_{\text{crit}}(z) = \frac{1.686}{D(z)}, \tag{3.20}$$

where we set $D(z=0) = 1$.

Even a slight violation of the exact symmetry of the initial perturbation can prevent the top-hat from collapsing to a point. Instead, the halo reaches a state of virial equilibrium by violent relaxation (phase mixing). Using the virial theorem $U = -2K$ to relate the potential energy U to the kinetic energy K in the final state (implying that the virial radius is half the turnaround radius — where the kinetic energy vanishes), the final overdensity relative to the critical density at the collapse redshift is $\Delta_c = 18\pi^2 \simeq 178$ in the Einstein–de Sitter model, modified in a Universe with $\Omega_m + \Omega_\Lambda = 1$ to the fitting formula

$$\Delta_c = 18\pi^2 + 82d - 39d^2, \tag{3.21}$$

where $d \equiv \Omega_m^z - 1$ is evaluated at the collapse redshift, so that

$$\Omega_m^z = \frac{\Omega_m(1+z)^3}{\Omega_m(1+z)^3 + \Omega_\Lambda + \Omega_k(1+z)^2}. \tag{3.22}$$

A halo of mass M collapsing at redshift z thus has a virial radius

$$r_{\rm vir} = 0.784 \left(\frac{M}{10^8\, h^{-1}\, M_\odot}\right)^{1/3} \left[\frac{\Omega_m}{\Omega_m^z} \frac{\Delta_c}{18\pi^2}\right]^{-1/3} \left(\frac{1+z}{10}\right)^{-1} h^{-1}\,{\rm kpc}, \tag{3.23}$$

and a corresponding circular velocity,

$$V_c = \left(\frac{GM}{r_{\rm vir}}\right)^{1/2} = 23.4 \left(\frac{M}{10^8\, h^{-1}\, M_\odot}\right)^{1/3} \left[\frac{\Omega_m}{\Omega_m^z} \frac{\Delta_c}{18\pi^2}\right]^{1/6}$$

$$\times \left(\frac{1+z}{10}\right)^{1/2} {\rm km\,s^{-1}}. \tag{3.24}$$

In these expressions we have assumed a present Hubble constant written in the form $H_0 = 100\, h\,{\rm km\,s^{-1}\,Mpc^{-1}}$. We may also define a virial temperature

$$T_{\rm vir} = \frac{\mu m_p V_c^2}{2k} = 1.98 \times 10^4 \left(\frac{\mu}{0.6}\right) \left(\frac{M}{10^8\, h^{-1}\, M_\odot}\right)^{2/3} \left[\frac{\Omega_m}{\Omega_m^z} \frac{\Delta_c}{18\pi^2}\right]^{1/3}$$

$$\times \left(\frac{1+z}{10}\right) {\rm K}, \tag{3.25}$$

where μ is the mean molecular weight and m_p is the proton mass. Note that the value of μ depends on the ionization fraction of the gas; for a fully ionized primordial gas $\mu = 0.59$, while a gas with ionized hydrogen but only singly-ionized helium has $\mu = 0.61$. The binding energy of the halo is approximately[1]

$$E_b = \frac{1}{2}\frac{GM^2}{r_\text{vir}} = 5.45 \times 10^{53} \left(\frac{M}{10^8\, h^{-1}\, M_\odot}\right)^{5/3} \left[\frac{\Omega_m}{\Omega_m^z}\frac{\Delta_c}{18\pi^2}\right]^{1/3}$$
$$\times \left(\frac{1+z}{10}\right) h^{-1}\, \text{erg}. \qquad (3.26)$$

Note that the binding energy of the baryons is smaller by a factor equal to the baryon fraction Ω_b/Ω_m.

Although spherical collapse captures some of the physics governing the formation of halos, structure formation in cold dark matter models proceeds hierarchically. At early times, most of the dark matter is in low-mass halos, and these halos continuously accrete and merge to form high-mass halos (see Fig. 3.4). Numerical simulations of hierarchical halo formation indicate a roughly universal spherically-averaged density profile for the resulting halos, though with considerable scatter among different halos. The typical profile has the form

$$\rho(r) = \frac{3H_0^2}{8\pi G}(1+z)^3 \frac{\Omega_m}{\Omega_m^z}\frac{\delta_c}{c_\text{N} x (1+c_\text{N} x)^2}, \qquad (3.27)$$

where $x = r/r_\text{vir}$, and the characteristic density δ_c is related to the concentration parameter c_N by

$$\delta_c = \frac{\Delta_c}{3}\frac{c_\text{N}^3}{\ln(1+c_\text{N}) - c_\text{N}/(1+c_\text{N})}. \qquad (3.28)$$

The concentration parameter itself depends on the halo mass M, at a given redshift z.

3.2.3 Formation of the first stars

Theoretical expectations for the properties of the first galaxies are based on the standard cosmological model outlined in Sec. 1. The formation of the first bound objects marked the central milestone in the transition

---
[1] The coefficient of $1/2$ in Eq. (3.26) would be exact for a singular isothermal sphere with $\rho(r) \propto 1/r^2$.

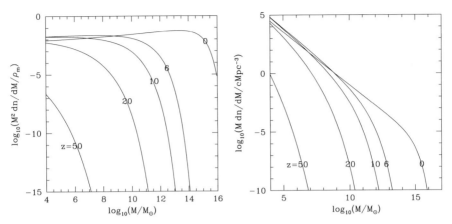

Fig. 3.4 Left panel: The mass fraction incorporated into dark matter halos per logarithmic bin of halo mass $(M^2 dn/dM)/\rho_m$, as a function of M at different redshifts z. Here $\rho_m = \Omega_m \rho_c$ is the present-day matter density, and $n(M)dM$ is the comoving density of halos with masses between M and $M + dM$. The halo mass distribution was calculated based on an improved version of the Press–Schechter formalism for ellipsoidal collapse (R. K. Sheth and G. Tormen, 2002) that fits better numerical simulations. Right panel: Number density of halos per logarithmic bin of halo mass, $M dn/dM$ (in units of comoving Mpc^{-3}), at various redshifts. **Figure credit:** A. Loeb, *How Did the First Stars and Galaxies Form?* (Princeton University Press, 2010).

from the initial simplicity (discussed in the previous subsection) to the present-day complexity. Stars and accreting black holes output copious radiation and also produced explosions and outflows that brought into the IGM chemical products from stellar nucleosynthesis and enhanced magnetic fields. However, the formation of the very first stars, in a universe that had not yet suffered such feedback, remains a well-specified problem for theorists.

Stars form when large amounts of matter collapse to high densities. However, the process can be stopped if the pressure exerted by the hot intergalactic gas prevents outlying gas from falling into dark matter concentrations. As the gas falls into a dark matter halo, it forms shocks due to converging supersonic flows and in the process heats up and can only collapse further by first radiating its energy away. This restricts such a process of collapse to very large clumps of dark matter that are around 100 000 times the mass of the Sun. Inside these clumps, the shocked gas loses energy by emitting radiation from excited molecular hydrogen that formed naturally within the primordial gas mixture of hydrogen and helium.

The first stars are expected to have been quite different from the stars that form today in the Milky Way. The higher pressure within the primordial gas due to the presence of fewer cooling agents suggests that fragmentation only occurred into relatively large units, in which gravity could overcome the pressure. Due to the lack of carbon, nitrogen, and oxygen — elements that would normally dominate the nuclear energy production in modern massive stars — the first stars must have condensed to extremely high densities and temperatures before nuclear reactions were able to heat the gas and balance gravity. These unusually massive stars produced high luminosities of UV photons, but their nuclear fuel was exhausted after 2–3 million years, resulting in a huge supernova or in collapse to a black hole. The heavy elements which were dispersed by the first supernovae in the surrounding gas, enabled the enriched gas to cool more effectively and fragment into lower mass stars. Simple calculations indicate that a carbon or oxygen enrichment of merely $< 10^{-3}$ of the solar abundance is sufficient to allow solar mass stars to form. These second-generation "low-metallicity" stars are long-lived and could in principle be discovered in the halo of the Milky Way galaxy, providing fossil record of the earliest star formation episode in our cosmic environment.

Advances in computing power have made possible detailed numerical simulations of how the first stars formed. These simulations begin in the early universe, in which dark matter and gas are distributed uniformly, apart from tiny variations in density and temperature that are statistically distributed according to the patterns observed in the CMB. In order to span the vast range of scales needed to simulate an individual star within a cosmological context, the adopted codes zoom in repeatedly on the densest part of the first collapsing cloud that is found within the simulated volume. The simulation follows gravity, hydrodynamics, and chemical processes in the primordial gas, and resolves a scale that is > 10 orders of magnitudes smaller than that of the simulated box. In state-of-the-art simulations, the resolved scale is approaching the scale of the protostar. The simulations have established that the first stars formed within halos containing $\sim 10^5\,M_\odot$ in total mass, and indicate that the first stars most likely weighed tens to hundreds of solar masses each.

To estimate *when* the first stars formed we must remember that the first 100 000 solar mass halos collapsed in regions that happened to have a particularly high density enhancement very early on. There was initially only a small abundance of such regions in the entire universe, so a simulation that is limited to a small volume is unlikely to find such halos until much later. Simulating the entire universe is well beyond the capabilities of current simulations, but analytical models predict that the first observable star in the universe probably formed 30 million years after the Big Bang, less than a quarter of one percent of the Universe's total age of 13.7 billion years.

Although stars were extremely rare at first, gravitational collapse increased the abundance of galactic halos and star formation sites with time (Fig. 3.2). Radiation from the first stars is expected to have eventually dissociated all the molecular hydrogen in the intergalactic medium, leading to the domination of a second generation of larger galaxies where the gas cooled via radiative transitions in atomic hydrogen and helium. Atomic cooling occurred in halos of mass above $\sim 10^8\,M_\odot$, in which the in-falling gas was heated above 10 000 K and became ionized. The first galaxies to form through atomic cooling are expected to have formed around redshift 45, and such galaxies were likely the main sites of star formation by the time reionization began in earnest. As the IGM was heated above 10 000 K by reionization, its pressure jumped and prevented the gas from accreting into newly forming halos below $\sim 10^9\,M_\odot$. The first Milky Way-sized halo $M = 10^{12}\,M_\odot$ is predicted to have formed 400 million years after the Big Bang, but such halos have become typical galactic hosts only in the last five billion years.

Hydrogen is the most abundant element in the Universe, The prominent Lyman-α spectral line of hydrogen (corresponding to a transition from its first excited level to its ground state) provides an important probe of the condensation of primordial gas into the first galaxies. Existing searches for Lyman-α emission have discovered galaxies robustly out to a redshift $z \sim 7$ with some unconfirmed candidate galaxies out to $z \sim 10$. The spectral break

Fig. 3.5 A full-scale model of the James Webb Space Telescope (JWST), the successor to the Hubble Space Telescope. JWST includes a primary mirror 6.5 m in diameter, and offers instrument sensitivity across the infrared wavelength range of 0.6–28 μm which will allow detection of the first galaxies. The size of the Sun shield (the large flat screen in the image) is 22 m × 10 m (72 ft × 29 ft). The telescope will orbit 1.5 million kilometers from Earth at the Lagrange L2 point. **Image credit:** JWST/NASA (http://www.jwst.nasa.gov/).

owing to Lyman-α absorption by the IGM allows to identify high-redshift galaxies photometrically. Existing observations provide only a preliminary glimpse into the formation of the first galaxies.

Within the next decade, NASA plans to launch an infrared space telescope (*JWST*; Fig. 3.5) that will image some of the earliest sources of light (stars and black holes) in the Universe. In parallel, there are several initiatives to construct large-aperture infrared telescopes on the ground with the same goal in mind.

The next generation of ground-based telescopes will have a diameter of 20–30 m (Fig. 3.6). Together with *JWST* (which will not be affected by the atmospheric background) they will be able to image and make spectral studies of the early galaxies. Given that these galaxies also create the ionized bubbles around them by their UV emission, during reionization the locations of galaxies should correlate with bubbles within

Fig. 3.6 Artist's conception of the designs for three future giant telescopes that will be able to probe the first generation of galaxies from the ground: the European Extremely Large Telescope (EELT, left), the Giant Magellan Telescope (GMT, middle), and the Thirty Meter Telescope (TMT, right). **Image credits:** http://www.eso.org/sci/facilities/eelt/, http://www.gmto.org/, and http://www.tmt.org/.

the neutral hydrogen. Within a decade it should be possible to explore the environmental influence of individual galaxies by using these telescopes in combination with 21-cm probes of reionization.

3.2.4 *Gamma-ray bursts: probing the first stars one star at a time*

So far, to learn about diffuse IGM gas pervading the space outside and between galaxies, astronomers routinely study its absorption signatures in the spectra of distant quasars, the brightest long-lived astronomical objects. Quasars' great luminosities are believed to be powered by accretion of gas onto black holes weighing up to a few billion times the mass of the Sun that are situated in the centers of massive galaxies. As the surrounding gas spirals in toward the black hole sink, the viscous dissipation of heat makes the gas glow brightly into space, creating a luminous source visible from afar.

Over the past decade, an alternative population of bright sources at cosmological distances was discovered, the so-called afterglows of *Gamma-Ray Bursts* (GRBs). These events are characterized by a flash of high-energy ($>0.1\,\mathrm{MeV}$) photons, typically lasting 0.1–100 s, which is followed by an afterglow of lower-energy photons over much longer timescales. The afterglow peaks at X-ray, UV, optical and eventually radio wavelengths on timescales of minutes, hours, days, and months, respectively. The central engines of GRBs are believed to be associated with the compact remnants (neutron stars or stellar-mass black holes) of massive stars. Their high luminosities make them detectable out to the edge of the visible Universe. GRBs offer the opportunity to detect the most distant (and hence earliest) population of massive stars, the so-called Population III (or Pop III), one star at a time. In the hierarchical assembly process of halos that are dominated by cold dark matter (CDM), the first galaxies should have had lower masses (and lower stellar luminosities) than their more recent counterparts. Consequently, the characteristic luminosity of galaxies or quasars is expected to decline with increasing redshift. GRB afterglows, which already produce a peak flux comparable to that of quasars or starburst galaxies at $z \sim 1$–2, are therefore expected to outshine any competing source at the highest redshifts, when the first dwarf galaxies formed in the Universe.

GRBs, the electromagnetically-brightest explosions in the Universe, should be detectable out to redshifts $z > 10$. High-redshift GRBs can be identified through infrared photometry, based on the Lyman-α break induced by absorption of their spectrum at wavelengths below $1.216\,\mu$m $[(1+z)/10]$. Follow-up spectroscopy of high-redshift candidates can then be performed on a 10-meter-class telescope. GRB afterglows offer the opportunity to detect stars as well as to probe the metal enrichment level of the intervening IGM. Recently, the *Swift* satellite has detected a GRB originating at $z \simeq 8.2$, thus demonstrating the viability of GRBs as probes of the early Universe.

Another advantage of GRBs is that the GRB afterglow flux at a given observed time lag after the γ-ray trigger is not expected to fade significantly with increasing redshift, since higher redshifts translate to earlier times in the source frame, during which the afterglow is intrinsically brighter. For standard afterglow lightcurves and spectra, the increase in the luminosity distance with redshift is compensated by this cosmological time-stretching effect as shown in Fig. 3.7.

GRB afterglows have smooth (broken power-law) continuum spectra unlike quasars which show strong spectral features (such as broad emission lines or the so-called "blue bump") that complicate the extraction of IGM absorption features. In particular, the extrapolation into the spectral regime marked by the IGM Lyman-α absorption during the epoch of reionization is

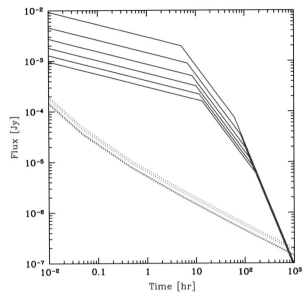

Fig. 3.7 GRB afterglow flux as a function of time since the γ-ray trigger in the observer frame. The flux (solid curves) is calculated at the redshifted Lyman-α wavelength. The dotted curves show the planned detection threshold for the *James Webb Space Telescope* (*JWST*), assuming a spectral resolution $R = 5000$ with the near infrared spectrometer, a signal-to-noise ratio of 5 per spectral resolution element, and an exposure time equal to 20% of the time since the GRB explosion. Each set of curves shows a sequence of redshifts, namely $z = 5, 7, 9, 11, 13,$ and 15, respectively, from top to bottom. **Figure credit:** R. Barkana and A. Loeb, *Astrophys. J.* **601**, 64 (2004).

much more straightforward for the smooth UV spectra of GRB afterglows than for quasars with an underlying broad Lyman-α emission line. However, the interpretation may be complicated by the presence of damped Lyman-α absorption by dense neutral hydrogen in the immediate environment of the GRB within its host galaxy. Since GRBs originate from the dense environment of active star formation, such damped absorption is expected and indeed has been seen, including in the most distant GRB at $z = 8.2$.

3.2.5 *Supermassive black holes*

The fossil record in the present-day Universe indicates that every bulged galaxy hosts a supermassive black hole (BH) at its center. This conclusion is derived from a variety of techniques which probe the dynamics of stars and gas in galactic nuclei. The inferred BHs are dormant or faint most of the time, but occasionally flash in a short burst of radiation that lasts for a small fraction of the age of the Universe. The short duty cycle accounts for the fact that bright quasars are much less abundant than their host galaxies, but it begs the more fundamental question: *"Why is the quasar activity so brief?"* A natural explanation is that quasars are suicidal, namely the energy output from the BHs regulates their own growth.

Supermassive BHs make up a small fraction, $< 10^{-3}$, of the total mass in their host galaxies, and so their direct dynamical impact is limited to the central star distribution where their gravitational influence dominates. Dynamical friction on the background stars keeps the BH close to the center. Random fluctuations in the distribution of stars induces a Brownian motion of the BH. This motion can be described by the same Langevin equation that captures the motion of a massive dust particle as it responds to random kicks from the much lighter molecules of air around it. The characteristic speed by which the BH wanders around the center is small, $\sim (m_\star/M_{\rm BH})^{1/2} \sigma_\star$, where $m_\star$ and $M_{\rm BH}$ are the masses of a single star and the BH, respectively, and $\sigma_\star$ is the stellar velocity dispersion. Since the random force fluctuates on a dynamical time, the BH wanders across a region that is smaller by a factor of $\sim (m_\star/M_{\rm BH})^{1/2}$ than the region traversed by the stars inducing the fluctuating force on it.

The dynamical insignificance of the BH on the global galactic scale is misleading. The gravitational binding energy per rest-mass energy of galaxies is of order $\sim (\sigma_\star/c)^2 < 10^{-6}$. Since BH are relativistic objects, the gravitational binding energy of material that feeds them amounts to a substantial fraction of its rest-mass energy. Even if the BH mass amounts to a fraction as small as $\sim 10^{-4}$ of the baryonic mass in a galaxy, and only a percent of the accreted rest-mass energy is deposited into the gaseous environment of the BH, this slight deposition can unbind the entire gas reservoir of the host galaxy. This order-of-magnitude estimate explains why quasars may be short lived. As soon as the central BH accretes large quantities of gas so as to significantly increase its mass, it releases large amounts of energy and momentum that could suppress further accretion onto it. In short, the BH growth might be *self-regulated*.

The principle of *self-regulation* naturally leads to a correlation between the final BH mass, $M_{\rm BH}$, and the depth of the gravitational potential well to which the surrounding gas is confined. The latter can be characterized by the velocity dispersion of the associated stars, $\sim \sigma_\star^2$. Indeed a correlation between $M_{\rm BH}$ and $\sigma_\star^4$ is observed in the present-day Universe. If quasars shine near their Eddington limit as suggested by observations of low- and high-redshift quasars, then a fraction of $\sim 5-10\%$ of the energy released by the quasar over a galactic dynamical time needs to be captured in the surrounding galactic gas in order for the BH growth to be self-regulated. With this interpretation, the $M_{\rm BH}-\sigma_\star$ relation reflects the limit introduced to the BH mass by self-regulation; deviations from this relation are inevitable during episodes of BH growth or as a result of mergers of galaxies that have no cold gas in them. A physical scatter around this upper envelope could also result from variations in the efficiency by which the released BH energy couples to the surrounding gas.

Various prescriptions for self-regulation were sketched in the literature. These involve either energy or momentum-driven winds, with the latter type being a factor of $\sim v_c/c$ less efficient. The quasar remains active during the dynamical time of the initial gas reservoir, $\sim 10^7$ years, and fades afterward due to the dilution of this reservoir. The BH growth may resume if the cold gas reservoir is replenished through a new merger. Following early analytic work, extensive numerical simulations demonstrated that galaxy mergers do produce the observed correlations between black hole mass and spheroid properties. Because of the limited resolution near the galaxy nucleus, these simulations adopt a simple prescription for the accretion flow that feeds the black hole. The actual feedback in reality may depend crucially on the geometry of this flow and the physical mechanism that couples the energy or momentum output of the quasar to the surrounding gas.

The inflow of cold gas toward galaxy centers during the growth phase of the BH would naturally be accompanied by a burst of star formation. The fraction of gas that is not consumed by stars or ejected by supernova-driven winds, will continue to feed the BH. It is therefore not surprising that quasar and starburst activities co-exist in Ultra Luminous Infrared Galaxies, and that all quasars show broad metal lines indicating pre-enrichment of the surrounding gas with heavy elements.

The upper mass of galaxies may also be regulated by the energy output from quasar activity. This would account for the fact that cooling flows are suppressed in present-day X-ray clusters, and that massive BHs and stars in galactic bulges were already formed at $z \sim 2$. In the cores of cooling X-ray clusters, there is often an active central BH that supplies sufficient energy to compensate for the cooling of the gas. The primary physical process by which this energy couples to the gas is still unknown.

The quasars discovered so far at the $z \sim 6$ mark the early growth of the most massive BHs and galactic spheroids. The BHs powering these bright quasars possess a mass of a few billion solar masses. A quasar radiating at its Eddington limiting luminosity, $L_E = 1.4 \times 10^{47} \, \text{erg s}^{-1} (M_{\text{BH}}/10^9 \, M_\odot)$, with a radiative efficiency, $\epsilon_{\text{rad}} = L_E/\dot{M}c^2$, for converting accreted mass into radiation, would grow exponentially in mass as a function of time t, $M_{\text{BH}} = M_{\text{seed}} \exp\{t/t_E\}$ from its initial seed mass M_{seed}, on a timescale, $t_E = 4.1 \times 10^7 \, \text{yr} \, (\epsilon_{\text{rad}}/0.1)$. Thus, the required growth time in units of the Hubble time $t_{\text{Hubble}} = 10^9 \, \text{yr}[(1+z)/7]^{-3/2}$ is

$$\frac{t_{\text{growth}}}{t_{\text{Hubble}}} = 0.7 \left(\frac{\epsilon_{\text{rad}}}{10\%}\right) \left(\frac{1+z}{7}\right)^{3/2} \ln\left(\frac{M_{\text{BH}}/10^9 \, M_\odot}{M_{\text{seed}}/100 \, M_\odot}\right). \qquad (3.29)$$

The age of the Universe at $z \sim 6$ provides just sufficient time to grow a BH with $M_{\text{BH}} \sim 10^9 \, M_\odot$ out of a stellar mass seed with $\epsilon_{\text{rad}} = 10\%$. The growth time is shorter for smaller radiative efficiencies or a higher seed mass.

3.2.6 *The epoch of reionization*

Given the understanding described above of how many galaxies formed at various times, the course of reionization can be determined Universe-wide by counting photons from all sources of light. Both stars and black holes contribute ionizing photons, but the early universe is dominated by small galaxies which in the local universe have central black holes that are disproportionately small, and indeed quasars are rare above redshift 6. Thus, stars most likely dominated the production of ionizing UV photons during the reionization epoch (although high-redshift galaxies should have also emitted X-rays from accreting black holes and accelerated particles in collisionless shocks). Since most stellar ionizing photons are only slightly more energetic than the 13.6 eV ionization threshold of hydrogen, they are absorbed efficiently once they reach a region with substantial neutral hydrogen. This makes the IGM during reionization a two-phase medium characterized by highly ionized regions separated from neutral regions by sharp ionization fronts.

We can obtain a first estimate of the requirements of reionization by demanding one stellar ionizing photon for each hydrogen atom in the IGM. If we conservatively assume that stars within the early galaxies were similar to those observed locally, then each star produced $\sim 4\,000$ ionizing photons per baryon. Star formation is observed today to be an inefficient process, but even if stars in galaxies formed out of only $\sim 10\%$ of the available gas, it would still be sufficient to accumulate a small fraction (of order 0.1%) of the total baryonic mass in the universe into galaxies in order to ionize the entire IGM. More accurate estimates of the actual required fraction account for the formation of some primordial stars (which were massive, efficient ionizers, as discussed above), and for recombinations of hydrogen atoms at high redshifts and in dense regions.

From studies of quasar absorption lines at $z \sim 6$ we know that the IGM is highly ionized a billion years after the big bang. There are hints, however, that some large neutral hydrogen regions persist at these early times and so this suggests that we may not need to go to much higher redshifts to begin to see the epoch of reionization. We now know that the universe could not have fully reionized earlier than an age of 300 million years, since WMAP observed the effect of the freshly created plasma at reionization on the large-scale polarization anisotropies of the CMB and this limits the reionization redshift; an earlier reionization, when the universe was denser, would have created a stronger scattering signature that would be inconsistent with the WMAP observations. In any case, the redshift at which reionization ended only constrains the overall cosmic efficiency of ionizing photon production. In comparison, a detailed picture of reionization, as it happens, will teach us a great deal about the population of young galaxies that produced this cosmic phase transition. A key point is that the spatial distribution of ionized bubbles is determined by clustered groups of galaxies and not by individual galaxies. At such early times galaxies were strongly clustered even on very large scales (up to tens of Mpc), and these scales therefore dominate the structure of reionization. The basic idea is simple. At high redshift, galactic halos are rare and correspond to rare, high density peaks. As an analogy, imagine searching on Earth for mountain peaks above 5000 m. The 200 such peaks are not at all distributed uniformly but instead are found in a few distinct clusters on top of large mountain ranges. Given the large-scale boost provided by a mountain range, a small-scale crest need only provide a small additional rise in order to become a 5000 m peak. The same crest, if it formed within a valley, would not come anywhere near 5000 m in total height. Similarly, in order to find the early galaxies, one must first locate a region with a large-scale density enhancement, and then galaxies will be found there in abundance.

The ionizing radiation emitted from the stars in each galaxy initially produces an isolated ionized bubble. However, in a region dense with galaxies the bubbles quickly overlap into one large bubble, completing reionization in this region while the rest of the universe is still mostly neutral. Most importantly, since the abundance of rare density peaks is very sensitive to small changes in the density threshold, even a large-scale region with a small enhanced density (say, 10% above the mean density of the universe) can have a much larger concentration of galaxies than in other regions (e.g. a 50% enhancement). On the other hand, reionization is harder to achieve in dense regions, since the protons and electrons collide and recombine more often in such regions, and newly-formed hydrogen atoms need to be reionized again by additional ionizing photons. However, the overdense regions end up reionizing first since the number of ionizing sources in these regions is increased so strongly. The large-scale topology of reionization is therefore inside-out, with underdense voids reionizing only at the very end of reionization, with the help of extra ionizing photons coming in from their surroundings (which have a higher density of galaxies than the voids themselves). This is a key prediction awaiting observational testing.

Detailed analytical models that account for large-scale variations in the abundance of galaxies confirm that the typical bubble size starts well below an Mpc early in reionization, as expected for an individual galaxy, rises to 5–10 Mpc during the central phase (i.e. when the universe is half ionized), and then by another factor of ~ 5 toward the end of reionization. These scales are given in comoving units that scale with the expansion of the universe, so that the actual sizes at a redshift z were smaller than these numbers by a factor of $(1+z)$. Numerical simulations have only recently begun to reach the enormous scales needed to capture this evolution. Accounting precisely for gravitational evolution on a wide range of scales but still crudely for gas dynamics, star formation, and the radiative transfer of ionizing photons, the simulations confirm that the large-scale topology of reionization is inside-out, and that this topology can be used to study the abundance and clustering of the ionizing sources.

The characteristic observable size of the ionized bubbles at the end of reionization can be calculated based on simple considerations that only depend on the power-spectrum of density fluctuations and the redshift. As the size of an ionized bubble increases, the time it takes for a 21-cm photon emitted by hydrogen to traverse gets longer. At the same time, the variation in the time at which different regions reionize becomes smaller as the regions grow larger. Thus, there is a maximum size above which the photon crossing time is longer than the cosmic variance in ionization time. Regions bigger than this size will be ionized at their near side by the time a 21-cm photon will cross them toward the observer from their far side. They would appear to the observer as one-sided, and hence signal the end of reionization. These considerations imply a characteristic size for the ionized bubbles of ~ 10 physical Mpc at $z \sim 6$ (equivalent to 70 Mpc today). This result implies that future radio experiments should be tuned to a characteristic angular scale of tens of arcminutes for an optimal detection of 21-cm brightness fluctuations near the end of reionization (see Sec. 3.3.2).

Existing data on the polarization anisotropies of the CMB as well as the Lyman-α forest can be used to derive a probability distribution for the

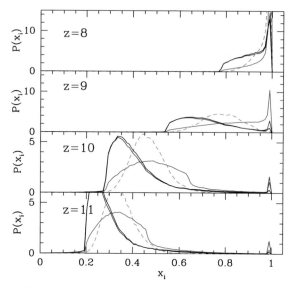

Fig. 3.8 Probability distribution of x_i at redshifts $z = 8$, 9, 10, and 11, based on existing data on the CMB polarization anisotropies and the Lyman-α forest. The different lines correspond to different combinations of data sets. **Figure credit:** J. R. Pritchard, A. Loeb and S. Wyithe, *Mon. Not. R. Astron. Soc.*, in press (2010); http://arxiv.org/abs/0908.3891.

hydrogen ionization fraction (x_i) as a function of redshift. Figure 3.8 shows this likelihood distribution in four redshift bins of interest to upcoming observations. Although there is considerable uncertainty in x_i at each redshift, it is evident from existing data that hydrogen is highly ionized by $z = 8$ (at least to $x_i > 0.8$).

To produce one ionizing photon per baryon requires a minimum comoving density of Milky Way (so-called Population II) stars of,

$$\rho_\star \approx 1.7 \times 10^6 f_{\rm esc}^{-1} \, M_\odot \, {\rm Mpc}^{-3}, \qquad (3.30)$$

or equivalently, a cosmological density parameter in stars of $\Omega_\star \sim 1.25 \times 10^{-5} f_{\rm esc}^{-1}$. More typically, the threshold for reionization involves at least a few ionizing photons per proton (with the right-hand side being $\sim 10^{-6}\,{\rm cm}^{-3}$), since the recombination time at the mean density is comparable to the age of the Universe at $z \sim 10$.

For the local mass function of (Population II) stars at solar metallicity, the star formation rate per unit comoving volume that is required for balancing recombinations in an already ionized IGM, is given by

$$\dot{\rho}_\star \approx 2 \times 10^{-3} f_{\rm esc}^{-1} C \left(\frac{1+z}{10}\right)^3 M_\odot \, {\rm yr}^{-1} \, {\rm Mpc}^{-3}, \qquad (3.31)$$

where $C = \langle n_e^2 \rangle / \langle n_H \rangle^2$ is the volume-averaged clumpiness factor of the electron density up to some threshold overdensity of gas which remains neutral. Current state-of-the-art surveys (HST WFC3/IR) are only sensitive to the bright end of the luminosity function of galaxies at $z > 6$ and hence provide a lower limit on the production rate of ionizing photons during reionization.

3.2.7 *Post-reionization suppression of low-mass galaxies*

After the ionized bubbles overlapped in each region, the ionizing background increased sharply, and the IGM was heated by the ionizing radiation to a temperature $T_{\rm IGM} > 10^4$ K. Due to the substantial increase in the IGM pressure, the smallest mass scale into which the cosmic gas could fragment, the so-called Jeans mass, increased dramatically, changing the minimum mass of forming galaxies.

Gas in-fall depends sensitively on the Jeans mass. When a halo more massive than the Jeans mass begins to form, the gravity of its dark matter overcomes the gas pressure. Even in halos below the Jeans mass, although the gas is initially held up by pressure, once the dark matter collapses its increased gravity pulls in some gas. Thus, the Jeans mass is generally higher than the actual limiting mass for accretion. Before reionization, the IGM is cold and neutral, and the Jeans mass plays a secondary role in limiting galaxy formation compared to cooling. After reionization, the Jeans mass is increased by several orders of magnitude due to the photoionization heating of the IGM, and hence begins to play a dominant role in limiting the formation of stars. Gas in-fall in a reionized and heated Universe has been investigated in a number of numerical simulations. Three-dimensional numerical simulations found a significant suppression of gas in-fall in even larger halos ($V_c \sim 75 \, {\rm km \, s}^{-1}$), but this was mostly due to a suppression of late in-fall at $z < 2$.

When a volume of the IGM is ionized by stars, the gas is heated to a temperature $T_{\rm IGM} \sim 10^4$ K. If quasars dominate the UV background at reionization, their harder photon spectrum leads to $T_{\text{IGM}} > 2 \times 10^4$ K. Including the effects of dark matter, a given temperature results in a linear Jeans mass corresponding to a halo circular velocity of

$$V_J \approx 80 \left(\frac{T_{\text{IGM}}}{1.5 \times 10^4 \, \text{K}} \right)^{1/2} \, \text{km s}^{-1}. \tag{3.32}$$

In halos with a circular velocity well above V_J, the gas fraction in in-falling gas equals the universal mean of Ω_b/Ω_m, but gas in-fall is suppressed in smaller halos. A simple estimate of the limiting circular velocity, below which halos have essentially no gas in-fall, is obtained by substituting the virial overdensity for the mean density in the definition of the Jeans mass. The resulting estimate is

$$V_{\text{lim}} = 34 \left(\frac{T_{\text{IGM}}}{1.5 \times 10^4 \, \text{K}} \right)^{1/2} \, \text{km s}^{-1}. \tag{3.33}$$

This value is in rough agreement with the numerical simulations mentioned before.

Although the Jeans mass is closely related to the rate of gas in-fall at a given time, it does not directly yield the total gas residing in halos at a given time. The latter quantity depends on the entire history of gas accretion onto halos, as well as on the merger histories of halos, and an accurate description must involve a time-averaged Jeans mass. The gas content of halos in simulations is well-fit by an expression which depends on the filtering mass, a particular time-averaged Jeans mass.

The reionization process was not perfectly synchronized throughout the Universe. Large-scale regions with a higher density than the mean tended to form galaxies first and reionized earlier than underdense regions. The suppression of low-mass galaxies by reionization is therefore modulated by the fluctuations in the timing of reionization. Inhomogeneous reionization imprint a signature on the power-spectrum of low-mass galaxies. Future high-redshift galaxy surveys hoping to constrain inflationary parameters must properly model the effects of reionization; conversely, they will also place new constraints on the thermal history of the IGM during reionization.

3.3 Probing the Diffuse Intergalactic Hydrogen

3.3.1 *Lyman-alpha absorption*

Resonant Lyman-α absorption has thus far proved to be the best probe of the state of the IGM. The optical depth to absorption by a uniform intergalactic medium is

$$\tau_s = \frac{\pi e^2 f_\alpha \lambda_\alpha n_{\text{H-I}}(z)}{m_e c H(z)} \quad (3.34)$$

$$\approx 6.45 \times 10^5 x_{\text{H-I}} \left(\frac{\Omega_b h}{0.0315}\right) \left(\frac{\Omega_m}{0.3}\right)^{-1/2} \left(\frac{1+z}{10}\right)^{3/2},$$

where $H \approx 100\, h\, \text{km}\,\text{s}^{-1}\,\text{Mpc}^{-1}\, \Omega_m^{1/2}(1+z)^{3/2}$ is the Hubble parameter at redshift z; $f_\alpha = 0.4162$ and $\lambda_\alpha = 1216\,\text{Å}$ are the oscillator strength and the wavelength of the Lyman-α transition; $n_{\text{H-I}}(z)$ is the neutral hydrogen density at z (assuming primordial abundances); Ω_m and Ω_b are the present-day density parameters of all matter and of baryons, respectively; and $x_{\text{H-I}}$ is the average fraction of neutral hydrogen. In the second equality we have implicitly considered high redshifts.

Lyman-α absorption is thus highly sensitive to the presence of even trace amounts of neutral hydrogen. The lack of full absorption in quasar spectra then implies that the IGM has been very highly ionized during much of the history of the Universe, from at most a billion years after the big bang to the present time. At redshifts approaching 6, however, the optical depth increases, and the observed absorption becomes very strong. The difference between the unabsorbed expectation and the actual observed spectrum can be used to measure the amount of absorption, and thus to infer the atomic hydrogen density.

Several quasars beyond $z \sim 6.1$ show in their spectra a strong (so-called "Gunn–Peterson") trough, a blank spectral region at wavelengths shorter than Ly-α at the quasar redshift (Fig. 3.9). The detection of Gunn–Peterson troughs indicates a rapid change in the neutral content of the IGM at $z \sim 6$, and hence a rapid change in the intensity of the background ionizing flux. However, even a small atomic hydrogen fraction of $\sim 10^{-3}$ would still produce nearly complete Ly-α absorption.

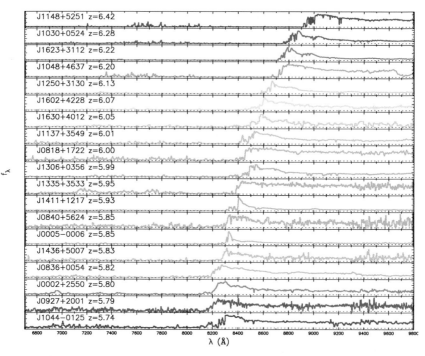

Fig. 3.9 Spectra of 19 quasars with redshifts $5.74 < z < 6.42$ from the *Sloan Digital Sky Survey*. For some of the highest-redshift quasars, the spectrum shows no transmitted flux shortward of the Lyman-α wavelength at the quasar redshift (the so-called "Gunn–Peterson trough"), indicating a non-negligible neutral fraction in the IGM. **Figure credit:** X. Fan *et al.*, *Astron. J.* **125**, 1649 (2005).

While only resonant Ly-α absorption is important at moderate redshifts, the damping wing of the Ly-α line plays a significant role when neutral fractions of order unity are considered at $z > 6$. The scattering cross-section of the Ly-α resonance line by neutral hydrogen is given by

$$\sigma_\alpha(\nu) = \frac{3\lambda_\alpha^2 \Lambda_\alpha^2}{8\pi} \frac{(\nu/\nu_\alpha)^4}{4\pi^2(\nu - \nu_\alpha)^2 + (\Lambda_\alpha^2/4)(\nu/\nu_\alpha)^6}, \qquad (3.35)$$

where $\Lambda_\alpha = (8\pi^2 e^2 f_\alpha / 3 m_e c \lambda_\alpha^2) = 6.25 \times 10^8 \, \text{s}^{-1}$ is the Ly-α ($2p \to 1s$) decay rate, $f_\alpha = 0.4162$ is the oscillator strength, and $\lambda_\alpha = 1216 \, \text{Å}$ and $\nu_\alpha = (c/\lambda_\alpha) = 2.47 \times 10^{15} \, \text{Hz}$ are the wavelength and frequency of the Ly-α line. The term in the numerator is responsible for the classical Rayleigh scattering.

Although reionization is an inhomogeneous process, we consider here a simple illustrative case of instantaneous reionization. Consider a source at a redshift z_s beyond the redshift of reionization, z_{reion}, and the corresponding scattering optical depth of a uniform, neutral IGM of hydrogen density $n_{\text{H},0}(1+z)^3$ between the source and the reionization redshift. The optical depth is a function of the observed wavelength λ_{obs},

$$\tau(\lambda_{\text{obs}}) = \int_{z_{\text{reion}}}^{z_s} dz \, \frac{cdt}{dz} \, n_{\text{H},0}(1+z)^3 \sigma_\alpha[\nu_{\text{obs}}(1+z)], \quad (3.36)$$

where $\nu_{\text{obs}} = c/\lambda_{\text{obs}}$ and for a flat Universe with $(\Omega_m + \Omega_\Lambda) = 1$,

$$\frac{dt}{dz} = [(1+z)H(z)]^{-1} = H_0^{-1} \times [\Omega_m(1+z)^5 + \Omega_\Lambda(1+z)^2]^{-1/2}. \quad (3.37)$$

At wavelengths longer than Ly-α at the source, the optical depth obtains a small value; these photons redshift away from the line center along its red wing and never resonate with the line core on their way to the observer. Considering only the regime in which $|\nu - \nu_\alpha| \gg \Lambda_\alpha$, we may ignore the second term in the denominator of Eq. (3.35). This leads to an analytical result for the red damping wing of the Gunn–Peterson trough,

$$\tau(\lambda_{\text{obs}}) = \tau_s \left(\frac{\Lambda}{4\pi^2 \nu_\alpha}\right) \tilde{\lambda}_{\text{obs}}^{3/2}[I(\tilde{\lambda}_{\text{obs}}^{-1}) - I([(1+z_{\text{reion}})/(1+z_s)]\tilde{\lambda}_{\text{obs}}^{-1})], \quad (3.38)$$

an expression valid for $\tilde{\lambda}_{\text{obs}} \geq 1$, where τ_s is given in Eq. (3.34), and we also define

$$\tilde{\lambda}_{\text{obs}} \equiv \frac{\lambda_{\text{obs}}}{(1+z_s)\lambda_\alpha} \quad (3.39)$$

and

$$I(x) \equiv \frac{x^{9/2}}{1-x} + \frac{9}{7}x^{7/2} + \frac{9}{5}x^{5/2} + 3x^{3/2} + 9x^{1/2} - \frac{9}{2}\ln\left[\frac{1+x^{1/2}}{1-x^{1/2}}\right]. \quad (3.40)$$

3.3.2 21-cm absorption or emission

3.3.2.1 The spin temperature of the 21-cm transition of hydrogen

The ground state of hydrogen exhibits hyperfine splitting owing to the possibility of two relative alignments of the spins of the proton and the electron. The state with parallel spins (the triplet state) has a slightly higher energy than the state with anti-parallel spins (the singlet state). The 21-cm line associated with the spin-flip transition from the triplet to the singlet state is often used to detect neutral hydrogen in the local universe. At high redshift, the occurrence of a neutral pre-reionization IGM offers the prospect of detecting the first sources of radiation and probing the reionization era by mapping the 21-cm emission from neutral regions. While its energy density is estimated to be only a 1% correction to that of the CMB, the redshifted 21-cm emission should display angular structure as well as frequency structure due to inhomogeneities in the gas density field, hydrogen ionized fraction, and spin temperature. Indeed, a full mapping of the distribution of H-I as a function of redshift is possible in principle.

The basic physics of the hydrogen spin transition is determined as follows. The ground-state hyperfine levels of hydrogen tend to thermalize with the CMB background, making the IGM unobservable. If other processes shift the hyperfine level populations away from thermal equilibrium, then the gas becomes observable against the CMB in emission or in absorption. The relative occupancy of the spin levels is usually described in terms of the hydrogen spin temperature T_S, defined by

$$\frac{n_1}{n_0} = 3\exp\left\{-\frac{T_*}{T_S}\right\}, \qquad (3.41)$$

where n_0 and n_1 refer respectively to the singlet and triplet hyperfine levels in the atomic ground state ($n=1$), and $T_* = 0.068\,\text{K}$ is defined by $k_B T_* = E_{21}$, where the energy of the 21-cm transition is $E_{21} = 5.9 \times 10^{-6}\,\text{eV}$, corresponding to a frequency of 1420 MHz. In the presence of the CMB alone, the spin states reach thermal equilibrium with $T_S = T_{\text{CMB}} = 2.725(1+z)\,\text{K}$ on a timescale of $T_*/(T_{\text{CMB}} A_{10}) \simeq 3 \times 10^5 (1+z)^{-1}\,\text{yr}$, where $A_{10} = 2.87 \times 10^{-15}\,\text{s}^{-1}$ is the spontaneous decay rate of the hyperfine transition. This timescale is much shorter than the age of the Universe at all redshifts after cosmological recombination.

The IGM is observable when the kinetic temperature T_k of the gas differs from T_{CMB} and an effective mechanism couples T_S to T_k. Collisional de-excitation of the triplet level dominates at very high redshift, when the gas density (and thus the collision rate) is still high, but once a significant galaxy population forms in the universe, the spin temperature is affected also by an indirect mechanism that acts through the scattering of Lyman-α photons. Continuum UV photons produced by early radiation sources redshift by the Hubble expansion into the local Lyman-α line at a lower redshift. These photons mix the spin states via the Wouthuysen–Field process whereby an atom initially in the $n = 1$ state absorbs a Lyman-α photon, and the spontaneous decay which returns it from $n = 2$ to $n = 1$ can result in a different final spin state from the initial one. Since the neutral IGM is highly opaque to resonant scattering, and the Lyman-α photons receive Doppler kicks in each scattering, the shape of the radiation spectrum near Lyman-α is determined by T_k, and the resulting spin temperature (assuming $T_S \gg T_*$) is then a weighted average of T_k and $T_{\rm CMB}$:

$$T_S = \frac{T_{\rm CMB} T_k (1 + x_{\rm tot})}{T_k + T_{\rm CMB} x_{\rm tot}}, \qquad (3.42)$$

where $x_{\rm tot} = x_\alpha + x_c$ is the sum of the radiative and collisional threshold parameters. These parameters are

$$x_\alpha = \frac{P_{10} T_\star}{A_{10} T_{\rm CMB}}, \qquad (3.43)$$

and

$$x_c = \frac{4\kappa_{1-0}(T_k) n_H T_\star}{3 A_{10} T_{\rm CMB}}, \qquad (3.44)$$

where P_{10} is the indirect de-excitation rate of the triplet $n = 1$ state via the Wouthuysen–Field process, related to the total scattering rate P_α of Lyman-α photons by $P_{10} = 4 P_\alpha / 27$. Also, the atomic coefficient $\kappa_{1-0}(T_k)$ is tabulated as a function of T_k. The coupling of the spin temperature to the gas temperature becomes substantial when $x_{\rm tot} > 1$; in particular, $x_\alpha = 1$ defines the thermalization rate of P_α:

$$P_{\rm th} \equiv \frac{27 A_{10} T_{\rm CMB}}{4 T_*} \simeq 7.6 \times 10^{-12} \left(\frac{1+z}{10}\right) \, {\rm s}^{-1}. \qquad (3.45)$$

A patch of neutral hydrogen at the mean density and with a uniform T_S produces (after correcting for stimulated emission) an optical depth at a present-day (observed) wavelength of $21(1+z)$ cm,

$$\tau(z) = 9.0 \times 10^{-3} \left(\frac{T_{\text{CMB}}}{T_S}\right) \left(\frac{\Omega_b h}{0.03}\right) \left(\frac{\Omega_m}{0.3}\right)^{-1/2} \left(\frac{1+z}{10}\right)^{1/2}, \quad (3.46)$$

assuming a high redshift $z \gg 1$. The observed spectral intensity I_ν relative to the CMB at a frequency ν is measured by radio astronomers as an effective brightness temperature T_b of blackbody emission at this frequency, defined using the Rayleigh–Jeans limit of the Planck radiation formula: $I_\nu \equiv 2k_B T_b \nu^2/c^2$.

The brightness temperature through the IGM is $T_b = T_{\text{CMB}} e^{-\tau} + T_S(1-e^{-\tau})$, so the observed differential antenna temperature of this region relative to the CMB is

$$T_b = (1+z)^{-1}(T_S - T_{\text{CMB}})(1-e^{-\tau})$$

$$\simeq 28 \,\text{mK} \left(\frac{\Omega_b h}{0.033}\right) \left(\frac{\Omega_m}{0.27}\right)^{-1/2} \left(\frac{1+z}{10}\right)^{1/2} \left(\frac{T_S - T_{\text{CMB}}}{T_S}\right), \quad (3.47)$$

where $\tau \ll 1$ is assumed and T_b has been redshifted to redshift zero. Note that the combination that appears in T_b is

$$\frac{T_S - T_{\text{CMB}}}{T_S} = \frac{x_{\text{tot}}}{1 + x_{\text{tot}}} \left(1 - \frac{T_{\text{CMB}}}{T_k}\right). \quad (3.48)$$

In overdense regions, the observed T_b is proportional to the overdensity, and in partially ionized regions T_b is proportional to the neutral fraction. Also, if $T_S \gg T_{\text{CMB}}$ then the IGM is observed in emission at a level that is independent of T_S. On the other hand, if $T_S \ll T_{\text{CMB}}$ then the IGM is observed in absorption at a level that is enhanced by a factor of T_{CMB}/T_S. As a result, a number of cosmic events are expected to leave observable signatures in the redshifted 21-cm line, as discussed below in further detail.

Figure 3.10 illustrates the mean IGM evolution for three examples in which reionization is completed at different redshifts, namely $z = 6.47$ (thin curves), $z = 9.76$ (medium curves), and $z = 11.76$ (thick curves). The top panel shows the global evolution of the CMB temperature T_{CMB} (dotted curve), the gas kinetic temperature T_k (dashed curve), and the spin temperature T_S (solid curve). The middle panel shows the evolution

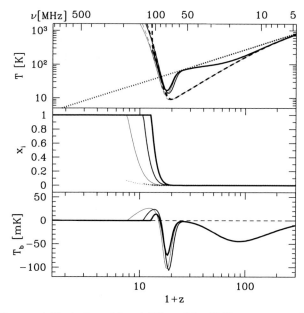

Fig. 3.10 Top panel: Evolution with redshift z of the CMB temperature $T_{\rm CMB}$ (dotted curve), the gas kinetic temperature T_k (dashed curve), and the spin temperature T_S (solid curve). Middle panel: Evolution of the gas fraction in ionized regions x_i (solid curve) and the ionized fraction outside these regions (due to diffuse X-rays) x_e (dotted curve). Bottom panel: Evolution of mean 21 cm brightness temperature T_b. The horizontal axis at the top provides the observed photon frequency at the different redshifts shown at the bottom. Each panel shows curves for three models in which reionization is completed at different redshifts, namely $z = 6.47$ (thin curves), $z = 9.76$ (medium curves), and $z = 11.76$ (thick curves). **Figure credit:** J. Pritchard and A. Loeb, *Phys. Rev. D* **78**, 3511 (2008).

of the ionized gas fraction and the bottom panel presents the mean 21 cm brightness temperature, T_b.

3.3.2.2 *A handy tool for studying cosmic reionization*

The prospect of studying reionization by mapping the distribution of atomic hydrogen across the universe using its prominent 21-cm spectral line has motivated several teams to design and construct arrays of low-frequency radio telescopes; the Low Frequency Array (http://www.lofar.org/), the Murchison Wide-Field Array (*http://www.mwatelescope.org/*), PAPER (*http://arxiv.org/abs/0904.1181*), GMRT (*http://arxiv.org/abs/0807.1056*), 21CMA (*http://21cma.bao.ac.cn/*), and ultimately the Square

Kilometer Array (*http://www.skatelescope.org*) will search over the next decade for 21-cm emission or absorption from $z \sim 6.5-15$, redshifted and observed today at relatively low frequencies which correspond to wavelengths of 1.5 to 4 m.

The idea is to use the resonance associated with the hyperfine splitting in the ground state of hydrogen. While the CMB spectrum peaks at a wavelength of 2 mm, it provides a still-measurable intensity at meter wavelengths that can be used as the bright background source against which we can see the expected 1% absorption by neutral hydrogen along the line of sight. The hydrogen gas produces 21-cm absorption if its spin temperature is colder than the CMB and excess emission if it is hotter. Since the CMB covers the entire sky, a complete three-dimensional map of neutral hydrogen can in principle be made from the sky position of each absorbing gas cloud together with its redshift z. Different observed wavelengths slice the Universe at different redshifts, and ionized regions are expected to appear as cavities in the hydrogen distribution, similar to holes in Swiss cheese. Because the smallest angular size resolvable by a telescope is proportional to the observed wavelength, radio astronomy at wavelengths as large as a meter has remained relatively undeveloped. Producing resolved images even of large sources such as cosmological ionized bubbles requires telescopes which have a kilometer scale. It is much more cost-effective to use a large array of thousands of simple antennas distributed over several kilometers, and to use computers to cross-correlate the measurements of the individual antennas and combine them effectively into a single large telescope. The new experiments are being placed mostly in remote sites, because the cosmic wavelength region overlaps with more mundane terrestrial telecommunications.

In approaching redshifted 21-cm observations, although the first inkling might be to consider the mean emission signal in the bottom panel of Fig. 3.10, the signal is orders of magnitude fainter than foreground synchrotron emission from relativistic electrons in the magnetic field of our own Milky Way as well as other galaxies (see Fig. 3.12). Thus cosmologists have focused on the expected characteristic variations in T_b, both with position on the sky and especially with frequency, which signifies redshift for the cosmic signal. The synchrotron foreground is expected to have a smooth frequency spectrum, and so it is possible to isolate the cosmological signal by taking the difference in the sky brightness fluctuations at slightly different frequencies (as long as the frequency separation corresponds to the characteristic size of ionized bubbles). The 21-cm brightness temperature depends on the density of neutral hydrogen. As explained in the previous subsection, large-scale patterns in the reionization are driven by spatial variations in the abundance of galaxies; the 21-cm fluctuations reach $\sim 5\,\mathrm{mK}$ (root mean square) in brightness temperature on a scale of 10 comoving Mpc. While detailed maps will be difficult to extract due to the foreground emission, a statistical detection of these fluctuations is expected to be well within the capabilities of the first-generation experiments now being built. Current work suggests that the key information on the topology and timing of reionization can be extracted statistically.

While numerical simulations of reionization are now reaching the cosmological box sizes needed to predict the large-scale topology of the ionized bubbles, they do this at the price of limited small-scale resolution (see Fig. 3.11). These simulations cannot yet follow in any detail the

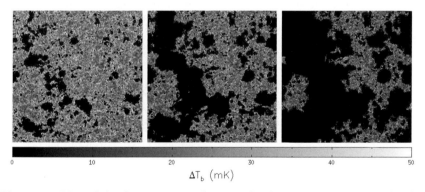

Fig. 3.11 Map of the fluctuations in the 21 cm brightness temperature on the sky, ΔT_b (mK), based on a numerical simulation which follows from the dynamics of dark matter and gas in the IGM as well as the radiative transfer of ionizing photons from galaxies. The panels show the evolution of the signal in a slice of 140 comoving Mpc on a side, in three snapshots corresponding to the simulated volume being 25, 50, and 75% ionized. Since neutral regions correspond to strong emission (i.e. a high T_b), the 21-cm maps illustrate the global progress of reionization and the substantial large-scale spatial fluctuations in the reionization history. **Figure credit:** H. Trac, R. Cen and A. Loeb, *Astrophys. J.* **689**, L81 (2009).

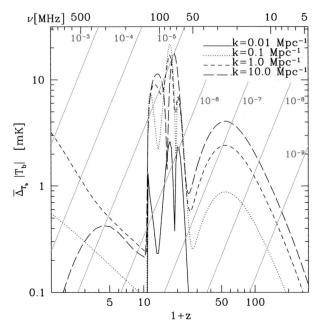

Fig. 3.12 Predicted redshift evolution of the angle-averaged amplitude of the 21-cm power spectrum ($|\bar{\Delta}_{T_b}| = [k^3 P_{21-\rm cm}(k)/2\pi^2]^{1/2}$) at comoving wavenumbers $k = 0.01$ (solid curve), 0.1 (dotted curve), 1.0 (short dashed curve), 10.0 (long dashed curve), and 100.0 Mpc^{-1} (dot-dashed curve). In the model shown, reionization is completed at $z = 9.76$. The horizontal axis at the top shows the observed photon frequency at the different redshifts. The diagonal straight (red) lines show various factors of suppression for the synchrotron Galactic foreground, necessary to reveal the 21-cm signal. **Figure credit:** J. R. Pritchard and A. Loeb, *Phys. Rev. D* **78**, 3511 (2008).

formation of individual stars within galaxies, or the feedback that stars produce on the surrounding gas, such as photo-heating or the hydrodynamic and chemical impact of supernovae, which blow hot bubbles of gas enriched with the chemical products of stellar nucleosynthesis. Thus, the simulations cannot directly predict whether the stars that form during reionization are similar to the stars in the Milky Way and nearby galaxies or to the primordial 100 $M_\odot$ stars. They also cannot determine whether feedback prevents low-mass dark matter halos from forming stars. Thus, models are needed, that make it possible to vary all these astrophysical parameters of the ionizing sources and to study the effect on the 21-cm observations.

The theoretical expectations presented here for reionization and for the 21-cm signal are based on rather large extrapolations from observed galaxies to deduce the properties of much smaller galaxies that formed at an earlier cosmic epoch. Considerable surprises are thus possible, such as an early population of quasars or even unstable exotic particles that emitted ionizing radiation as they decayed. In any case, the forthcoming observational data in 21-cm cosmology should make the next few years a very exciting time.

At high redshifts prior to reionization, spatial perturbations in the thermodynamic gas properties are linear and can be predicted precisely (see Sec. 3.2.1). Thus, if the gas is probed with the 21-cm technique then it becomes a promising tool of fundamental, precision cosmology, able to probe the primordial power spectrum of density fluctuations imprinted in the very early Universe, perhaps in an era of cosmic inflation. The 21-cm fluctuations can be measured down to the smallest scales where the baryon pressure suppresses gas fluctuations, while the CMB anisotropies are damped on small scales (through the so-called Silk damping). This difference in damping scales can be seen by comparing the baryon-density and photon-temperature power spectra. Since the 21-cm technique is also three-dimensional (while the CMB yields a single sky map), there is a much large potential number of independent modes probed by the 21-cm signal: $N_{21-cm} \sim 3 \times 10^{16}$ compared to $N_{cmb} \sim 2 \times 10^7$. This larger number should provide a measure of non-Gaussian deviations to a level of $\sim N_{21\,cm}^{-1/2}$, constituting a test of the inflationary origin of the primordial inhomogeneities which are expected to possess non-Gaussian deviations $> 10^{-6}$.

The 21-cm fluctuations are expected to simply trace the primordial power-spectrum of matter density perturbations (which is shaped by the initial conditions from inflation and the dark matter) either before the first population of galaxies had formed (at redshifts $z > 25$) or after reionization ($z < 6$) — when only dense pockets of self-shielded hydrogen (such as damped Lyman-α systems) survive. During the epoch of reionization, the fluctuations are mainly shaped by the topology of ionized regions, and thus depend on uncertain astrophysical details involving star formation.

However, even during this epoch, the imprint of peculiar velocities (which are induced gravitationally by density fluctuations) can in principle be used to separate the implications for fundamental physics from the astrophysics.

Peculiar velocities imprint a particular form of anisotropy in the 21-cm fluctuations that is caused by gas motions along the line of sight. This anisotropy, expected in any measurement of density that is based on a spectral resonance or on redshift measurements, results from velocity compression. Consider a photon traveling along the line of sight that resonates with absorbing atoms at a particular point. In a uniform, expanding universe, the absorption optical depth encountered by this photon probes only a narrow strip of atoms, since the expansion of the universe makes all other atoms move with a relative velocity that takes them outside the narrow frequency width of the resonance line. If there is a density peak, however, near the resonating position, the increased gravity will reduce the expansion velocities around this point and bring more gas into the resonating velocity width. This effect is sensitive only to the line-of-sight component of the velocity gradient of the gas, and thus causes an observed anisotropy in the power spectrum even when all physical causes of the fluctuations are statistically isotropic. This anisotropy is particularly important in the case of 21-cm fluctuations. When all fluctuations are linear, the 21-cm power spectrum takes the form

$$P_{21\text{-cm}}(\mathbf{k}) = \mu^4 P_\rho(k) + 2\mu^2 P_{\rho\text{-iso}}(k) + P_{\text{iso}}, \qquad (3.49)$$

where $\mu = \cos\theta$ in terms of the angle θ between the wave vector $\mathbf{k}$ of a given Fourier mode and the line of sight, P_{iso} is the isotropic power spectrum that would result from all sources of 21-cm fluctuations without velocity compression, $P_\rho(k)$ is the 21-cm power spectrum from gas density fluctuations alone, and $P_{\rho\text{-iso}}(k)$ is the Fourier transform of the cross-correlation between the density and all sources of 21-cm fluctuations. The three power spectra can also be denoted $P_{\mu^4}(k)$, $P_{\mu^2}(k)$, and $P_{\mu^0}(k)$, according to the power of μ that multiplies each term. At these redshifts, the 21-cm fluctuations probe the in-fall of the baryons into the dark matter potential wells. The power spectrum shows remnants of the photon-baryon acoustic oscillations on large scales, and of the baryon pressure suppression on small scales.

Once stellar radiation becomes significant, many processes can contribute to the 21-cm fluctuations. The contributions include fluctuations in gas density, temperature, ionized fraction, and Ly-α flux. These processes can be divided into two broad categories: The first, related to *"physics"*, consists of probes of fundamental, precision cosmology; and the second, related to *"astrophysics"*, consists of probes of stars. Both categories are interesting — the first for precision measures of cosmological parameters and studies of processes in the early universe, and the second for studies of the properties of the first galaxies. However, the astrophysics depends on complex nonlinear processes (collapse of dark matter halos, star formation, supernova feedback), and must be cleanly separated from the physics contribution, in order to allow precision measurements of the latter. As long as all the fluctuations are linear, the anisotropy noted above allows precisely this separation of the *fundamental physics* from the *astrophysics* of the 21-cm fluctuations. In particular, the $P_{\mu^4}(k)$ is independent of the effects of stellar radiation, and is a clean probe of the gas density fluctuations. Once nonlinear terms become important, there arises a significant mixing of the different terms; in particular, this occurs on the scale of the ionizing bubbles during reionization.

The 21-cm fluctuations are affected by fluctuations in the Lyman-α flux from stars, a result that yields an indirect method to detect and study the early population of galaxies at $z \sim 20$. The fluctuations are caused by biased inhomogeneities in the density of galaxies, along with Poisson fluctuations in the number of galaxies. Observing the power-spectra of these two sources would probe the number density of the earliest galaxies and the typical mass of their host dark matter halos. Furthermore, the enhanced amplitude of the 21-cm fluctuations from the era of Ly-α coupling improves considerably the practical prospects for their detection. Precise predictions account for the detailed properties of all possible cascades of a hydrogen atom after it absorbs a photon. Around the same time, X-rays may also start to heat the cosmic gas, producing strong 21-cm fluctuations due to fluctuations in the X-ray flux.

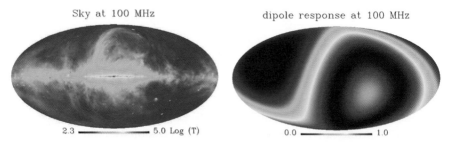

Fig. 3.13 Left panel: Radio map of the sky at 100 MHz. Right panel: Ideal dipole response averaged over 24 hours. **Figure credits:** J. Pritchard and A. Loeb, *Phys. Rev. D*, in press (2010); A. de Oliveira-Costa *et al.*, *Mon. Not. R. Astron. Soc.* **388**, 247 (2008).

Different from interferometric arrays, single dipole experiments which integrate over most of the sky can search for the global (spectral) 21-cm signal shown in Fig. 3.10. Examples of such experiments are CoRE or EDGES (*http://www.haystack.mit.edu/ast/arrays/Edges/*). Rapid reionization histories which span a redshift range $\Delta z < 2$ can be constrained, provided that local foregrounds (see Fig. 3.13) can be well modeled by low-order polynomials in frequency. Observations in the frequency range 50–100 MHz can potentially constrain the Lyman-α and X-ray emissivity of the first stars forming at redshifts $z \sim 15-25$, as illustrated in Fig. 3.14.

3.4 Epilog

The initial conditions of our Universe can be summarized on a single sheet of paper. Yet the Universe is full of complex structures today, such as stars, galaxies and groups of galaxies. This chapter discussed the standard theoretical model for how complexity emerged from the simple initial state of the Universe through the action of gravity. In order to test and inform the related theoretical calculations, large-aperture telescopes and arrays of radio antennae are currently being designed and constructed.

The actual transition from simplicity to complexity has not been observed as of yet. The simple initial conditions were already traced in maps of the microwave background radiation, but the challenge of detecting the first generation of galaxies defines one of the exciting frontiers in the future

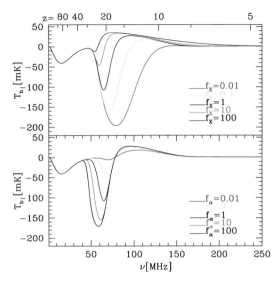

Fig. 3.14 Dependence of global 21-cm signal on the X-ray (top panel) and Lyman-α (bottom panel) emissivity of stars. Each case depicts examples with the characteristic emissivity reduced or increased by a factor of up to 100. **Figure credit:** J. Pritchard and A. Loeb, *Phys. Rev. D*, in press (2010).

of cosmology. Once at hand, the missing images of the infant Universe might potentially surprise us and revise our current ideas.

Acknowledgments

I thank my collaborators on the topics covered by this chapter: Dan Babich, Rennan Barkana, Volker Bromm, Steve Furlanetto, Zoltan Haiman, Joey Munoz, Jonathan Pritchard, Hy Trac, Stuart Wyithe, and Matias Zaldarriaga.

Further Reading

A. Loeb, *How Did the First Stars and Galaxies Form?* (Princeton University Press, 2010).

References

Barkana, R. and Loeb, A. (2004). *Astrophys. J.* **601**, p. 64.
Barkana, R. and Loeb, A. (2007). *Rep. Prog. Phys.* **70**, p. 627.
de Oliveira-Costa, A. *et al.* (2008). *Mon. Not. R. Astron. Soc.* **388**, p. 247.

Fan, X. *et al.* (2005). *Astron. J.* **125**, p. 1649.
Loeb, A. (2010). *How Did the First Stars and Galaxies Form?* Princeton University Press.
Pritchard, J. and Loeb, A. (2008). *Phys. Rev. D* **78**, p. 3511.
Pritchard, J. and Loeb, A. (2010). *Phys. Rev. D*, in press.
Pritchard, J. R., Loeb, A. and Wyithe, S. (2010). *Mon. Not. R. Astron. Soc.*, in press.
Sheth, R. K. and Tormen, G. (2002) *Mon. Not. R. Astron. Soc.* **329**, p. 61.
Trac, H., Cen, R. and Loeb, A. (2009). *Astrophys. J.* **689**, p. L81.

CHAPTER 4

CLUSTERS OF GALAXIES

ELENA PIERPAOLI
Department of Physics and Astronomy
University of Southern California
Los Angeles, CA 90089, USA

4.1 What are Galaxy Clusters? Why are They Interesting?

With a mass of about 10^{14}–10^{15} solar masses (one solar mass being 1.99×10^{30} kg), they are the biggest gravitationally bound objects in our Universe. The vast majority of their mass (about 80%) is believed to be dark matter, which we cannot directly observe, but whose gravitational effect impacts the behavior of the galaxies within the cluster (about 5% in mass, of the order of a few hundred in number) and of the diffuse ionized gas (about 15% in mass). Clusters are particularly interesting to astrophysicists and cosmologists because they formed relatively recently, when the Universe was already more than 4 billions years old. As a consequence, their properties quite closely reflect the ones of initial density fluctuations, so that we *believe* we can interpret them easily. For this reason, clusters have been extensively used to constrain cosmology: their number density is particularly sensitive to the amount of matter in the Universe and the amplitude of initial fluctuations set by mechanisms occuring in the Universe's first instants (e.g. inflation). Because they started to form at a time when the Universe's expansion was likely driven by dark energy, their abundance also gives specific information on the nature of this very mysterious component. In addition, clusters can provide information on the growth of structures during cosmic times, which in turn allows to constrain fundamental theories of particles physics through, for example, the measurement of neutrino masses. Being the biggest objects in the Universe, clusters also allow to verify the law of gravity at very large scales, potentially challenging Einstein's prediction and opening new windows for the discovery of physical laws.

Apart from their relevance in Cosmology, clusters are also very important for understanding and interpreting structure formation at smaller scales, such as galaxies. Since a cluster is a region of large overdensity, galaxies within it tend to interact more frequently, merge, disrupt and give rise to bigger galaxies with different morphological characteristics. These violent events can lead to the production of supermassive black holes and active galaxies where substantial energy is released in the intra-cluster medium as material falls on a central supermassive black hole. Jets are produced which input relativistic particles in the intra-cluster medium, disrupting the balance between gravity and pressure and modifying the way we see clusters. Such violent events can deeply affect other important astrophysical mechanisms, like star formation, and impact galaxy evolution in general.

It is important to realize that, while the dark matter is the dominant component of the clusters, the behavior of the galaxy and the gas needs to be well understood in order to make correct cosmological interpretations. This is because, given a cosmological model, we are able to make accurate predictions on the number of objects with a given mass; however, we are typically not able to observe the mass directly: we need to *guess* it by looking at the components we do see. We then infer the mass from the observed quantities: for example we expect a more massive object to contain more galaxies, and to be more luminous at all observational frequencies.

Correct cosmological interpretations are therefore only possible if we have adequately understood the galaxies and gas characteristics for given mass clusters at various cosmic times.

4.2 Structure Formation

In its very early days, the Universe was very smooth; with just tiny ripples in its energy density. These ripples, which came in different sizes, all followed a similar evolution. In the beginning, when density fluctuations were small, their radius expanded with the Universe itself — just at a slightly different rate. Overdense regions, however, expand more slowly, and eventually stopped growing and collapsed. From this time on, the internal dynamics of these formed objects was little affected from what happened in the outside world. For a given object, the collapse stops when the internal pressure is sufficient to oppose the effects of gravity. Typically, at a given time, there is one size objects that starts forming in appreciable quantity. According to our current understanding, small objects form first, and larger objects form at a later time. Clusters are the latest comers to this game: they started to form relatively recently, at a "redshift" of about 2 (when the Universe was about a third of the current size), and are still forming at the present time. The fact that they formed so recently makes them even more interesting to study. The evolution of bound structures becomes more and more complicated the larger the overdensity involved is. Observations of bound objects that just formed are easier to interpret because these objects did not have much time to change their characteristics (e.g. chemical composition, distribution of matter on small scale, luminosities and colors) with respect to the rest of the Universe. The amount of "baryons" (intended as protons and electrons) with respect to the dark matter and the relative abundances of the heavy elements with respect to hydrogen in clusters are therefore expected to trace primordial values. When structure evolves, merging events, cooling and heating processes, as well as star formation may disrupt this ideal situation. In summary, clusters are so special because they are nearby and easily observable, and yet clearly predictable.

Scientist nowadays make use of sophisticated numerical simulations to understand structure formation. Millions of particles are placed in three-dimensional boxes with a spatial distribution that reflects the statistical characteristics of the inhomogeneities expected in a given cosmological model. This artificial reproduction of the Universe is let evolve

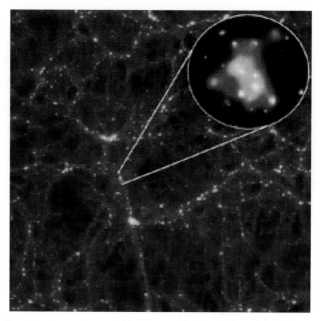

Fig. 4.1 Computer simulation of a large volume of the Universe showing voids and filamentary structures. An XMM-Newton X-ray image of a real galaxy cluster is superimposed to illustrate the formation of galaxy clusters in the densest parts of the Universe. **Figure credit:** Virgo consortium; Chandra cluster gallery.

under the laws of gravity up to the present time. Such simulations show that matter displays itself in a spider-web structure which surrounds large voids (see Fig. 4.1). Clusters are typically located at the crossing points of this spider-web, and they accrete material along the filaments departing from their outskirts. So, while it is largely possible to simply interpret clusters as if they formed from an initial density fluctuation that collapses under the effect of its own gravity, the real picture is more complicated — involving both collapse and accretion. The spider-web structure found in numerical simulations is also observed in nature: with the advent of surveys like the SDSS, extending on a quarter of the sky and observing objects out to half of the Universe's history, it is possible to study galaxies and clusters' distribution (see Fig. 4.2). The spider-web structure is present in both galaxy and cluster maps.

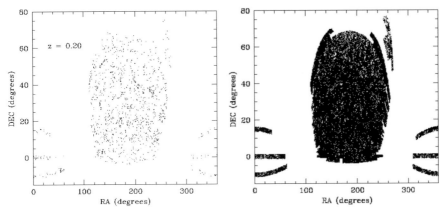

Fig. 4.2 Left panel: distribution of AMF clusters in the SDSS DR6 data. The axes represent right ascension and declination (i.e. location on the sky). Each point represents one cluster. The spider-web structure they reside in, with filaments and voids, is clearly visible (Szabo et al., 2010). Right panel: the whole set of SDSS AMF clusters on the sky. There are about 70 000 of them in 95 000 square degrees.

4.3 How do We Observe Clusters?

Clusters have been first observed with the naked eye. In the 1960s George Abell, at the time a Caltech PhD student, visually inspected images of the Palomar Sky Survey in search for groups of sufficiently bright galaxies placed in a compact configuration. He found about three thousand objects that way. The Abell Catalog is still used nowadays for the sake of comparison. Current optical catalogs (Wen et al., 2009; Szabo et al., 2010; Hao et al., 2010; Koester et al., 2007, Fig. 4.3), constructed on the basis of multi-band observations and by means of sophisticated computer codes, contain about 20 times the number of Abell clusters: it is still quite impressive what Abell could achieve with the naked eye!

Optical observations not only allowed the first detection of galaxy clusters, but also contributed to the understanding that the galaxies were not the only component. By measuring the Doppler effect in spectral lines of clusters' galaxies, scientists could infer the typical galaxies' velocity. At the same time, they could also estimate the total mass of the observed cluster by counting the galaxies and assigning them a reasonable mass, as derived

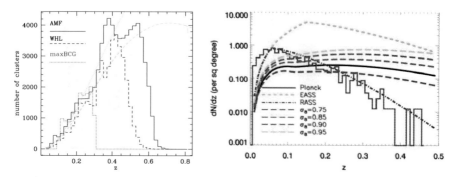

Fig. 4.3 Left: Redshift distribution of optical clusters in different SDSS catalogs: AMF (solid line, Szabo et al., 2010), (dashed-dot line, Wen et al., 2009) maxBCG (dotted line, Koester et al., 2007). The area covered by SDSS is about one quarter of the sky. The green lines represent the expected number of clusters for the amplitude of the matter power spectrum $\sigma_8 = 0.8$ (dashed) and 0.9 (dotted) and a cluster mass threshold of $4 \times 10^{13}\,h^{-1} M_{\text{solar}}$. The value of the reduced Hubble parameter $h = 0.70$. Right panel: the redshift distribution clusters in all-sky X-ray and SZ surveys from the ROSAT satellite (actual data as solid line, model dot-dashed), future SZ observations with Planck (black line as reference, long-dashed refer to different possible matter power spectrum normalizations) and the future X-ray satellite Erosita (green, short dashed).

from local galaxy studies. By comparing the inferred mass with the velocity measured, they soon realized that the former was too small to hold together galaxies with the estimated velocity. Something else needed to be present, even though not yet visible.

With the advent of X-ray astronomy, the puzzle gained an additional element. X-ray observations showed that the space within the galaxies was filled with diffuse gas at very high temperatures (millions of Kelvins, see Fig. 4.4). This was interpreted as *bremsstrahlung* emission of ionized plasma filling the space between clusters. Estimates of its density lead to the conclusion that more mass was in the form of gas than in galaxies; at the same time the sum of these two components was not yet sufficient to explain either galaxies' velocity or the high temperatures of the intra-cluster medium. There was another evidence that something else needed to be present. Since nobody at the time questioned the reliability of Einstein's theory of gravity, these observations were taken as a demonstration of dark matter's existence. It was already evident at those times that dark matter needed to be the most prominent component in the cluster for all galaxies and gas to hold together the way it was observed.

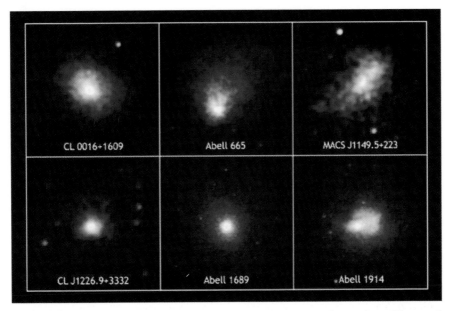

Fig. 4.4 Clusters' gas component, as visible through X-ray observations. The panel presents a set of X-ray clusters with different morphologies as observed by the Chandra satellite. The brightest spot typically indicates the region with the biggest gas density. Small blobs may be in or around the cluster. **Figure credit:** Chandra clusters gallery.

In the 90s, the advent of adaptive optics and optical space astronomy opened new perspectives in the study of optical clusters. Imaging became so precise that it was now possible to measure galaxies' shapes and study the distribution of matter through measurements of the gravitational lensing effect. The light from background galaxies aligned with a cluster would need to pass through or near the cluster before reaching us. The cluster's gravitational potential would bend such light, and, as a result, the galaxies' images would get distorted. Of course, since we do not have any other independent way to estimate the shape of a given background galaxy, it is not possible to derive information on the intervening gravitational potential from each individual observation. However, by looking at the shapes of many galaxies around a known cluster, a specific circular pattern can be seen (Fig. 4.5). The images of background galaxies form visible arcs around the cluster. Studies of the gravitational lensing effect allow for an independent estimate of the cluster's mass, and constitute the most

Fig. 4.5 Cluster Abell 2218 as observed in the optical band. The dominant central galaxy is clearly visible, as well as the distorted shapes of background galaxies whose light is deviated by the gravitational potential of the cluster. **Figure credit:** NASA/JPL.

direct method to probe the existence and distribution of dark matter in clusters.

Clusters' observations experienced a further major era of excitement in the last ten years, as a completely new observational strategies are started to be pursued: the study of the Sunyaev–Zeldovich (SZ) effect. In the 70s, after the discovery of the cosmic microwave background (CMB) radiation, Sunyaev and Zeldovich pointed out that CMB photons traveling through galaxy clusters would interact with the hot intra-cluster gas. Because the CMB has a current temperature of just a few Kelvin (and not much higher at the time of initial cluster's formation), when CMB photons interact with electrons within the cluster they gain energy through inverse Compton scattering. Since no additional photons are generated in the process, the net result is a distortion of the CMB spectrum in the direction of galaxy clusters. This is a new, independent way to probe electron density and temperature of the intra-cluster medium, complementary to what is achievable with X-ray observations. At difference with X-rays, the SZ effect

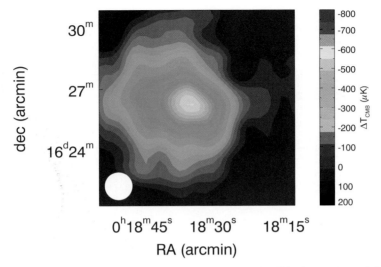

Fig. 4.6 Clusters' gas component, as visible through SZ observations. The SZ observation of Cl0016 (the same as in the upper left X-ray Chandra image) performed by the Bolocam instrument (Sayers et al., 2010). This cluster is at a redshift of 0.56.

is expected to be redshift-independent: for a given cluster's electron density and temperature, the observed distortion does not depend on the distance of the cluster. The opportunity of observing distant clusters as if they were nearby is a clear advantage of SZ measurements over optical and X-ray ones. While initial observation of a few tens of clusters are already reported in the literature (see, e.g. Fig. 4.6), the scientific community is looking forward to the results of the Planck satellite as well as ground-based experiments which collectively are expected to discover thousands of objects.

4.4 Clusters in Cosmology

The most straightforward and exploited way to derive cosmological constraints from clusters consists in counting them. The number of objects per given mass, solid angle on the sky and within a given distance can be predicted under the assumption that clusters form through the simple process of gravitational collapse described before. Needless to say, such number strongly depends on the initial amplitude of density perturbations set by processes in the very early Universe: the bigger the perturbations, the

Fig. 4.7 The cluster 1E0657-56 "bullet cluster": the result of the collision between two clusters. The pink area shows the distribution of the gas as observed by Chandra. The blue areas represent the distribution of the (dark) matter as reconstructed using the gravitational lensing effect. Galaxies, as observed by the Hubble space telescope, are also displayed in orange and white.

earlier the structures start to form and the more we observe today. So in a way by looking at clusters we can indirectly investigate aspects of physical processes that set up the initial conditions for structures to form. Along with early Universe processes, the general cosmological setting, that is the content and evolution of the Universe, comes into place in determining clusters' abundance. This is because of two reasons: first because it sets the rate of the Universe's expansion; second because it affects the actual growth rate of individual structures. The observed volume is determined by the maximum distance of objects we are considering and the geometry of the Universe around us. The way we determine such distance is by looking at the "redshift" of the object, which informs us of how much smaller the Universe was when the light we are receiving was actually emitted. Since light travels at a specific speed, we can determine how distant the object is from us, given an underlying model for expansion. The actual growth rate of individual structures is affected by the matter content and the Universe's expansion rate; for example, structures grow more slowly in a low density

Universe. In the 90s, When the first cluster samples with well defined selection properties was constructed from observations, it became clear that the number of objects observed and its evolution in time supported evidence for a low-density Universe. This was in striking contrast with the most popular cosmological model at the time according to which the matter density was close to its critical value (that is there was enough matter to create a flat space, with no need for other major contributions from other sources of energy). These findings were amongst the first indications that we live in a low density Universe, and forced cosmologists to think of new paradigms for the Universe's content and evolution.

Beyond measuring the total matter abundance, number counts can, in some cases, also shed light on the actual type of particles that compose this mysterious component of the Universe. This is because the specific kind of dark matter present in the Universe can influence the growth of structures. If particles possess relatively high velocities, they may be able to climb out of the potential well they are in and escape. By doing so, their distribution tends to become more homogeneous than the other components', subtracting part of the source for the gravitational potential and reducing the amplitude of perturbations. The typical example of such particles are light neutrinos: we do know they exist and have a (small) mass, so that they certainly constitute part of the dark matter. We know quite precisely the physics of weak interactions and can evaluate when they decoupled from the CMB photons and therefore can accurately predict their temperature at any given time. Because they are light, they remain relativistic for a relevant fraction of the Universe's history (the specific span depending on the actual neutrino mass value). Their damping effect on density fluctuation then acquires time to add up. As a result, if we compare the fluctuation amplitude as derived from galaxy cluster number counts with the one we can probe at higher redshifts (bigger distances) by observing the cosmic microwave background, we can derive the amount of damping occurred and infer the value of the neutrino mass. This is in practice the strategy adopted by cosmologists to constrain the neutrino mass. Tightest cosmological constraints on neutrino mass are expected to come not from cluster number counts but rather from measurements of fluctuations on smaller scales where the neutrinos' damping effect is even more efficient. Clusters, however, when combined with other probes, can currently constrain the sum of the three neutrino masses to below 0.33 eV at 95% C.L., which is much lower than the value obtained from current particle physics experiments (Mantz et al., 2010). Next generation of clusters' surveys should improve on this upper limit thanks to better clusters' statistics.

One of the most intriguing and sensational discoveries of the past 15 years in Cosmology is the fact that the Universe is currently dominated by a mysterious source of energy (the so called *Dark Energy* (DE)). In lack of better theoretical understanding, DE is phenomenologically described through an equation of state that relates its pressure and density. In its simplest form, such equation gets specified by only one proportionality parameter (w); but describing its possible variation in time may require a few more. For a given dark energy density, the specific form of the equation of state parameter alters both the volumes observed out to a certain redshift and the growth of structures.

It is believed that dark energy prevailed over standard matter in leading the Universe's expansion right about when galaxy clusters started to form in relevant numbers. This fact makes clusters an ideal probe of dark energy. Cluster abundances can be used to constrain both the dark energy density and its equation of state. The former gets indirectly constrained by measurements of the matter content and considerations on the Universe's geometry. CMB observations have shown that the Universe's geometry is consistent with flatness, implying that the total energy density in our Universe should be close to its critical value. Clusters are able to probe the matter energy density, implying a dark energy content equal to the remainder. In addition, cluster's redshift distribution conveys information on the value of the equation of state parameter (Mantz et al., 2008).

Another fundamental property of our Universe that cluster number counts can contribute to determine is the statistical properties of density fluctuations. Early Universe processes which created the initial perturbations had a quantum mechanical nature, and determined the statistical characteristics of density fluctuations on each given physical scale. They did not dictate however, specific locations where over- and under-dense regions should be, nor the specific value of the density in a given place. The statistical properties of fluctuations throughout the history of the Universe therefore reflect the ones of the quantum field (or fields) which played a major role in the very first instant of the Universe's history.

Simplest early Universe models, like single field inflation, predict such fluctuations to be very close to Gaussian, implying that they can be described with only one number for each scale: their variance. In such scenarios density perturbations also follow a Gaussian statistic, as long as they remain small and their evolution is in the linear regime. The study of CMB anisotropies, reflecting the properties of fluctuations when the Universe was about 4×10^5 years old, has confirmed that initial fluctuations were Gaussian, or close to it. As structures evolve, they become progressively nonlinear and their statistics, while it still retains memory of the initial conditions, becomes more complicated to interpret. Clusters, which started to form relatively recently, are considered to be the smallest objects whose statistics can be reliably predicted from given initial conditions using linear theory only. Along with constraining cosmological parameters, cluster number counts can therefore be used to constrain the statistical properties of fluctuations on the range of physical scales they represent. When fluctuations are Gaussian, the predicted number of objects at a given time has an exponential cutoff in the high mass range. The mass at which cutoff occurs increases with time, so that, according to Gaussian initial condition predictions it is quite unlikely to find very massive objects at large distances. One of the simplest ways to test the statistics of fluctuations consists in looking at very distant objects, and try to infer if their masses are consistent with what expected from the Gaussian hypothesis. This is now possible thanks to long exposure X-ray observations, as well as to the new SZ measurements which naturally discover objects at the time when the Universe was about half or a third of its current size. Just a few clusters have been observed at such great distances, and some preliminary studies seem to suggest that they may be more massive than expected from Gaussian initial fluctuations. Such results, however, strongly depend on mass estimates, which can be carried out with different strategies. While various mass estimates seem to be compatible with each other, the level of uncertainty is still quite big. We eagerly wait for more data on high-redshift clusters to confirm or discard these findings.

Clusters' number counts typically provide information about fluctuations on scales that are associated with their masses, given the average density of the Universe at the present time. Such scale is about 10 Mpc (Megaparsec, that is 3×10^{22} m), and should not be confused with the actual typical size of a cluster (about 1 Mpc). However, cosmological information is contained in fluctuations on all kinds of scales, as they are all generated in the early Universe. Clusters can not inform us about scales smaller than the ones probed with number counts, but they can be used as tracers and exploited to describe larger scale perturbations. This can be achieved in a few different ways.

With the advent of large surveys, it is possible to study the distribution of clusters. Clusters form in region of the Universe where the overdensity is above a certain threshold. If a cluster-scale perturbation is located in a region where there is also an overdensity on a bigger scale, the total overdensity may pass such threshold and the object forms. If, on the contrary, the same cluster-scale perturbation is located in a larger underdense region, the total overdensity may be too low for the object to form. As a result, a higher number of clusters are expected to be found in correspondence of larger scale overdensities, so that galaxy clusters are supposed to be clustered themselves. The physical scales probed by such strategy largely overlap with the ones that are probed by CMB studies. However it is still worthwhile to study them as the CMB conveys information about a totally different period in the history of the Universe: namely when it was about 10^{-3} its current size.

Finally, clusters can also be used as tracers of velocities. The inhomogeneous distribution of matter in the Universe sources peculiar velocity fields on all scales. These velocities add on top of the general expansion which is in common for all locations at a given time. As long as density fluctuations are small, the statistical properties of peculiar velocities are implied by the density initial conditions and how they were generated. While it is impossible to observe the transverse velocity of objects of cosmological interest such as galaxies or galaxy clusters, it is possible however to infer their radial velocity. This can be done, for instance, by looking at the *kinetic* SZ signal. If the cluster is moving toward or away from us, so are, on average, the electrons in the diffuse gas component. As a consequence, the coherent velocity of the electrons in the clusters leaves an imprint in the SZ signal through the Doppler effect. Measurements of this effect are quite difficult: the signature of the kinetic SZ signal is much weaker and more difficult to identify than the thermal SZ distortion as it presents the same frequency dependence of the CMB signal it is embedded in. Nevertheless, attempts to measure the coherent signal in a set of known X-ray galaxy clusters have been made (Kashlinsky *et al.*, 2009). The result was obtained by combining knowledge of the clusters' location, temperature and gas density from X-ray observations for a set of massive clusters distributed in a symmetric way and covering about 80% of the sky, with recent observations of the CMB on the same area provided by the WMAP result.

Such measurements indicated that the sphere around us with a radius of a few hundred Mpc is coherently moving with a velocity of a few hundred km/s. This velocity would be well above and inconsistent with what is predicted by the leading cosmological model and inflationary scenario. Current research in this area is still underway, and such claim have been challenged by other results showing no discrepancy between current data and "standard" cosmology predictions (Keisler, 2009; Osborne *et al.*, 2010). With the results of the new Planck Satellite, which is observing the CMB with higher spatial resolution and lower instrumental noise, it will be possible to reduce the current upper limits on the bulk velocity by an order of magnitude (Mak, Pierpaoli and Osborne, 2011). New, deeper cluster catalogs, constructed both from the Planck satellite itself and from the planned X-ray new satellite eRosita, will allow to reach even bigger distances, where standard cosmology predictions for bulk velocities are even smaller. Any departure from the standard scenario will be put in evidence more easily.

There are other ways in which galaxy clusters can be used to constrain large scale energy density perturbations. They can in fact be used to overcome the most relevant limitation we have in probing large-scale perturbations: the existence of cosmic variance (Kamionkowski and Loeb, 1997). The largest observable perturbations have the size of our present horizon, current constraints on such large-scale fluctuations are derived from the observation of the largest-scale CMB anisotropies: specifically the dipole and the quadrupole of the CMB temperature anisotropies. As density fluctuations are statistical in nature, we would like to take several independent measurements of them. On the largest scales, however, we only have a handful of independent samples for such measurements when we probe fluctuations through CMB observations on the celestial sphere. We are therefore limited in constraining the underlying theoretical model simply because we cannot get enough experimental samples. Clusters may provide a way to overcome such limitation by offering another "point of observation" in the Universe. This can be achieved, in particular, by looking at the polarization signal in the SZ effect.

If the incoming CMB radiation the off-scattering of clusters' electrons has a quadrupole component in the electrons' rest frame, the scattered light is polarized. By looking at such polarization, we can therefore derive information on the density perturbations producing a quadrupole pattern around the cluster at the time when the scattering occurred. The combined information from several distant clusters may reduce the theoretical limitation presented by cosmic variance. In practice, since clusters form relatively recently, the quadrupole observed by most clusters is not dramatically different from the one we observe, and the effectiveness of such "solution" to the cosmic variance problem is low. However, there is an additional possible use of SZ polarization measurements in inferring constraints on fluctuations of large scales. When we observe the quadrupole of the CMB fluctuations on our sky, we in fact get information from a mixture of three-dimensional fluctuations projected on the sphere centered on us and extending to the dust-scattering surface. Certainly a given spherical mode is dominated by the signal around a particular physical scale, but what we observe is a mixture. If we take the SZ polarization measurements for several clusters, we can study the pattern of such polarization signal on the sky. The large-scale signal of this polarization pattern for clusters that are quite distant from us is a much sharper probe of a given physical mode, allowing better testing of the initial conditions imposed by early-Universe mechanisms like inflation (Seto and Pierpaoli, 2005). This is particularly relevant because CMB observations have shown a peculiarity on very large scale: the measured quadrupole in the CMB angular power spectrum is much lower than expected in the best-fit cosmological model for all scales. This is a puzzling fact that could be explained through some non-standard mechanism in the early Universe which only affects one particular scale or range of scales. More precise information on large-scale fluctuations may help in resolving this puzzle. Unfortunately, the predicted polarization signal in SZ clusters is relatively low. As the first (intensity) observations of SZ clusters just started, we are anyway quite far from having a map of SZ cluster signal on the whole sky. It will still take some years for this method to be exploited.

4.5 Dark Matter or Modified Gravity?

Since the vast majority of the clusters' mass is believed to be dark matter, it is natural to look at them when it comes to probe its nature. Aside from its gravitational interaction, dark matter particles can also be detected in other ways. They are, in fact, not completely dark: some of the most accredited candidates of dark matter are weakly interacting massive particles (WIMPs), which exist in several well-motivated theoretical extensions of the Particle Physics Standard Model. WIMPs can be detected in astronomical observations because they may annihilate or decay. Consequences of such processes can be seen at various frequencies ranging from the radio to gamma-rays. Recent observations from the Fermi satellite in the direction of galaxy clusters, for instance, have shown the lack of a prominent gamma-ray signal. This fact allowed to put the most stringent limits available to-date on the lifetime of dark matter particles for a wide range of particle masses and decay final states (Dugger *et al.*, 2010). The determination of WIMP properties from dark matter observations, however, is not immune from theoretical uncertainties. The annihilation rate , for instance, is proportional to the square of the dark matter density and therefore strongly depends on the cuspiness of the cluster's internal profile. Precise predictions require numerical simulations, and the details of the central density may in fact depend on the simulation's resolution. As computers improve performances and observations in various bands become available, we expect more definitive results on dark matter properties from clusters' observations.

Despite the numerous evidences in its favor, dark matter is perceived by many in an uneasy way. As it calls for mysterious new particles and to solve a discrepancy between theory and observations, evidence for the need of dark matter mainly comes from relatively big scales, ranging from galactic sizes and up. Rotation curves of galaxies, for instance, showed that the galaxy's outskirts were moving with a higher velocity than expected had all and only matter been what was visible. More specifically, the rotation velocity which, at a radius whose corresponding sphere contains the whole galaxy, was expected to decrease as $1/r$, was in fact flat up to where observations could reach. Motivated mainly by these findings, which clearly called for some new element in the theoretical picture provided by general relativity, in the 80s and 90s some scientists tried to develop models that could reproduce galaxy observations without invoking the existence of dark matter. These new models, generally referred to as MOND (MOdified Newtonian Dynamics), postulate that the the law of gravity would not follow the familiar $1/r^2$ form when accelerations are lower than a minimum value g_0. The relevant parameters are then tuned to reproduce the observed rotation galaxy curves. If MOND manifests itself at galactic scale, it should also show its effect at galaxy cluster's one. Such simple MOND description, however, when applied to galaxy clusters does not seem to be sustained by the data. In general, MOND predicts a shallower slope of the mass profile than what the data show, irrespective of the critical acceleration parameter g_0.

Perhaps the most striking cluster-based evidence in favor of dark matter is constituted by the famous "bullet cluster" (the nickname for cluster 1E 0657-56). The bullet cluster has first been observed in the X-rays where it showed a prominent shock front. This fact brought 1E 0657-56 to the attention of the community as one of the clusters that are not relaxed and presumably experienced recent major merging. A following analysis of optical images allowed to reconstruct the mass distribution within the cluster from the lensing effect. The combination of the optical and X-ray information revealed interesting information: the findings were consistent with 1E 0657-56 being a merger between two large structures, but the gas distribution was different from the one of the total matter (Clowe et al., 2006). Overlapped images showed that gas is lagging behind the dark matter after the collision, which is interpreted as an effect of pressure acting on it. Since 1E 0657-56 was discovered, a number of similar objects have been systematically studied, further confirming these results (Takahashi and Chiba, 2007).

Despite the flows of MOND, modified gravity theories in general, however, continued to be pursued in order to solve the other major puzzle of our era: the cause for acceleration of the Universe. A major advancement in this field was achieved with the covariant formulation of a relativistic MOND inspired theory proposed by (Bekenstein, 2004) (TeVeS), which was then followed by a number of new approaches to modified gravity (e.g. DGP and f(R)).

Such theories can also be tested with galaxy clusters number counts and power spectra, as they predict a different expansion history and growth factor than the simplest dark energy models. Dedicated numerical simulations have recently been carried out to investigate the effect of such modifications on clusters' observable quantities. Attempts to constrain these models via X-ray clusters number counts do not seem to show any departure from standard general relativity (Rapetti et al., 2010), and so far allow to put upper limits on modified gravity parameters (Schmidt et al., 2009).

4.6 Gas, Galaxies and Their Evolution

Galaxy clusters also present a peculiar galaxy population. On average, galaxies within clusters are more elliptical in shape and redder in color than the general case, which is an indication of a more evolved population. It is very common for clusters to have a central dominant galaxy (as is also visible in Fig. 4.5). This is most often a giant elliptical or lenticular galaxy three to ten times more luminous than the typical galaxy in a

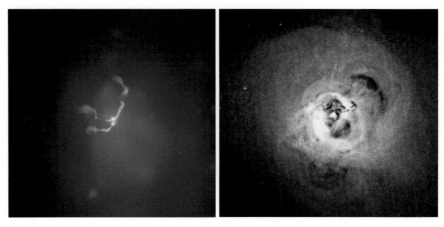

Fig. 4.8 Left panel: Cluster Abell 400 as seen in the radio (pink) and X-rays (blue). The radio emission shows jets departing from two supermassive black holes associated with two large galaxies in the process of merging. Right panel: the effect of the black hole emission on the gas in the cluster's center. **Figure credit:** Chandra image gallery.

cluster. A sizable fraction of clusters (about 20%) also show that the central galaxy is associated with an unusually high radio emission. It is believed that such radio emission is the consequence of mass accretion on a central supermassive black hole at the center of the galaxy. While material falls into the black hole, energetic jets are emitted to large distances. This is visible in Fig. 4.8 (left panel) which displays jets from two galaxies in the process of merging.

The presence of such peculiar galaxies in the cluster's center also have repercussion on the gas central properties. Figure 4.8 (right panel) shows the loops and ripples in the X-ray emission of the Perseus cluster. These features are testimony to the interaction between the central back hole (the white spot at the center) and the gas: the jets have formed two cavities around the black holes and sound waves are expanding from the center. While the presence of supermassive black holes at galaxy centers is now well established, how common it is and the details of galaxy evolution are not yet well understood.

Radio-loud central galaxies also seem associated with the presence of a cooling flow in a cluster. Cool-core clusters display a higher central density and lower temperature than expected for their total mass. The galaxies at their center display a more prominent blue light, indicative of active star formation. Overall, galaxy evolution in clusters is a very complex process, strictly related to the evolution of the gas within the cluster. New observations in the optical, X-ray and radio band yielding thousands of cluster candidates spanning large distances will help to shed light on this process.

Acknowledgments

Elena Pierpaoli is supported by NSF ADVANCE grant AST-0649899, NASA grant NNX07AH59G and JPL-Planck subcontract 1290790.

References

Sayers, J. *et al.* (2010). ArXiv:1010.1798.
Wen, Z. L., Han J. L. and Liu, F. S. (2009). *Astrophys. J. Suppl. Ser.* **183**, p. 197.
Szabo, T. *et al.* (2010). ArXiv:1011.0249v1.
Hao, J. *et al.* (2010). *Astrophys. J. Suppl.* **191**, pp. 254–274.
Koester, B. P. *et al.* (2007). *ApJ* **660**, p. 239.
Mantz, A., Allen, S. W. and D. Rapetti, (2010). *Mon. Not. R. Astron. Soc.* **406**, p. 1805.
Mantz, A. *et al.*, (2008). *Mon. Not. R. Astron. Soc.* **387**, p. 1179.
Osburne, S. *et al.*, (2010). ArXiv:1011.278v1
Kamionkowski, M. and Loeb, A. (1997). *Phys. Rev. D* **56**, p. 4511.
Seto, N. and Pierpaoli, E. (2005) *Phys. Rev. Lett.* **95**, p. 101302.
Dugger, L., Jeltema, T. E. and Profumo, S. (2010). *JCAP* **12**, p. 015.
Clowe, D. *et al.* (2006). *Astrophys. J. Lett.* **648**, p. L109.
Takahashi, R. and Chiba, T. (2007). *Astrophys. J.* **671**, p. 45.
Bekenstein, J. D. (2004). *Phys. Rev. D* **70**, p. 083509.
Rapetti, D. *et al.* (2010). *Mon. Not. R. Astron. Soc.* **406**, p. 1796.
Schmidt, F., Vikhlinin, A. and W. Hu, (2009). *Phys. Rev. D* **80**, p. 083505.
Kashlinsky, A. *et al.* (2009). *Astrophys. J. Lett.* **686**, p. L49.
Mak, D. S. Y., Pierpaoli, E. and Osborne, S. J. (2011). or Xiv: 1101.1581.
Keisler, R. (2009). *Astrophys. J. Lett.* **707**, p. L42.

CHAPTER 5

REIONIZING THE UNIVERSE WITH THE FIRST SOURCES OF LIGHT

VOLKER BROMM
University of Texas, Austin, TX 78712, USA
vbromm@astro.as.utexas.edu

Recently, cosmologists have succeeded in determining the fundamental parameters of our cosmic world model with extremely high precision. There remains, however, one crucial gap in our understanding: How and when did the first stars and galaxies form? Their emergence during the first billion years had dramatic consequences for the history of the Universe, leading to a phase-transition from a cold, neutral medium to an almost completely ionized, hot one. With the help of cutting-edge observatories and supercomputer simulations, we are about to lift the veil that is still shrouding the epoch of cosmic dawn.

The last decade has seen a breakthrough in our understanding of the Universe. We now have answers to the questions of the ages, thanks to NASA's *Wilkinson Microwave Anisotropy Probe* (WMAP). We now know that the age of the Universe is 13.7 billion years, that space is not curved, and that the cosmos consists of 73% dark energy, 23% dark matter and only 4% ordinary, baryonic matter (Turner, 2007; Weinberg, 2008). After reaching this milestone, the focus shifts to the emergence of structure inside this now well-understood background model (Mo *et al.*, 2010).

In particular, we wish to unravel the processes that governed the formation of the first stars and galaxies (Barkana and Loeb, 2001; Bromm et al., 2009; Loeb, 2010). Here, *WMAP* has provided us with vital clues as well. In measuring the temperature distribution of the cosmic microwave background (CMB), it took a snapshot of the Universe 400 000 years after the Big Bang. By this time, the Universe had sufficiently cooled to enable free electrons to recombine with protons, leading to a gas of almost completely neutral hydrogen and helium. The CMB temperature distribution contains tiny fluctuations of $\Delta T/T \sim 10^{-5}$, resulting from fluctuations in density that were in turn imprinted through quantum processes in the very early Universe (Weinberg, 2008). Without these primordial ripples on the underlying smooth cosmic sea, there would have been no stars, galaxies, nor any other structures today, including life and intelligent beings.

Gravity is ultimately responsible for establishing a Universe with a complex hierarchy of structures, from cosmological scales of hundreds of Mpc[1] to stellar scales of a few million km. It slowly amplified the initially very small density fluctuations until the self-gravity of select regions of space had grown strong enough to be able to overwhelm cosmic expansion, thus leading to the collapse of the region. Such regions, called *halos*, contained dark matter and normal, baryonic gas made of hydrogen and helium. Their shapes were close to spherical, and as a result of the gravitational collapse, a rough equilibrium between potential and kinetic energy was established. This equilibrium can be described by the virial theorem, connecting the total kinetic energy of random motions to the gravitational potential energy: $2E_{\rm kin} = -E_{\rm pot}$. The process of halo formation is therefore also often termed *virialization*. The disordered kinetic energy is transferred to the gas as heat through adiabatic compression, or through virialization shocks. If the conditions were right, the first stars were able to form inside such halos via continued runaway compression of the gas.

The currently-favored framework of cosmological structure formation is the Λ-dominated cold dark matter model (ΛCDM) (Mo et al., 2010). The

---

[1] Astronomers measure distances in units of parsec, where $1\,{\rm pc} = 3.09 \times 10^{18}$ cm.

"cold" here refers to the slow speed of the dark matter particles compared to the speed of light, and the prefix indicates that cosmic expansion is dominated by dark energy, possibly Einstein's cosmological constant. The CDM model predicts that structures formed in a bottom-up, hierarchical way: Regions of smaller mass had statistically larger density fluctuations than more massive ones. Gravity therefore amplified smaller regions more rapidly to the point where virialization could occur. The smaller systems would then grow later on through mergers or accretion of matter from the background medium. On the basis of CDM, cosmologists had realized already in the 1980s that the Universe after recombination, the epoch when the first neutral hydrogen atoms were created $\sim 400\,000$ years after the Big Bang, must have been completely dark. This darkness would pertain to a hypothetical human observer who would be around at these early times, since it refers to the fact that the photons of the CMB at recombination had been redshifted into the (near-) infrared part of the electromagnetic spectrum, invisible to us. This leads us to studying the end of this mysterious *cosmic dark ages* (Barkana and Loeb, 2001).

5.1 The End of The Dark Ages

The dark ages were the ultimate *tabula rasa*: a time without stars, galaxies, or indeed any visible change. The CDM model predicts that this epoch ended a few 100 million years after the Big Bang. At this point, the conditions in select dark matter halos admitted star formation to commence for the first time (Loeb, 2010; Tegmark *et al.*, 1997). The so-called minihalos had total masses[2] (gas and dark matter) of $\sim 10^6\,M_\odot$, and extended over $\sim 100\,\text{pc}$. What kind of stars formed inside these primordial sites? Big Bang nucleosynthesis did not create any heavy chemical elements, so that the first stars must have formed out of a pure hydrogen–helium gas. These pristine stars are termed *Population III*, distinguished from the Population I stars such as our Sun, that form in the present-day Universe. The latter stars already contain heavy elements beyond the primordial H/He, *metals* in the terminology employed by astronomers. Metal-poor stars, such as

---

[2]Astronomers measure mass in units of the solar mass ($M_\odot = 2 \times 10^{33}$ g).

those found in the outer regions (halo) of the Milky Way or in globular clusters, belong to the intermediate category of Population II. This sequence of stellar populations simply reflects the continuous enrichment of the interstellar medium with metals over many billions of years from supernova explosions and stellar winds.

Observations have shown that stars in the present-day Universe (Population I) form with typical masses close to that of the Sun. The relative number of stars rapidly declines with increasing mass, approximately described by the initial mass function (IMF) proposed by Edwin Salpeter in 1955: $dN/dm \approx Am^{-2.35}$, where m is the mass of a star, N the number of stars with a given mass, and A a normalization constant (Salpeter, 1955; Stahler and Palla, 2004). The first stars formed under very different circumstances, in particular out of gas without the heavy-element cooling agents, such as C, O, Si, and Fe that dominate present-day star formation. Their IMF was therefore likely also quite different. What are the physical consequences implied by the absence of heavy elements for the first stars? Qualitatively, one can argue as follows: With no heavy element present to cool the gas, the primordial cloud would be hotter than gas clouds today. Gravity, and consequently stellar masses, would then need to be larger to overwhelm the opposing thermal pressure. Only at the end of the 1990s had computer technology sufficiently advanced to allow astrophysicists to seriously address the first star-formation process (Abel *et al.*, 2002; Bromm *et al.*, 2002).

In this endeavor, we benefit from the much simpler conditions in the early Universe, where the cooling and radiation transport do not have to consider the complications brought in by the ions, molecules and dust grains made of heavy elements. Dust in particular, consisting of minerals such as graphite or silicate, presents one of the outstanding challenges in the theory of present-day star formation (Stahler and Palla, 2004). In addition, the early Universe likely did not yet contain dynamically significant magnetic fields. These are important actors in Population I star formation, due to the complex magneto-hydrodynamic (MHD) coupling to the partially ionized gas. Finally, we have a precise understanding of the initial conditions (ICs) for Population III star formation, given by the now-standard ΛCDM model (Komatsu et al., 2009). According to this model, the density fluctuations, which eventually led to the emergence of minihalos, had initially very small amplitudes: $(\rho - \rho_0)/\rho_0 \ll 1$, where ρ_0 is the mean density of the Universe. Fluctuations were thus initially still linear, so that they can be expressed as a Fourier series. The Fourier modes in turn can then be accurately implemented in the initial conditions fed into the computer. Initial conditions in present-day star forming regions, on the other hand, are already nonlinear and thus extremely complex. This can be seen in radioastronomical images of giant molecular clouds (GMCs), indicating a highly structured, possibly fractal, medium that is pervaded by supersonically-turbulent MHD flows. Again, all of these complications were absent prior to the formation of the first stars.

In developing our computer models, we had to take into account the chemistry of molecular hydrogen formation and destruction, together with the corresponding H_2 cooling processes (Galli and Palla, 1998). In the temperature range of a few thousand K, which is typical for the gas in minihalos, metal ions and molecules, together with dust grains, provide very efficient coolants in present-day gas clouds. In the primordial case, H_2 is the only viable cooling agent, albeit not a very good one. In general, cooling happens when a thermal particle, here a hydrogen atom, collides with the coolant, here H_2, leaving it in an excited state. For the hydrogen molecule, these are rovibrational levels, where every vibrational quantum state is split up further, when rotational motions are superimposed on its vibrations. If decay occurs through the spontaneous emission of a photon, and if the photon can escape the cloud, the initial kinetic (thermal) energy of the collision is carried away, thus cooling the cloud.

5.2 Formation of a Population III Star

The computer simulations yielded a very interesting result: Independent of small variations in the ICs, the primordial gas settles into the center of the dark matter minihalos, acquiring temperatures of $\sim 200\,\mathrm{K}$ and number densities of $\sim 10^4\,\mathrm{cm}^{-3}$. After having reached this state, the gas could initially not condense any further. Up to this point, the gas simply followed the gravitational drag of the dark matter during the virialization process. The collapse into the minihalo center required that the gas could radiate away (dissipate) the heat generated through adiabatic compression. In the absence of metals, cooling had to rely on the presence of H_2. The calculations told us that the gas cloud consisted mostly of neutral H and He atoms, with only a trace amount of molecular hydrogen, of order 10^{-3} by number. Initially, even this small molecule fraction sufficed to radiate away the compressional heat energy. To recap the basic physics: After the excitation of rovibrational quantum states of H_2 through collisions with H atoms, spontaneous radiative decay leads to the emission of photons in the radio- and infrared wavebands. And these photons carry away the random kinetic (thermal) energy of the colliding particles, thus effectively cooling the gas. The H_2 cooling channel, however, becomes ineffective at temperatures below $\sim 200\,\mathrm{K}$, since then the kinetic energy of the impacting atoms will no longer suffice to excite even the lowest-lying rotational transition in H_2. Technically, this corresponds to the $J = 0 \to 2$ transition, where J is the rotational quantum number. Here, the $\Delta J = 2$ is enforced by the quantum-mechanical selection rules, as applied to the hydrogen molecule with its high degree of symmetry (Galli and Palla, 1998). We thus have an explanation for the characteristic temperature mentioned above. How can we understand the corresponding characteristic value for gas number density?

At low densities, every collisional excitation is followed by a radiative decay, resulting in cooling. Toward higher densities, however, collisions begin to compete in de-exciting the upper quantum state, with a zero net effect on cooling. The critical density where collisions compete with radiative processes is exactly the $\sim 10^4\,\mathrm{cm}^{-3}$ value encountered in the simulations. Above the critical density, cooling effectively saturates, and the gas experiences only slow, quasi-hydrostatic contraction. At the same time, additional gas is falling onto the central gas cloud. Once the central gas mass exceeds a few $100\,M_\odot$, the cloud's self-gravity is able to overwhelm the opposing thermal pressure (Bromm *et al.*, 2002). An exponentially accelerating collapse follows, eventually leading to central densities that are encountered in stars, $n \sim 10^{22}\,\mathrm{cm}^{-3}$. At this point, a primordial protostar has formed (Omukai and Nishi, 1998; Yoshida et al., 2008). The critical mass of a few $100\,M_\odot$, where gravitational instability sets in, is termed *Jeans instability* in astrophysics. A simple way to understand this key instability is as follows. Consider a cloud of density ρ_0 and temperature T_0, where a region of size L is compressed by some chance event, such as converging turbulent flows. This region will continue to collapse if its now-enhanced gravity overwhelms the thermal pressure inside of it. The strength of gravity can be measured by the free-fall time, $t_{\rm ff} \sim (G\rho_0)^{-1/2}$, where G is Newton's constant, in the sense that stronger gravity corresponds to a shorter free-fall time. Similarly, the effect of the opposing thermal pressure can be estimated with the sound-crossing time, $t_{\rm s} \sim L/c_{\rm s}$, where $c_{\rm s} \sim T_0^{1/2}$ is the sound speed. Again, strong pressure implies that sound-waves can travel faster. Jeans instability then occurs for: $t_{\rm ff} < t_{\rm s}$. This defines a critical length, which a perturbation needs to exceed for instability to set in, the so-called Jeans length: $L > L_{\rm J} \sim c_{\rm s}/(G\rho_0)^{1/2}$. The intuition behind this expression is that a compressed region can continue to collapse, if sound-waves fail to wipe out the density inhomogeneity quickly enough. It is straightforward to express the same physics in terms of a critical mass, the Jeans mass: $M_{\rm J} \sim \rho_0 L_{\rm J}^3 \sim T_0^{3/2} \rho_0^{-1/2}$. Inserting the characteristic values for primordial gas (see above), one finds: $M_{\rm J} \sim 200\,M_\odot$, in good agreement with the simulation results (see Fig. 5.1).

The Population III star is assembled by the accretion of gas from the Jeans-unstable cloud onto the hydrostatic core in the center. The build-up process thus proceeds in an inside-out fashion. Initially, the core has a mass of only $0.01\,M_\odot$, similar to initial protostellar masses in the present-day Universe (Stahler and Palla, 2004). The final mass of the first stars, therefore, depends on how and when the accretion process is shut off. This question is still not completely solved, but simulations are approaching the degree of sophistication necessary to provide answers. What is needed are three-dimensional calculations, initialized on cosmological (Mpc) scales, and the ability to zoom-in on the tiny (sub-AU) scale of the protostar itself. Note that whereas the natural unit to measure cosmological scales is the Megaparsec ($1\,{\rm Mpc} \sim 10^{24}\,{\rm cm}$), the physics of star formation unfolds on scales of the *Astronomical Unit*, i.e. the distance between the Earth

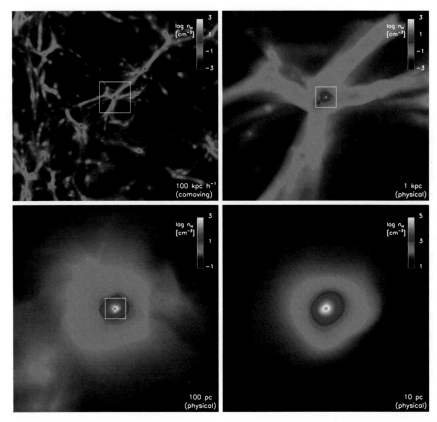

Fig. 5.1 Cosmological initial conditions. A Population III star forms in the center of a minihalo, approximately 300 million years after the Big Bang. The simulation shows how the dark matter is condensing the primordial gas to the point where gravitational instability sets in. The successive zooms show the situation from cosmological to protostellar scales, where gas density is color-coded as indicated. Adopted from (Stacy et al., 2010).

and the Sun ($1\,\mathrm{AU} \sim 10^{13}$ cm), or even smaller, eventually reaching stellar dimensions, of order the solar radius ($1\,R_\odot \sim 10^{11}$ cm). This comparison of units re-emphasizes the huge range of scales, extending over more than 12 orders of magnitude, that has to be addressed in primordial star formation. There is a correspondingly large range in the timescales involved, from the expansion age of the Universe at redshifts $z \sim 20$ (10^8 yr), to the time needed to complete a given chemical reaction (~ 1 s or less) during the formation or destruction of molecular hydrogen. This is the main reason why progress in the field relies on the availability of extremely large-scale supercomputing facilities, operating in a massively parallel mode, together with the development of fast but accurate algorithms.

In addition, the problem requires a coupled treatment of the gas flow and radiation transport, technically a radiation-hydrodynamics (RHD) approach (Castor, 2004). At lower densities, encountered during the earlier stages of star formation, the radiation emitted by the H_2 can easily escape the cloud. The hydrodynamics of the collapsing gas can then be solved without the need to follow the detailed transport of the radiation. Such a situation is called *optically thin*. At increasingly high densities, however, the interaction of the photons with the gas particles (scattering, absorption, and re-emission processes) can no longer be neglected. The mean-free path for such an interaction to occur becomes smaller than the size of the cloud, and the problem enters the *optically thick* phase. Now, the equations of hydrodynamics have to be solved simultaneously with those of radiation transport, thus hugely boosting the complexity. Even the most powerful computers currently in existence cannot yet handle this complexity in its full generality. The art of carrying out such RHD simulations consists in devising clever approximate schemes. A commonly employed strategy assumes that the photons are bottled up so completely that their transport resembles a diffusive random walk. Ignoring the formal limits of where diffusion applies, the computational schemes assume that it can be used everywhere in the simulation box, in particular close to the cloud boundary. Near the boundary, it is obvious that photons will not slowly diffuse, but instead escape with an increasing probability. If one wrongly treats such a situation as a diffusion process, radiative fluxes can become unphysically large. One then adds an ad hoc fudge by imposing an upper limit to the flux. The resulting scheme is then called *flux-limited diffusion*. With this approximate treatment of the RHD, computations can then even afford to consider the full three-dimensional flow.

Such diffusion algorithms have recently been applied to the problem of understanding the formation of massive stars in the present-day Milky Way

(Krumholz et al., 2009). The situation is somewhat less advanced, as far as full three-dimensional RHD treatments are concerned, for primordial star formation. The main reason is the necessity to consider the huge dynamic range of length and time scales described above. The computational cost involved in handling this disparity of scales compounds what is incurred in the RHD calculations. However, we are approaching the point, where such all-encompassing simulations will become feasible. In the meantime, we have obtained intriguing hints for how the primordial star grows by accretion by simulating the first few thousand years of accretion, after the initial protostellar core had formed (Stacy et al., 2010; Clark et al., 2011). During this early phase of accretion, the radiation emitted by the central protostar is not yet so strong as to dominate the flow. One can then still work in the optically thin limit. From these simulations (see Fig. 5.2), a number of key features have already become apparent: (i) Population III protostars can reach masses in excess of 10 $M_\odot$ early on; (ii) a circumstellar disk assembles around the initial protostar, when material with non-negligible angular momentum falls in; (iii) this disk is unstable to further fragmentation, leading to a small, multiple group of stars, possibly dominated by a massive binary; and (iv) there will be a range of masses, dominated by massive stars, but with a possible extension to lower masses as well. The final shape of the primordial IMF will depend on how the complex interplay of these elements will evolve over the $\sim 10^5$ years where accretion is expected to take place (see Fig. 5.3). Incidentally, the recent results of widespread disk fragmentation in present-day star formation, and the key role played by angular momentum, render the primordial case much more similar to the present-day case than traditionally assumed (Stahler and Palla, 2004; Krumholz et al., 2009).

5.3 Feedback in the Early Universe

After the first stars had appeared on the cosmic scene, a number of feedback processes fundamentally changed the early Universe (Ciardi and Ferrara, 2005). These come in different flavors: The high-energy radiation emitted by massive Population III stars leads to radiative feedback, whereas supernova

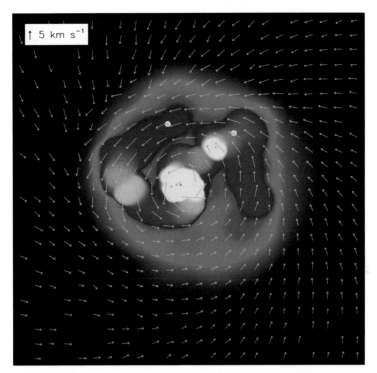

Fig. 5.2 Accretion disk around the first star. Shown are the central 5000 AU of the minihalo in Fig. 5.1. Arrows indicate velocity vectors, and symbols mark the location of individual protostars. The disk structure and the resulting fragmentation are evident. This simulation suggests that the first stars formed as members of small stellar groups, and not in isolation as previously thought. **Figure credit:** Stacy *et al.* (2010).

explosions result in mechanical and chemical feedback. Let us discuss them in turn, beginning with the radiative one.

A fully developed Population III star has an extremely hot atmosphere, characterized by an effective temperature of $T_{\text{eff}} \sim 10^5$ K (compared to our Sun's 5800 K). There are two main reasons for such high temperatures (Bromm *et al.*, 2001). First, the strong gravity associated with the Population III star's high mass requires an equally strong counteracting thermal pressure, which in turn implies high temperatures. Second, the absence of any heavy elements in the outer layers of the star results in a significantly reduced resistance to the flow of radiation, technically the *opacity* of the stellar material. The radiation that escapes from the surface

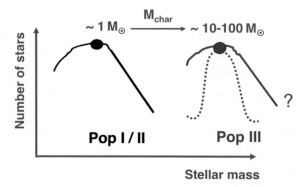

Fig. 5.3 The initial stellar mass function for various stellar generations. We show schematically the number of stars that form with a given mass versus this mass. In the present-day universe (Population I), most stars form with masses close to the solar value. The distribution towards higher masses is approximately described by the Salpeter power-law. For Population III, the peak of the distribution is predicted to be shifted by a factor of 10–100. Whether the primordial IMF is simply a scaled-up version of the present-day case is not yet known.

thus originates in deeper layers of the star, where temperatures are higher. In the present-day case of metal-enriched stars, opacities are much larger, and radiation therefore emanates from shallower, and thus colder, regions. Furthermore, the first stars exhibited extreme luminosities (i.e. the power of the emitted radiation), up to 10^6–10^7 times the solar luminosity. With such high values, the first stars approach the theoretical upper limit for any stellar luminosity, the so-called Eddington luminosity ($L_{\rm Edd}$). This limit describes the situation where the outward force exerted by radiation pressure, which is proportional to the number of absorbed photons per unit time, exactly equals the inward force of gravity. A star with $L > L_{\rm Edd}$ would thus be very short-lived, as it would quickly lose a fraction of its mass to a strong stellar wind, akin to what is observed in the famous η Carinae system.

Such a hot and luminous star produces copious amounts of radiation capable of ionizing the initially neutral H and He in its vicinity. A Population III star of mass $\sim 100\,M_\odot$ produces of order $\sim 10^{50}$ H-ionizing photons per second (Bromm et al., 2001). Despite its large mass, and the correspondingly large stock of nuclear fuel, a primordial star of mass M only lives for an astronomically short time: $t \sim 0.007\,Mc^2/L_{\rm Edd} \sim 3 \times 10^6$

years. The factor 0.007 is the efficiency of hydrogen burning, the most long-lived of all the nuclear burning stages. During its brief existence, the star creates a bubble of $\sim 20\,000$ K hot ionized gas around it, called H II region by astronomers (Greif et al., 2009). At the end of the star's life, the bubble has reached an extent of ~ 5 kpc. The emergence of the first H II regions initiated the long process of cosmic reionization. The neutral intergalactic medium (IGM) that filled the early Universe during the dark ages is thereby transformed again into an almost completely ionized state. The term *reionization* is appropriately chosen, since the universe started out in an ionized state briefly after the Big Bang, lasting until the epoch of recombination $\sim 400\,000$ years later when the cosmic dark ages set in. The bubbles around the first stars initially were like isolated holes in a Swiss cheese, but eventually the entire IGM must have been reionized again. We know this from observations of very distant quasars detected within the Sloan Digital Sky Survey (SDSS). Their spectra show non-zero flux at wavelengths shortward of the Lyman-α resonance line (Fan et al., 2006). Such flux would be very effectively absorbed by neutral hydrogen atoms between the quasar and us (the so-called Gunn–Peterson effect). To see this, remember that the wavelength of photons is redshifted because of cosmic expansion. Consider a photon that was emitted by the quasar with a wavelength shorter than 121.6 nm (the rest wavelength of Lyman-α), and let us also assume for the moment that the quasar is surrounded by a still neutral IGM. For a while, the photon can travel unimpeded by any interaction with the gas. At some point, however, the photon's wavelength is redshifted to 121.6 nm. Due to the resonance character of the transition, the probability for absorbing this photon is extremely high, so that even a very small fraction of neutral H absorbers (10^{-5} by number) would suffice to eliminate it from the line of sight. The H atom will re-emit another Lyman-α photon shortly afterwards, but since this emission is isotropic, the photon is effectively lost from the original quasar spectrum. From the absence of the Gunn–Peterson trough one can therefore infer that the intervening IGM must be ionized to a very high degree. More specifically, the quasar observations indicate that reionization must have been virtually completed 800 million years after the Big Bang.

5.4 Brief History of Reionization

During the cosmic dark ages, the IGM was cold and extremely homogeneous. Roughly 50 million years after the Big Bang, the Universe even cooled below the temperature of the CMB. Before that, gas and photons were still thermally coupled via Thomson scattering. With the continuing Hubble expansion, densities dropped below the threshold needed to maintain effective gas–photon coupling. In a perfectly homogeneous Universe, this "ice age" would have lasted forever. The Universe contained, however, tiny density fluctuations, even in its infancy. Crucially for the history of the Universe, and more specifically that of reionization, the density perturbations were not distributed smoothly, but instead occurred in groups and clusters (Mo *et al.*, 2010). Cosmologists call this phenomenon *biasing*. The basic idea is very simple, best understood in terms of a terrestrial analogy: If you are searching for tall mountains, you should focus on mountain ranges like the Himalayas or Andes, where indeed the number of such giant peaks is greatly enhanced. The tallest summits are not uniformly distributed over the globe after all. Where one already has an underlying mountain range, only a small additional elevation is required in order to reach extreme heights. The presence of such a range, therefore, boosts the probability that additional smaller heaps on top of the underlying range add up to a very tall mountain. Within our analogy, the height of a mountain corresponds to the density of a given fluctuation. The mountain range corresponds to a Fourier mode with a long wavelength, on top of which additional Fourier modes with shorter wavelengths are added, resulting again in biased density peaks. High-density peaks correspond to halos that virialize first, because gravity needs to amplify them less before their overdensity has grown to the critical threshold needed for collapse. The first stars therefore form in the most highly biased halos, emerging in regions with the largest overdensities, and these are thus strongly clustered.

Consequently, the end of the cosmic dark ages had an "insular" character (see Fig. 5.4). The sites where the first stars and galaxies formed were initially still very rare. The first stellar archipelagos were also the sources for the beginning of reionization. Because of the clustered nature of

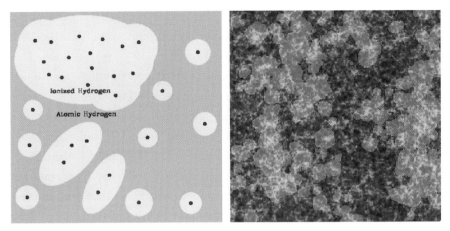

Fig. 5.4 Topology of reionization. Left panel: Early in cosmic history, galaxies are strongly clustered. The ionized regions (yellow) that form around them are thus able to merge into vast "super bubbles". On the other hand, galaxies, and therefore ionized bubbles, are rare in underdense regions. Right panel: Numerical simulations confirm this picture. Notice how the ionized regions (here shown in orange) encompass an increasing fraction of the neutral IGM (shown in green). **Figure credit:** Barkana and Loeb (2007) and Mellema *et al.* (2006).

early structure formation, the H II regions around individual Population III stars could merge into larger ionized regions (super bubbles). The patches of space, where locally the IGM was already reionized, at first only comprised a tiny fraction of the volume of the Universe, and they were separated by large domains of still cold, neutral, gas. With time, however, gravity led to the collapse of an increasing fraction of the mass, and the ionized regions steadily grew. The underdense, cold and neutral zones became smaller and less frequent, until they disappeared completely at the end of the reionization process. This must have occurred at redshifts around $z \sim 6$. This particular sequence of events is sometimes described as *inside-out reionization*, implying that regions with the highest densities were reionized first (Barkana and Loeb, 2007; Mellema *et al.*, 2006).

The global sequence of reionization could have happened very differently though. An implicit assumption in the theory of inside-out reionization is that the sources of the ionizing radiation were stars. The spectra of massive stars peak in the UV, and the resulting H II regions had sharp boundaries, or very thin ionization fronts. The basic reason is that UV photons have energies, or alternatively frequencies ν, close to the ionization threshold for hydrogen. The probability for a photon to be absorbed, measured by the corresponding cross section of $\sigma \sim \nu^{-3}$, is thus huge. This feature is ultimately responsible for the Swiss-cheese topology mentioned above. If on the other hand the first sources of light had been miniquasars — accreting black holes with not too high masses — the bulk of the photons would have been emitted as X-rays. Different from UV radiation, such photons have a much smaller cross section for the photo-ionization of hydrogen, corresponding to their much higher frequency. The resulting photon mean-free path was then also much larger, so that the ionizing photons were not bottled up in the immediate vicinity of the source. They could have escaped into the general IGM, preferentially ionizing underdense regions, the so-called voids. This is a consequence of the long timescales for recombination at low densities: $t_{\rm rec} \sim 1/n$. At high densities, an atom needs to be ionized multiple times in order to offset the repeated recombinations. In regions of low-density, e.g. in the voids, on the other hand, a single ionizing photon per hydrogen atom suffices to keep it permanently ionized. The global sequence of reionization might then have been outside-in (low-density regions first). Theoretically, one can argue for both cases, although the current standard model favors the inside-out, stellar driven, model.

The density dependence of the recombination timescale opens up the prospect for another fascinating possibility for the sequence of reionization (Cen, 2003). In 2003, Renyue Cen at Princeton University proposed the model of Double Reionization. Cen assumed that the first stars were very massive Population III stars, which were very efficient producers of ionizing radiation, as discussed above. These stars caused a first episode of at least partial reionization. Massive Population III stars, however, were also very short-lived, and could only form early on, since later generations of stars formed out of already metal-enriched material. This second burst of star formation could have occurred only after the gas that was heated by the first stars had sufficiently cooled down again. Theory posits that such second-generation stars that already contained heavy elements, the so-called Population II, were less massive. The reason is that metals can cool the gas to lower temperatures, which in turn lead to a smaller Jeans mass. Population II stars therefore also have lower surface temperatures, and consequently emit fewer ionizing photons. The key now is that, in a sense, the Universe could have "forgotten" about such an initial epoch of reionization. At high redshifts, cosmic densities were still very high, so that the recombination timescale would have been correspondingly short. More specifically, one can calculate that t_{rec} was shorter than the age of the Universe (the Hubble time), as long as $z > 8$. The first reionization epoch would then have been followed up by a period of decreased star formation and ionizing photon production that lasted for a few 100 million years. The Universe would again have become almost completely neutral and cold. Eventually, with the onset of second-generation star formation, a second epoch of reionization would have occurred, this time encompassing the entire Universe and lasting permanently. Cen's model received a lot of attention when it was first proposed, but one of course asks how it can be tested with observations, and more generally, how one can empirically decide between the different scenarios. Let us consider this issue next.

5.5 Empirical Probes for Reionization

We have direct observational hints for the end of reionization, provided by the Gunn–Peterson effect, but also, more recently, for its beginning. We again owe this clue to the *WMAP* satellite, which has, as we have discussed above, very successfully measured the temperature fluctuations in the CMB, thereby determining the fundamental parameters of our world model with unprecedented precision. In addition, *WMAP* has measured how strongly the photons of the CMB were scattered by free electrons encountered along their journey toward us. This Thomson scattering also generates a small degree of polarization, imparted to the CMB. The stronger the polarization, the more frequent the interaction with free electrons. One of the surprises from the *WMAP* mission was that the measured polarization signal, and therefore also the importance of Thomson scattering, was stronger than expected (Komatsu *et al.*, 2009). Technically, the quantity measured was the *optical depth* to Thomson scattering: $\tau_T \sim 0.09 \pm 0.03$, where optical depth, roughly, corresponds to the ratio of total distance traveled to the photon mean path ($\tau \sim L/\lambda_{\mathrm{mfp}}$). In order to have free electrons, ionizing photons are required, which in turn are produced by stars. The *WMAP* result indicates that the Universe had to be, at least partially, ionized already at $z \sim 10$, possibly by Population III stars. On the other hand, reionization could not have begun too early, since the polarization signal would then have surpassed the measured value. Even tighter constraints are expected to come from the *Planck* mission, recently launched by the European Space Agency. This satellite has exquisite sensitivity to specifically measure CMB polarization. It may be even possible to approximately determine the history of when the scattering events took place, or equivalently the redshift dependence of the optical depth, $\tau(z)$, as opposed to just measuring the integral signal along the entire line-of-sight, as was done by *WMAP*.

Thomson scattering originates in the ionized phase of the IGM. There exists, however, another way to observe hydrogen at high redshifts, this time targeting the neutral gas, ideally complementary to the Thomson scattering signal (Furlanetto *et al.*, 2006). This is the famous 21 cm radiation, created by the spin-flip transition, where the electron spin switches from parallel to anti-parallel orientation relative to the spin of the proton. Neutral regions of the IGM are then observable in the redshifted 21 cm line, whereas already reionized stretches will appear dark on 21 cm maps. Currently, a number of experiments are engaged in the search for this reionization signature in the radio sky at meter wavelengths (the redshifted 21 cm radiation). The best chance for an early detection may belong to the Dutch-European LOFAR (Low-Frequency Array) experiment. The potential for 21 cm cosmology is immense, since it can, in principle, map the three-dimensional distribution of neutral hydrogen in the early Universe: Redshift-slice by redshift-slice, akin to the tomography carried out by diagnostic medicine. It should then also be possible to distinguish between the competing theoretical models for reionization. Interestingly, it may be even possible to detect the cumulative signal from all Population III stars in the redshifted 21 cm radiation, at least with planned next-generation facilities, such as the Square Kilometer Array (SKA) (Greif *et al.*, 2009). An individual Population III star, formed in the center of a minihalo, is surrounded by an extended H II region, as we have seen. After the star dies, the H II region does not immediately disappear again, but instead lingers on for a while as a "relic H II region". Such a relic now is already substantially neutral as a consequence of recombinations, but is still much warmer than the cold, neutral IGM that surrounds it. These are prime conditions enabling the emission of 21 cm radiation, and the radiation background resulting from the superposition of all the individual, relic H II regions around Population III stars would have a distinct spectral signature, together with an overall amplitude that renders it readily detectable with the SKA.

5.6 Explosions at Cosmic Dawn

When a Population III star, after 2–3 million years, has exhausted its nuclear fuel, it will encounter catastrophic, gravitational collapse. Depending on mass, either a black hole is created, or the star is completely disrupted in a gargantuan explosion. Let us first consider the second possibility. If the star had a mass in the range 140–260 $M_\odot$, the explosion that ended its life was fundamentally different from the conventional supernovae from stars with $M > 8\,M_\odot$. First, the explosion mechanism is very distinct. In general, massive stars beyond 100 solar masses balance the inward force of gravity with radiation pressure. Once the temperature in the stellar core exceeds $\sim 10^9$ K in the course of the inevitable, gradual contraction, the so-called pair-creation instability sets in, where two gamma-photons (γ) are converted into an electron-positron pair: $\gamma + \gamma \rightarrow e^+ + e^-$. As the $\gamma\gamma$-process reduces the number of photons, radiation pressure is also weakened, so that gravity remains unbalanced, and an implosion ensues. As a consequence, the stellar matter is adiabatically heated, so that the temperature suddenly jumps to very high values. All remaining nuclear fuel is then simultaneously ignited, triggering a colossal thermonuclear explosion, whose energy release suffices to completely disrupt the star. Such an explosion is called pair-instability supernova (PISN), according to its peculiar triggering mechanism (Barkat et al., 1967; Heger and Woosley, 2002). Compared to conventional core-collapse supernovae, where a neutron star or black hole is left behind, it has an explosion energy that is up to 100 times higher, with a maximum of $\sim 10^{53}$ erg. Crucially for the chemical evolution of the early Universe, a PISN leaves no compact remnant behind, so that it can release the entire complement of heavy elements produced in the progenitor star into the surrounding IGM. In a core-collapse explosion, on the other hand, a large fraction of the heavy-element production is permanently locked up in the central compact object. A first generation of PISN explosions could thus have provided a "bedrock" of metals briefly after the first stars began to form.

One of the important questions in modern astrophysics is whether these remarkable explosions really did occur in non-negligible numbers in the early Universe. That they were extremely rare in the present-day Universe has a simple reason: Metal-enriched stars experience strong mass loss through winds that are driven by photon pressure on the heavy element absorbers in the star's outer layers (Kudritzki, 2002; Lamers and Cassinelli, 1999). Even if a Population I star was born with a mass $>140\,M_\odot$, such winds would quickly reduce the mass again below the PISN threshold. There may thus have been a brief window of opportunity at the end of the dark ages, where conditions allowed PISN events to occur. An international group of supernova hunters made an important discovery in 2009, by detecting an ultra-luminous explosion whose properties seem to require that a huge amount of heavy elements were released (Gal-Yam et al., 2009). The most natural explanation is that SN2007bi, its official catalog number, was indeed a PISN. This interpretation is supported by the fact that the host galaxy was a metal-poor dwarf galaxy, and not one of the more enriched, Milky Way-sized galaxies. The plausibility for PISN explosions to exist is significantly strengthened, if they can occur even in the present-day Universe, where the physics of star formation and subsequent stellar evolution does not favor them.

A PISN is so extremely bright that the *James Webb Space Telescope* (JWST), the mission that will replace the Hubble Space Telescope around 2014 and that is designed to observe the first stars and galaxies in the near-infrared waveband, should be able to detect even a single explosion at very high redshifts (Gardner et al., 2006). A complementary approach is provided by what is often termed Stellar Archaeology, where astronomers scrutinize the chemical abundance patterns in extremely metal-poor stars in our Milky Way (Beers and Christlieb, 2005; Frebel et al., 2009). The basic idea is that these stars effectively display the fossil signature of the first supernova explosions, where supernova progenitor stars with different masses give rise to distinct explosions and nucleosynthetic patterns. One can thus indirectly infer the properties of the first stars. None of the observed metal-poor stars to-date shows the distinct abundance pattern expected for a PISN. Instead, the observations can naturally be explained if the Population III progenitors were somewhat less massive, having a few tens of solar masses, so that the resulting supernovae would be of the core-collapse type (Iwamoto et al., 2005). This probably indicates that PISN explosions were rare events, but the proper interpretation is confounded by the fact that surveys of metal-poor stars may have systematically overlooked stars that carry the signature of PISN enrichment (Karlsson et al., 2008). The problem may be that a single PISN already enriched its surroundings to quite high levels of metallicity, so that any next generation star that formed out of the cooled PISN debris would not have been that metal-poor at all. The surveys, on the other hand, select their candidate stars according to low metallicity. Future, extremely large-scale observational campaigns should be able to decide this controversy.

Recently, it has become possible to study the impact of the first supernovae on the chemical evolution of the Universe with large-scale supercomputer simulations (Wise and Abel, 2008; Greif et al., 2010). Remarkably, even a single explosion influences the IGM on cosmological scales, in particular if it is of the PISN type. We have learned that a PISN completely destroys its host system, the minihalos discussed above. The supernova blastwave travels into the surrounding IGM out to a distance of $\sim 2\,\mathrm{kpc}$, before losing its thermal energy to radiation and adiabatic expansion. Inside of these bubbles, the metallicity reaches order of 1% of the solar value (see Fig. 5.5). After the cosmic gas in the vicinity of the PISN has thus been enriched with heavy elements, the character of star formation fundamentally changed. The more efficient metal cooling enabled the formation of low mass stars, since the lower temperatures result in smaller Jeans masses (Bromm and Loeb, 2003a). The second generation of stars would then already have been quite similar to the stars formed today.

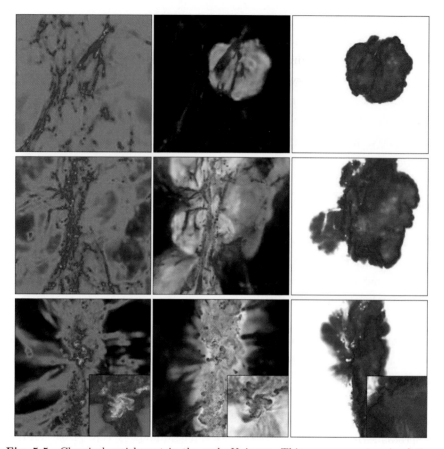

Fig. 5.5 Chemical enrichment in the early Universe. This supercomputer simulation shows how the metals produced in one of the first supernova explosions were distributed into the surrounding IGM, and how they were eventually re-assembled in the center of a primordial galaxy. From left to right, one can see the gas density, the temperature, and metallicity (the amount of heavy chemical elements), at three different moments: 15, 100, and 300 million years after the explosion. The IGM is heated through the supernova explosion, and through the light from neighboring Population III stars. The small panels in the bottom row are a zoom-in into the center of the emerging first galaxy. Here, the metallicity is already quite high, so that Population II stars are predicted to form there. **Figure credit:** Greif *et al.* (2010).

5.7 The First Black Holes

A massive Population III star, however, could have died in a very different fashion. If its mass was less than $140\,M_\odot$, or more massive than $260\,M_\odot$, the pair-instability would not occur, and the star would directly collapse into a black hole. The first black holes therefore were likely more massive than the $\sim 10\,M_\odot$ holes formed by the death of massive stars today. Possible massive black holes originating from Population III stars are candidates for the seed that would grow to the supermassive holes inferred to be present in the centers of most galaxies. Even more extreme cases are found in quasars, with central black hole masses of up to $\sim 10^9\,M_\odot$. Intriguingly, observations within the SDSS have turned up such luminous quasars already at redshifts $z \sim 6$. This means that there were less than 800 million years available to grow the required supermassive black holes from any seed that formed earlier. There is an upper limit to possible growth via accretion of infalling matter, imposed by conditions such as the Eddington luminosity or the spatial distribution of the surrounding gas. Simulations indicate that accretion was rather inefficient early on, mostly due to the strong radiative feedback involved. First, when the Population III progenitor is still alive, its UV radiation photo-heats the gas in its vicinity, and since it is located in the center of the shallow gravitational potential well of the minihalo, the heated gas is evaporated out into the IGM (Greif et al., 2009). The black hole, once formed, finds itself therefore in a near vacuum, with little potential for accretion. Second, even if dense gas is eventually funneled into the vicinity of the black hole, accretion is still inefficient, this time because of the outward force exerted by radiation pressure. In any case, accretion rates stay significantly below the maximum, Eddington, rate. It is therefore quite questionable whether the rapid growth required to explain the high-redshift SDSS quasars could really have happened.

To circumvent the problem of limited early growth, a number of alternative scenarios has been discussed. One model suggests that more massive, one million solar mass seed black holes could have formed by the direct collapse of a primordial gas cloud embedded in dark matter halos that were more massive than the minihalos (Bromm and Loeb, 2003b). If the halos were sufficiently massive to be able to heat the infalling gas to roughly $\sim 10^4\,\mathrm{K}$, cooling could occur through lines of *atomic* hydrogen, as opposed to the cooling via H_2 at lower temperatures. Contrary to H_2 molecules which can readily be destroyed by soft UV photons, atomic hydrogen is always present, and the corresponding cooling therefore cannot be shut down. The model now assumes that the cloud is embedded in a very strong background of soft UV radiation, capable of suppressing the formation of molecules. The temperature of gas during the collapse into the center of the dark matter halo is thus locked at $\sim 10^4$ K. The gas can therefore undergo overall collapse, but it cannot fragment into smaller pieces. Star formation during the collapse is consequently suppressed. This also implies that any negative feedback from such star formation that could act to prevent the gas from assembling in a compact central object would be suppressed as well. The atomic cooling model has problems as well. For example, one needs to explain how the required very strong background of soft UV radiation could have been produced. In summary, the problem of how the first supermassive black holes were seeded remains unsolved.

5.8 Toward the First Galaxies

The first stars formed in isolation, or as a member of a small multiple system. Despite their large luminosities, such single Population III stars cannot be observed even with the JWST. The only chance is to catch them at the moment of their death, when they undergo PISN explosions, which are predicted to be extremely bright. Our attention therefore shifts to the next stage in the hierarchy of cosmic structure formation. The minihalos that hosted the formation of the first stars were drawn together by gravity to merge into more massive structures. How and when did the next generation of stars form? Theory suggests $\sim 10^8\ M_\odot$ dark matter halos as candidates for hosting the second round of star formation (Bromm *et al.*, 2009). They would then also constitute the first bona fide galaxies, if we mean by a galaxy a system of long-lived stars that is embedded in a dark matter halo. Such halos emerged roughly 500 million years after the Big Bang, with gravitational potential wells that were deep enough to recollect the material that had previously been heated by the very first stars and been blown out into the IGM (see Fig. 5.6). The gas that is thus assembled into the first galaxies was likely already enriched with heavy elements to a level of 0.001 the solar value. The first galaxies are thus predicted to be already metal-enriched, giving rise to the formation of stars that included low-mass

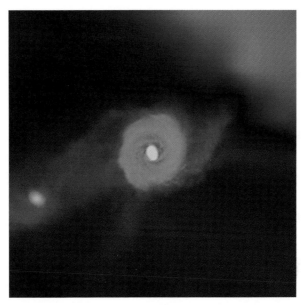

Fig. 5.6 Primordial galaxy. This supercomputer frame shows how the pristine, hydrogen and helium gas has assembled in the center of one of the first galaxies, 500 million years after the Big Bang. The prominent disk has a few kpc across, with a massive gas cloud in the very center. This cloud is expected to fragment and give rise to a massive cluster of stars, possibly bright enough to be observable with the JWST. **Figure credit:** Pawlik, Milosavljevic and Bromm (2011).

stars as well. Such low-mass stars would then survive virtually for the entire age of the Universe, and can still be probed as fossils of the dark ages in our local Milky Way neighborhood.

There is a second important factor that is likely crucial for star formation inside the first galaxies. In the aftermath of the gravitationally driven collapse of the dark matter/gas mixture, part of the gravitational potential energy is transformed into the kinetic energy of random, chaotic motion. Or, put differently: we witness the emergence of supersonic turbulence. Such supersonic turbulence endows the gas in present-day star forming clouds with a distribution of density fluctuations that encompass a spectrum of masses. Some of them become gravitationally unstable, and begin to collapse. Others are just temporary and bounce back after a short while. At the end of this complex process, a stellar cluster has emerged. We therefore expect that such clusters also formed in the center of the first galaxies, possibly with an IMF that already resembled the present-day mass spectrum. In extreme cases, some of them could even have grown to the size of globular clusters that consist of millions of stars, and could have survived to the present day.

The hunt for the first stars and galaxies is accelerating. We pursue them with the most powerful supercomputers, and the most technologically advanced telescopes and satellites. The next decade promises to be a golden age of discovery, where we will finally learn how the Universe emerged from its infancy at the end of the cosmic dark ages.

Acknowledgments

V.B. acknowledges support from the U.S. National Science Foundation through grants AST-0708795 and AST-1009928, as well as NASA through Astrophysics and Fundamental Physics Program grants NNX 08-AL43G and 09-AJ33G.

References

Abel, T., Bryan, G. L. and Norman, M. L. (2002). *Science* **295**, p. 93.
Barkana, R. and Loeb, A. (2001). *Phys. Rep.* **349**, p. 125.
Barkana, R. and Loeb, A. (2007). *Rep. Prog. Phys.* **70**, p. 627.
Barkat, Z., Rakavy, G. and Sack, N. (1967). *Phys. Rev. Lett.* **18**, p. 379.
Beers, T. C. and Christlieb, N. (2005). *Annual Rev. Astron. Astrophys.* **43**, p. 531.
Bromm, V., Kudritzki, R. P. and Loeb, A. (2001). *Astrophys. J.* **552**, p. 464.
Bromm, V., Coppi, P. S. and Larson, R. B. (2002). *Astrophys. J.* **564**, p. 23.
Bromm, V. and Loeb, A. (2003a). *Nature* **425**, p. 812.
Bromm, V. and Loeb, A. (2003b). *Astrophys. J.* **596**, p. 34.
Bromm, V. *et al.* (2009). *Nature* **459**, p. 49.
Castor, J. I. (2004). *Radiation Hydrodynamics*, Cambridge University Press, Cambridge.
Cen, R. (2003). *Astrophys. J.* **591**, p. 12.
Ciardi, B. and Ferrara, A. (2005). *Space Science Rev.* **116**, p. 625.
Clark, P. C. *et al.* (2011). *Sciene* **331**, p. 1040.
Fan, X., Carilli, C. L. and Keating, B. (2006). *Annual Rev. Astron. Astrophys.* **44**, p. 415.
Frebel, A., Johnson, J. L. and Bromm, V. (2009). *Mon. Not. R. Astron. Soc.* **392**, p. L50.
Furlanetto, S. R., Oh, S. P. and Briggs, F. H. (2006). *Phys. Rep.* **433**, p. 181.
Gal-Yam, A. *et al.* (2009). *Nature* **462**, p. 624.
Galli, D. and Palla, F. (1998). *Astron. Astrophys.* **335**, p. 403.

Gardner, J. P. *et al.* (2006). *Space Science Rev.* **123**, p. 485.
Greif, T. H. *et al.* (2009). *Mon. Not. R. Astron. Soc.* **399**, p. 639.
Greif, T. H. *et al.* (2010). *Astrophys. J.* **716**, p. 510.
Heger, A. and Woosley, S. E. (2002). *Astrophys. J.* **567**, p. 532.
Iwamoto, N. *et al.* (2005). *Science* **309**, p. 451.
Karlsson, T., Johnson, J. L. and Bromm, V. (2008). *Astrophys. J.* **679**, p. 6.
Komatsu, E. *et al.* (2009). *Astrophys. J. Supp.* **180**, p. 330.
Krumholz, M. R. *et al.* (2009). *Science* **323**, p. 754.
Kudritzki, R. P. (2002). *Astrophys, J.* **577**, p. 389.
Lamers, H. and Cassinelli, J. P. (1999). *Introduction to Stellar Winds*, Cambridge University Press, Cambridge.
Loeb, A. (2010). *How Did the First Stars and Galaxies Form?*, Princeton University Press, Princeton.
Mellema, G. *et al.* (2006). *Mon. Not. R. Astron. Soc.* **372**, p. 679.
Mo, H., van den Bosch, F. and White, S. (2010). *Galaxy Formation and Evolution*, Cambridge University Press, Cambridge.
Omukai, K. and Nishi, R. (1998). *Astrophys. J.* **508**, p. 141.
Pawlik, A. H., Milosavljevic, M. and Bromm, V. (2011). *Astrophys. J.* **731**, p. 54.
Salpeter, E. E. (1955). *Astrophys. J.* **121**, p. 161.
Stacy, A., Greif, T. H. and Bromm, V. (2010). *Mon. Not. R. Astron. Soc.* **403**, p. 45.
Stahler, S. W. and Palla, F. (2004). *The Formation of Stars*, Wiley-VCH, Weinheim.
Tegmark, M. *et al.* (1997). *Astrophys. J.* **474**, p. 1.
Turner, M. S. (2007). *Science* **315**, p. 59.
Weinberg, S. (2008). *Cosmology*, Oxford University Press, Oxford.
Wise, J. H. and Abel, T. (2008). *Astrophys. J.* **685**, p. 40.
Yoshida, N., Omukai, K. and Hernquist, L. (2008). *Science* **321**, p. 669.

CHAPTER 6

MAPPING THE COSMIC DAWN

STEVEN FURLANETTO

Department of Physics and Astronomy
University of California-Los Angeles
Los Angeles, CA 90095, USA

Deep in the bleak desolation of the Australian Outback, hundreds of kilometers from the nearest sizable town, sits a curious sight. Sprinkled around the desert, over a few hundred square meters, squat 32 "tiles", each composed of 16 small metal devices seated upon a coarse wire mesh. On each of these constructs, two crossed-metal "bowties" about a meter across radiate outward from a stout central tower, so that together the tiles resemble a small army of metal insects crawling across the red sand. Thick cables snake underground from each tile, connecting them to a computer station within the small headquarters building nearby.

A visitor would be forgiven for finding these queer sights out of place in a region known mostly for its kangaroos, sheep, and mines. But in fact these strange metal constructs are a *telescope* — though undoubtedly one of the oddest such instruments one will ever see. No dishes, mirrors, or lenses can be seen — not even anything that appears to point at the sky! No protective telescope domes dot the landscape, only half-hearted attempts to keep the Outback's wildlife at bay. Nevertheless, these squat bowties are indeed radio antennae "pointed" to the heavens, silently observing some of the most distant objects in the Universe, a crucial piece in a worldwide scientific quest to observe the first stars and black holes in the Universe. Giant but equally bizarre telescopes in the Netherlands, South Africa, and India have also joined the effort, racing each other to complete a decades-long effort.

How do these metal insects — collectively known as the *Murchison Wide-field Array*, or MWA — and their rivals observe the sky, and why did astronomers construct such a weird telescope? Why, of all places, did they place their army of metal insects in the desolate Australian Outback? And what do they hope to see?

6.1 A Brief History of Our Universe: From Soup to Galaxies

Many of the answers to this remarkable story begin with the even greater story of cosmology and humanity's drive to describe the history of our Universe: to understand how it evolved from remarkable simplicity, immediately after the Big Bang, to fascinating complexity today — the galaxies, stars, black holes, and planets that dot our night sky.

Over the past several decades, cosmologists have made enormous strides in disentangling the threads of this story. On the one hand, the increasing size and sophistication of conventional telescopes have allowed astronomers to observe objects at greater and greater distances from us — now up to a hundred million billion billion meters away! Because light travels at a finite (though large) speed, these enormous distances imply that we also see these distant objects as they were about 12 billion years ago. This seemingly simple property of light is in fact crucial to astronomers wishing to study the Universe's history, because it allows us to trace this history in detail by observing stars and galaxies at a variety of distances that correspond to different epochs in the history of our Universe.

Existing instruments probe galaxies as they were about 12 billion years ago — or about a billion years after the Big Bang. In many ways, these galaxies have surprised us. Even the brightest have only about one percent as many stars as our own Milky Way (which is a very typical galaxy in today's Universe). They are also much more violent: although most large galaxies today pirouette sedately through space while maintaining regular structures (usually either rotating disks like the Milky Way and

Andromeda, or else smooth ellipsoidal shapes), these early galaxies contain irregular knots and tendrils of stars that burst and smash into each other with abandon.

But, in many other ways, early galaxies are also very similar to their descendants that surround us today. We easily recognize the stars, which have chemical compositions similar to — though often more primitive than — today's galaxies, and their relative mix of stars, gas, and dust matches rapidly star-forming galaxies near us reasonably well. Many have supermassive black holes growing at prodigious rates, with properties remarkably similar to their nearby counterparts.

A second, independent method to map the history of the Universe is through radiation known the *cosmic microwave background*, which provides us a picture of the Universe's infancy.

Shortly after the Big Bang, the Universe was extraordinarily dense and hot — so hot that it contained a soup of particles and photons (or discrete packets of light). The matter particles constantly interacted, colliding and sticking to each other to form more massive particles — for example, protons and neutrons could stick together to form atomic nuclei, or protons and electrons could stick together to form atoms. But the particles also collided with photons; if these photons had enough energy — more than required to bind the new particle together — they could break that composite particle up into its constituent pieces. Thus, so long as the Universe remained hot, the continuous formation of larger groups of particles was opposed by their continuous destruction from photons.

However, as the Universe expanded after the Big Bang, it cooled. This happens to any gas — as an example, move your hand along the stream of air from a deflating balloon. The air (initially at room temperature inside the balloon) cools noticeably as it escapes and expands into the surrounding air. Similar "cooling" also applied to the photons. A photon's energy is inversely proportional to its wavelength, or the distance between two adjacent crests of the electromagnetic wave represented by the photon. As the photon travels through the expanding Universe, the distance between crests — just like every other distance — stretches. As this wavelength increases, the photon's energy also decreases, making the radiation "cooler".

Now consider how the soups of matter and photons interact as the Universe expanded. The photons "cooled" and eventually fell below the threshold energy needed to destroy composite particles: from that point forward, such composites stuck together rather than disintegrating after a photon collision. Thus, as time progressed, more and more weakly bound particles appeared: helium nuclei formed a few minutes after the Big Bang, and, a few hundred thousand years later, protons and electrons combined into hydrogen atoms (which are one hundred thousand times less tightly bound than nuclei). By this point, the photon energy had fallen far enough that they could no longer *ionize* the new atoms (or strip its electron from the proton), so hydrogen finally remained stable.

This *recombination era* marks a profound change in the relationship between the photons and matter in the Universe. Before it, the two soups constantly mixed and interacted, with photons ionizing atoms and bouncing, or *scattering*, off electrons. After it, with the electrons all bound into atoms that the photons could no longer ionize, the Universe suddenly became transparent to them. The photons traveled, (almost) unimpeded, for over 13 billion years.

Of course, while the photons traveled the Universe they continued to expand and their wavelengths stretched more and more. Today, these photons have become *microwaves* (with wavelengths in the millimeter range), and they are collectively known as the *cosmic microwave background* (or CMB). Because they have not scattered off matter since 400 000 years after the Big Bang, mapping their intensity provides us a "baby picture" of the Universe as it was at that moment — if today's Universe were a 40-year old person, the CMB would correspond to a snapshot just 10 hours after their birth!

As the only window into the infancy of our Universe, the CMB has provided an enormous amount of information about its composition and history since its discovery in 1969. (Amazingly, most readers have already "seen" this radiation, because about 10 percent of the static that appears when one tunes a television between broadcast stations comes from the CMB!) For our purposes here, its most important aspect is its remarkable uniformity: the CMB has the same intensity, to a few parts in 100 000, *everywhere* on the sky. This implies that the Universe was uniform to that same level a few hundred thousand years after the Big Bang: it was obviously an extraordinarily simple place during the recombination era.

Yet, clearly, by a billion or so years after the Big Bang — when the most distant galaxies we can now see lived — this simplicity had transformed itself into complexity comparable to the present day, with shockwaves smashing across galaxies, violent star formation, and black holes spewing high-energy radiation into the Universe around them. Thus a remarkable transformation occurred sometime during the "dark ages" in between the recombination era and the most distant galaxies we can see.

It is this transformation — known to astronomers as the *cosmic dawn*, during which the first black holes and galaxies formed — that is now one of the final, and most fundamental, frontiers of cosmology.

6.2   The Hidden Cosmic Dawn

So how can we study this transformation? Unfortunately, traditional astronomical tools have extraordinary difficulty. Theoretical models of these first galaxies and their immediate descendants predict that they are hundreds of thousands of times smaller — and hence fainter — than the Milky Way. Even the most ambitious new telescopes currently planned, such as the *James Webb Space Telescope* (NASA's successor to the *Hubble Space Telescope*) and the next generation of ground-based optical telescopes (with enormous mirrors up to 100 feet across focusing light to a pin-point) will only be able to scratch the surface of these early galaxies. Moreover, at these early times galaxies contain only a small fraction of the Universe's matter — nearly all actually still lies in the vast stretches between the galaxies, or the so-called *intergalactic medium*.

To focus in on the best approach to study the cosmic dawn, recall that the CMB appeared when protons and electrons combined into neutral hydrogen atoms. Because the expanding Universe had by then stretched the photons to low energies, these atoms were stable — and, because the CMB photons only stretched out more and more as they traveled, protons and electrons remained together for a long time.

In contrast, today the intergalactic gas is extremely highly ionized, with protons and electrons free from each other. The many hot stars and rapidly-accreting black holes (known as *quasars*) inside galaxies flood the Universe with radiation, which ionizes any atoms in the intergalactic medium. Observations of the most distant galaxies also show us that this gas was highly-ionized even 1 billion years after the Big Bang. Evidently the first galaxies, or their immediate descendants, must have *reionized* the intergalactic medium. In astronomers' eyes, this reionization era has become the hallmark event of the cosmic dawn, because it defines a radical shift in the composition of the intergalactic gas.

Because this reionization event occurred at roughly the same time as the most distant galaxies that we can currently see (or possibly slightly earlier, according to current data), the defining characteristic of the Universe during the cosmic dawn is that the intergalactic medium was composed of *neutral* hydrogen. To study this era, we should therefore focus on its properties.

Unfortunately, hydrogen is transparent to nearly all low-energy photons — the only tools we have to study the distant Universe. Quantum mechanics dictates that — unlike the planets in our solar system — electrons cannot orbit the central proton at just any distance; instead, only discrete orbits are allowed with very specific energies. This means that only those photons with precisely the correct energy to shift the atom's electron between its discrete quantum energy levels can interact with it; others simply pass through the atom as if it were invisible. In particular, a photon of the correct energy that strikes an atom can be absorbed and make the electron move to a wider orbit; the electron can then jump back on its own and emit a photon of that same energy. Usually, the most useful of these atomic orbit jumps correspond to optical or ultraviolet photons.

One might therefore hope to observe the radiation from these transitions during the cosmic dawn. However, there was so much neutral hydrogen in the intergalactic medium, and these transitions interact so strongly with the photons, that *all* of the photons would be completely absorbed as they traveled to us, leaving nothing for an observer to measure!

6.3 The Solution: Flipping Spins

Thus we must look beyond the usual suspects for a weaker, lower-energy transition. The best is the *hyperfine transition*. To understand its properties, it is helpful to picture the proton and electron as tiny spinning electric charges (this is not rigorously correct, but it is a fair approximation for our purposes). Any such spinning charge generates a magnetic field, so both the proton and electron act like tiny bar-magnets. If the north poles of these magnetic fields align, the configuration is unstable, and the electron will eventually flip its spin so that the north poles attain opposite orientations.

Just as with electrons moving between atomic orbits, this "spin-flip" releases energy in the form of a photon. The difference is its wavelength: 21 cm (which also corresponds to a frequency of 1420 MHz, making these radio waves). This means the photons are nearly 100 000 times "bigger" than the ultraviolet photons corresponding to the atomic orbits, or that their energies are 100 000 times smaller.

If a photon with 21 cm wavelength strikes an atom whose magnetic fields are anti-aligned, the atom *may* absorb the photon and flip its spins to the aligned state. However, because this transition is so improbable, such absorption turns out to be very unlikely — a photon traveling through the intergalactic medium has only about a 1% chance to be absorbed, even though it encounters an enormous number of atoms. As with the CMB, the relative transparency of the intergalactic gas to these photons lets us take pictures of the distant Universe with them.

Of course, the photons emitted by the spin-flip transition do not remain at this 21 cm wavelength, because they travel through the expanding Universe. Indeed, those emitted during the cosmic dawn travel many billions of light years before they reach Earth, expanding their wavelengths by a factor of six, ten, or even twenty (depending on the precise distance from which they were emitted).

Their final wavelengths are then in the range of about 1–10 m — from the size of a child to a large room — corresponding to frequencies of 30–200 MHz. This is a range familiar to many of us, for it is heavily used by humans — including, most importantly, FM radios at 90–110 MHz.

One might immediately wonder how easy it will be to see the cosmological signal, which is so intrinsically weak, through all of the radio broadcasts, TV stations, aircraft and satellite communications, and other electronic devices that fill our modern world. We will see that this is the primary reason the MWA telescope sits in the vast empty stretches of the Australian Outback. But these interfering signals constitute only one of the challenges confronting the efforts to see spin-flip radiation from the cosmic dawn, as we will see later.

6.4   The Spin-Flip Transition as an Astronomical Tool

Despite its intrinsic weakness, the spin-flip transition has a long and illustrious history in astronomy. After the initial discovery of cosmic radio emission from the center of the Milky Way in the 1930s, the eminent Dutch astronomer Jan Oort quickly realized that lines in the radio spectrum could provide crucial information about our galaxy — but, at that time, no such lines were known. At Oort's suggestion, the theoretical astrophysicist Hendrik van de Hulst examined a number of possibilities and eventually discovered the spin-flip transition in 1944 — although the line's extreme weakness originally led him to express pessimism over its utility.

In fact seven years of hard work did pass before the line was observed, first at Harvard by Harold Ewen and Edward Purcell (who later won a Nobel prize as co-discoverer of nuclear magnetic resonance technology). To detect the incredibly faint features, Ewen, a cyclotron engineer who spent his weekends working on the project as a PhD dissertation, introduced a novel "frequency switching" technique to astronomy that cancelled out the background noise.

Only after the line's detection did Ewen learn that van de Hulst himself was spending a sabbatical at Harvard, a mere 15 min walk from their telescope. After Ewen passed the news to van de Hulst and Oort, the Dutch group — who had in fact been pursuing the line for years, unbeknownst to Ewen and Purcell — built a similar radio receiver and detected the line themselves just a few weeks later. An Australian group followed suit later that summer.

The utility of the line for understanding our Galaxy was immediately obvious, especially to Oort, a leading expert on the Milky Way's structure. Just as in the intergalactic gas, the interstellar gas inside galaxies is primarily composed of hydrogen, and much of it is neutral. Thus, by mapping the intensity of 21-cm emission, one could map the structure of the Milky Way.

But the real beauty of the spin-flip line is that it overcomes one of astronomy's greatest difficulties: the measurement of distances. It is of course trivial to map objects' positions on the sky (such as the constellations). But it is very difficult to map structure in the third dimension (perpendicular to the sky, or directly outward from the observer). The spin-flip line surmounts this challenge thanks to its fixed 21 cm wavelength.

Imagine that an atom emitting spin-flip radiation is at rest relative to the Earth. Suppose a crest of the radiation wave leaves the atom and travels toward us. The next crest will leave only after a time-delay equal to the wavelength of the light (or distance between crests) divided by the speed of the wave (in this case the speed of light). This guarantees that precisely one wavelength separates the crests.

Now imagine that the atom moves away from us. The relative motion of the atom and the Earth makes no difference to processes within the atom, so the time period between the emission of crests remains unchanged. But, during this time, the atom has moved away from us — so, to the observer, the distance between successive crests appears slightly larger — or the wavelength grows. Thus so long as we know the initial wavelength, the *observed* value tells us how fast the atom was moving relative to Earth.

The variation of wavelength — or frequency — is known as the Doppler effect and is familiar from everyday life: it is the same reason that a siren receding from the observer appears to decrease its pitch (or, conversely, an approaching siren increases its pitch — this corresponds to a decreased wavelength when the atom approaches the observer).

The Doppler effect tells us how to determine the velocities of the emitting gas, but how does that give us a distance? The Milky Way is actually a rotating disk of gas and stars. Once we know this sense of rotation, any line of sight from Earth provides a fixed relation between velocity and distance. (This is a purely geometric effect, as you can see an analog with a phonograph record or any other spinning object.) Thus we can use the *velocities* measured by the observed *wavelengths* indirectly to determine the distance of gas, hence mapping the three-dimensional structure of the Milky Way.

Interestingly, this is completely analogous to our methods for mapping structure in the cosmic dawn. In that case, it is the expansion of the Universe that provides the stretching wavelength rather than relative motion, but the effect is the same: each observed wavelength corresponds to a different distance. So by mapping the spin-flip intensity at each individual wavelength, we can also map a slice of the Universe at a given distance from us. By arranging these slices consecutively, we can then build up a three-dimensional map of the spin-flip brightness (and, indirectly, of the hydrogen gas).

The spin-flip line has proven essential to mapping galaxies, and over the past 60 years it has provided the clearest maps of the distribution of gas (and, indirectly, stars) inside the Milky Way. It has also been detected in many nearby galaxies and used to measure their masses, rotation properties, and structure.

Nevertheless, its extreme weakness means that it has only been detected inside of relatively large or nearby galaxies — objects 10 times closer than, and thousands of times larger than, galaxies from the cosmic dawn. How can cosmologists hope to leapfrog from these local observations to see some of the most distant regions of the Universe?

6.5 Foiled!: Early Cosmology with the Spin-Flip Transition

In fact, efforts to find intergalactic gas via the 21-cm line date back almost to Ewen and Purcell. In 1959, George Field (also of Harvard) attempted to detect 21-cm absorption toward a nearby bright radio galaxy: he searched for the percent-level absorption that would have occurred if the intergalactic gas were neutral. This was a quietly revolutionary idea, and although it failed — because the nearby gas is so highly ionized, there is simply nothing to see — it presaged later techniques (still state-of-the-art) that used the normal electronic transitions of hydrogen to map the intergalactic gas at moderate distances from us.

In the early 1970s, after the detection of the CMB, cosmologists started to ask how structure formed throughout cosmic time. One theory, due to the brilliant Russian physicist Yakov Zel'dovich, suggested that the first structures to form would be enormous — a million billion times the mass of the Sun. These massive structures would then have fragmented into the smaller objects that surround us today. If this were true, the enormous primordial "pancakes" — for they resembled sheets of matter — would provide (relatively) strong 21-cm sources, despite their distance.

Dozens of radio astronomers set out to find these pancakes over the next 20 years, despite the formidable challenges they faced. These early searches focused, for the most part, on the era about two billion years after the Big Bang — then extraordinarily far away to astronomers, although now rather routinely probed by even modest telescopes.

Unfortunately, these efforts were doomed to failure — not only was the gas still highly ionized at this time (so that there could be no signal), but Zel'dovich's pancakes simply did not exist. We now understand that structure formed in the opposite direction — gravity collected matter into small galaxies first, which then accreted matter and merged with each other to form larger and larger objects all the way to the present day. (Interestingly, Craig Hogan and Martin Rees first predicted the 21-cm signal in such a model in 1979, but it was largely ignored since the observations would have been so difficult!) Led astray by theory, the early surveys could not succeed. Nevertheless, these early observational programs were crucial in spurring developments in radio astronomy, and many of their technical approaches will be used by the current generation of telescopes.

6.6 Spin-Flip Radiation Holds the Key to Observing the Cosmic Dawn

Only in the mid-1990s did our picture of structure formation mature enough to allow firm predictions of the spin-flip radiation from the cosmic dawn. The key insight was that, rather than attempt to identify individual objects, one should observe the diffuse emission from the intergalactic medium itself. This takes precisely the opposite approach from a naïve view of astronomy: it is trying to detect the "empty space" between the stars without seeing the stars themselves! But it is possible because the intergalactic gas contains an enormous mass of material — albeit incredibly tenuous gas.

If this gas were truly uniform, picking it out would be impossible — because each region would appear identical to every other one. Fortunately, though, the Universe is not truly uniform. Recall that, even 400 000 years after the Big Bang, the CMB contained tiny fluctuations, corresponding to differences in the Universe's density of one part in 100 000. Over the subsequent several hundred million years, gravity inexorably increased these differences: regions with slightly more matter than average exerted slightly more gravitational force on their surroundings, pulling material in and gradually increasing their density. Meanwhile, regions with slightly less matter than average exerted less gravity, so material flowed outward (into the higher-density surroundings) and the density decreased even further.

Thus, by the cosmic dawn, the intergalactic medium contained a "cosmic web" of high and low density regions. (In one important way, Zel'dovich's pancake model is correct, because this cosmic web was actually built from sheets or pancakes of matter; the "filaments" where two sheets intersect give the web its name.) The contrast between these regions is visible in the spin-flip line because more matter produces stronger radiation. Mapping the intensity of the emission provides a map of neutral hydrogen in the intergalactic medium during the cosmic dawn.

Compared to optical telescopes, which map the locations of discrete galaxies, this technique produces a very different picture of the Universe: one that contains no stars, but a great deal of matter that is heavily influenced by the stars, galaxies, and quasars in the early Universe. Conveniently, the spin-flip radiation therefore teaches us both about the objects that populate the Universe and the gas in between them.

The intensity of the spin-flip radiation depends on three essential factors: the underlying density of the gas, the fraction of the hydrogen that remains neutral (instead of ionized), and the gas temperature. The first allows us to map the distribution of matter in the Universe during the cosmic dawn, which is otherwise impossible. It turns out that this distribution depends on the fundamental cosmological parameters of the Universe — for example, the overall density of matter, the expansion rate of the Universe, and the properties of "cosmic inflation" that set the entire process of structure formation in motion.

In principle, the 21-cm background is a treasure trove of information on these questions. This is because it allows us to construct a three-dimensional map of the cosmic dawn: as described above, each *observed* wavelength corresponds to a different travel time through the Universe, and so a different distance from us. The three-dimensional information we receive about the density field (two dimensions along the sky and the third distance from us) allows us to probe — in principle — over *ten million billion* independent pieces of the Universe. (In contrast, the CMB is not a spectral line so it contains less than one hundred million pieces of information — yet it still provides the best constraints on cosmology to-date!) The statistical power of this sample is obviously enormous, although (as we will see) extracting it all will be prohibitively difficult.

The second factor, the neutral fraction of the gas, tells us when and how the first galaxies reionized the gas around them. Interestingly, this reionization event did not occur uniformly across the Universe; Fig. 6.1 shows some of the complex patterns that develop. Because only stars and black holes produce photons with enough energy to ionize the gas, reionization began with small ionized bubbles forming around the first such sources. As the galaxies grew and merged together, these individual bubbles rapidly grew and merged as well, becoming very large — tens of millions of light years across — until they filled the Universe. Obviously, the sizes of these bubbles depended on the brightness of each galaxy, and their locations correlated directly with the galaxies. If we could map these ionized bubbles by looking for regions where the spin-flip radiation vanished, we would learn about the galaxies creating them — even though most of these galaxies are so faint as to be invisible to conventional telescopes.

The third factor, the temperature, is the most complex (and potentially also the most interesting!). It determines the overall intensity of the

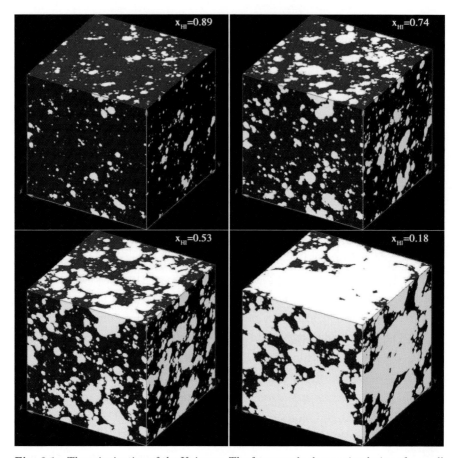

Fig. 6.1 The reionization of the Universe. The four panels show a simulation of a small part of the Universe (300 million light years across) during the cosmic dawn; each panel corresponds to a different time. As the first galaxies form, they launch photons into the gas around them, which produce small ionized bubbles around the galaxies (the white blobs in the upper left) embedded in a vast sea of neutral gas (black). As the galaxies grow and merge, these bubbles grow (upper right and lower left), eventually merging to fill space (bottom right); the labels in each panel tell how much of the gas remains neutral (from 89% at upper left to 18% at lower right). This event is the *reionization* of the intergalactic gas. **Figure credit:** A. Mesinger (Princeton University).

emission. In practice, the relevant comparison is to the temperature of the CMB: no matter which direction we look, we cannot escape the CMB, and we can only observe the spin-flip radiation against this backlight. If the gas lies at the same temperature as the CMB, the emission from both is precisely identical, so we cannot separate them — and the gas is effectively invisible! On the other hand, if the gas is hotter than the CMB, it radiates more strongly and we see *emission* against the CMB backlight. If the gas is colder than the CMB, it actually *absorbs* the CMB photons that pass through it and appears fainter.

6.7 The Spin-Flip Background: The First Stars

Thus the temperature of the hydrogen gas is crucial. Interestingly, it is actually set by the first galaxies to appear during the cosmic dawn — so by studying the temperature structure of the gas, we can learn about these sources, even though they are themselves invisible to us. Figure 6.2 shows an example spectrum of the spin-flip radiation background that outlines the important stages in this story. Here we look only at the *average* spin-flip signal across the entire Universe: in actuality, the signal fluctuates from point to point depending on the local physics (density, neutral fraction, and temperature), but this average presents a simplified measure of the Universe's evolution through time. A word of warning, though: the particular example in Fig. 6.2 is built from a simple theoretical model and the details are most certainly wrong — we await real measurements to understand the cosmic dawn!

The ordinate here shows the brightness of the spin-flip signal. It is reported in milliKelvin, a temperature unit, which is common for radio astronomy; the temperature represents the equivalent thermally-emitting object, here shown relative to the CMB (which we view as a backlight). The spin-flip signal is usually a fraction of a Kelvin, representing the exceptional weakness of this transition. Note also that this temperature can be *negative*: this means that the signal emits *less* than the CMB, or equivalently absorbs that background.

Immediately after the Universe became (mostly) neutral, the gas — although it was indeed almost completely neutral — was invisible to spin-flip radiation. This is because the small number of unattached electrons continue to scatter the CMB photons fast enough to keep their temperatures linked.

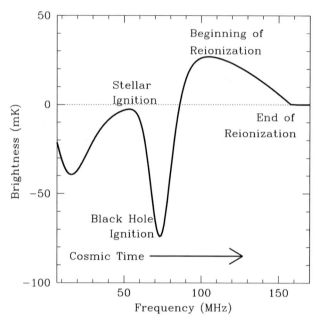

Fig. 6.2 The *average* spin-flip signal in a simple model of the cosmic dawn. The ordinate is the brightness of the line (conventionally measured in milliKelvin by radio astronomers). The abscissa is the observed frequency; because of the cosmological redshift this is a proxy for cosmic time, which increases from left to right. Four key cosmic epochs are marked in this model: the point at which the first stars appear, the point at which the first black holes begin to heat the hydrogen gas, and the beginning and end points of reionization, when early galaxies ionize all the hydrogen gas in the Universe. The horizontal dotted line corresponds to zero signal, when the spin-flip background has the same temperature as the cosmic microwave background. **Figure credit:** S. Furlanetto (UCLA).

About 5 million years later, though, this link began to break: photons and electrons stopped colliding, mostly because the photons had lost so much energy. At this point, the gas began to cool *faster* than the CMB, making it absorb the CMB behind it. Figure 6.2 shows this era as the trough at about 15 MHz.

This absorption continued for about 50 million years. But then the 21-cm background vanished again, because the expansion of the Universe continued to separate the hydrogen atoms. Eventually, they spread so far apart that collisions between them became very rare, so that the gas temperature stopped affecting the spinning charges. Instead, absorption of the CMB photons themselves eventually set the spin-flip line's effective temperature, and their intensities became equal. This, of course, poses a problem for attempts to study the cosmic dawn, because the gas was invisible. In Fig. 6.2, this happened at the peak near zero brightness at about 50 MHz.

Fortunately, at about this time (at least according to theoretical predictions), the first stars began to light up the Universe. Their radiation profoundly influenced the properties of the hydrogen gas. Not only did they produce ionizing photons (which created the bubbles in Fig. 6.1), but they produced lower-energy photons as well. Although these photons could not ionize the gas, they did cause electrons inside hydrogen to bounce between energy levels. During these jumps, the electron could also flip its spin: essentially, the atom makes a spin-flip by absorbing and re-emitting optical or ultraviolet photons. Once enough of these photons filled the Universe — generated by all the galaxies that existed — this process drove the temperature of the spin-flip transition far below that of the CMB, lighting the 21-cm background up in absorption again. This is analogous to the black lights employed by crime-scene investigators, which indirectly light up residual fluids otherwise invisible to the naked eye. Here the first stars play the role of the black light which renders the hydrogen gas — always there, but previously transparent — visible to the observer.

The crucial point is that, once these first stars appeared, the spin-flip transition became a useful gauge of the *collective* radiation field of all these galaxies. Again, even though we cannot see the galaxies themselves, we can indirectly infer their properties by mapping the 21-cm intensity. The turning point labeled *Stellar Ignition* in Fig. 6.2 marks the first appearance of this stellar background; the frequency at which the signal turns toward absorption tells us when these stars appeared, and the rate at which the absorption changes tells us their properties.

These first stars likely differed profoundly from those surrounding us today, because they formed through very different mechanisms: the primordial gas from which they coalesced lacked elements like carbon and iron that are crucial to star formation in the Milky Way. As a result, models show that they may have reached several hundred times the size of our Sun, which also made them extremely hot and luminous. But this formation mechanism is also extraordinarily fragile, and the very radiation background that renders the spin-flip background visible also chokes off the formation of these massive stars. The physics of these stars was extremely complex, but by observing the background can we begin to understand how these stars formed and how "normal" stars replaced them.

6.8 The Spin-Flip Background: The First Black Holes

However, spin-flip absorption persisted only as long as the gas remained cold. The most interesting heating mechanism was X-ray radiation from material falling onto black holes. These photons also knocked electrons off atoms, depositing a great deal of energy with the electron, which then rattled around the gas, heating it. Because the hydrogen was initially so cold, even a few X-rays sufficed to heat the gas significantly (marked in Fig. 6.2 by the point labeled *Black Hole Ignition*), eventually warming it above the CMB temperature. At this point the signal transformed from absorption to emission (at roughly 85 MHz in the figure).

The crucial piece of physics is *how* these black holes formed in the early Universe. Theoretical astrophysicists recognize a number of possibilities, all of them exotic. The least exceptional is that the first massive stars ended their lives with an enormous explosion, with their cores collapsing into black holes ten or so times as massive as the Sun — just like massive stars in our own Galaxy. These explosions — known as supernovae — were extremely powerful, flooding their environments with X-rays.

The black holes themselves could then continue to emit X-rays if gas fell onto them: the strong gravity of the black hole accelerated the gas to high velocities, and the rapidly-infalling streams then collided with each other, heating themselves to tens of millions of degrees and emitting X-rays. This accretion is common in the nearby Universe, but it requires a reservoir of gas to fall onto the black holes. During the cosmic dawn, the short-lived but violently powerful stars may have cleared out their surroundings by their intense radiation or winds.

On the other hand, such black holes may not even have formed. Stars a couple of hundred times as massive as our Sun do not explode in the conventional way. Instead, they become so hot that photons deep inside can spontaneously collide with each other to form electrons and positrons (the antimatter equivalent of electrons). These photons "waste" much of their energy forming these particles, so they provide less support against the star's enormous weight. The entire star then collapses, rebounds, and explodes *without* leaving a black hole. These so-called *pair-instability supernovae* represent genuinely new physics that may occur in the early Universe and would provide exciting constraints on the first stars.

In contrast, stars either somewhat less or somewhat more massive than those undergoing pair-instability supernovae completely implode without producing a supernova — simply swallowing all their material into a large black hole. Accretion onto the resulting black hole would still radiate X-rays, but their radiation properties would differ from those produced by pair-instability supernovae.

These different scenarios encapsulate both the excitement and uncertainty about the cosmic dawn. Theoretical models hint strongly that the first stars were very different from nearby stars, but the models are not yet sophisticated enough to pin down their properties. Only by observing them — or their impact on the surrounding gas — can we answer these questions and fill in this crucial gap in the history of our Universe. Unfortunately, these tiny, lonely stars are extraordinarily difficult to see directly, and none of the innovative telescopes on the horizon can even hope to do so. It is only through their collective effects on the 21-cm background that we can study these objects.

A completely different sort of black hole includes the so-called *supermassive black holes* (millions to billions of times as massive as our Sun) that live at the centers of galaxies. As gas falls onto these enormous objects, the collisions can generate a prodigious amount of heat, making some of these objects (then called *quasars*) the brightest sustained luminous sources in the Universe. Some of the strongest of these beasts, with estimated masses a billion times that of the Sun, already existed just a billion years after the

Big Bang. Presumably these objects began as smaller "seed" black holes that grew through rapid accretion of gas, which would have produced huge quantities of X-rays.

These are just a few of the many mechanisms to produce X-rays during the cosmic dawn. By studying the spin-flip background shown in Fig. 6.2, and especially the transition from absorption to emission, we will learn about the first black holes to appear in the Universe, as well as how they grew and evolved.

6.9 The Spin-Flip Background: The Epoch of Reionization

Figure 6.2 also shows the effects of reionization on the average spin-flip signal: in this simple model, it began in earnest at the point labeled *Beginning of Reionization* and ended when the signal vanishes at about 160 MHz. In this example, the neutral fraction (and hence brightness) decreased fairly steadily over this range, but that need not be true — the rate at which stars form, the types of stars that form, and the efficiency with which they are able to launch ionizing photons into the intergalactic gas are highly uncertain, and the physics of these processes has important implications for our understanding of galaxy formation.

As described above, the Universe's first stars (forming a few hundred million years after the Big Bang) were likely very different from those in the nearby Universe. Yet all of the galaxies we do see — including the most distant currently accessible — have stars much more similar to those around us. How did the transformation to this normal mode of star formation occur? Was it slow or rapid? Could the two kinds of stars co-exist inside the same galaxy or clusters of galaxies? Theoretical models suggest some answers, but the processes are so complex that no clear consensus has emerged.

One of the most important questions in all of astronomy is how efficiently galaxies can form stars. In galaxies like the Milky Way, stars form fairly sedately inside dense clouds of gas, but in other galaxies stars can form hundreds of times more efficiently — at least for brief intervals. These rapid bursts of star formation are often triggered by collisions with other galaxies. During the cosmic dawn, galaxies grew extremely rapidly, so these bursts may have been very common.

On the other hand, most of those galaxies were very small, and as such they were relatively fragile. An initial burst of star formation may even have blown them apart completely, halting the formation of any additional stars, when the shockwaves from exploding stars drove gas out of the galaxies.

Another question is how important the quasars were to the process of reionization, because they also produced huge numbers of ionizing photons. But these higher-energy photons did not produce well-defined ionized bubbles, as did the stars in Fig. 6.1: instead, the gas would have gradually transitioned from neutral to ionized. This would create very different patterns in the spin-flip signal and allow us to measure how these black holes grow beyond the heating era.

These issues provide just a sampling of the complex physics of the first galaxies, and they show the extreme difficulty — even at a qualitative level — of making robust theoretical predictions about their properties. The spin-flip background is so crucial because it will finally help to answer many of the mysteries of the cosmic dawn.

6.10 FM Radio Antennae as Cosmic Observatories

Obviously, this spin-flip background contains a tremendous amount of information, including fundamental cosmological measurements (from the density field), the astrophysics of the galaxies that reionized the intergalactic gas (from the neutral fraction), and the properties of the first stars and black holes in the Universe (from the temperature field).

As a result, several telescopes are racing to observe the signal in the near future. Most of these are, however, very different from the conventional "telescopes" that may immediately spring to mind. Remember that the spin-flip transition initially produces a photon with a frequency of 1420 MHz (or a wavelength of 21 cm). As the photon speeds through the expanding Universe, its wavelength stretches, and the frequency (which is simply the number of crests that pass an observer per second) decreases. Photons from the cosmic dawn increase their wavelength by factors of about 7 to

30 — yielding wavelengths in the range of 1.5 to 6 m, or frequencies from 200 MHz down to 40 MHz.

At these wavelengths, "telescopes" are very simple indeed: one just needs a metal rod a few meters across, approximately equal to the desired wavelength! Radio waves, like all light waves, cause electrons inside matter to oscillate across their wavelength, and a conducting rod of this size is perfectly matched to detect these oscillations as the wave travels by. These simple rods are known as "dipoles" in the telescope community.

In fact, such structures are very familiar to most of us, because analog television and FM radio antennae — which operate in the same wavelength range — are often constructed in just this way. (A portable radio's antenna is even simpler — it is the headphone cord itself, conveniently about a meter long!) In comparison to the typical dishes making up most radio telescopes (or mirrors in optical telescopes), these instruments *appear* much more primitive — although they are still perfectly-matched to the task at hand.

These telescopes can follow two basic strategies to observe the spin-flip background. The first is to measure the *average* signal (over the entire sky) as a function of frequency, just as Fig. 6.2 shows. The advantage of this approach is its modest telescope requirements: because one can utilize radiation from the entire sky, a single antenna a few meters across suffices to detect the signal. The disadvantage is that it provides only the most basic information, telling you nothing of the *structure* of the signal at any given time, just its average value. Thus it might reveal that the hydrogen gas was just starting to heat up, but it would not tell you where the heat sources themselves lived.

Two groups are now working hard to measure this sky-averaged spectrum. The *Experiment to Detect the Global "Epoch of Reionization" Signal*, or EDGES, is led by Judd Bowman at Arizona State University and Alan Rogers of the Haystack Observatory, while the *Cosmological Reionization Experiment*, or CoRE, is led by Ravi Subrahmanyan (of the Raman Research Institute in India). Figure 6.3 shows the EDGES experiment deployed in the field and a prototype of CoRE; note the simplicity of their antennae.

Fig. 6.3 The EDGES and CoRE experiments (left and right, respectively). For EDGES, the flat panel in the foreground is the antenna; the white object below contains the receiver electronics. For CoRE, the pair of objects in the foreground are the antennae, while the electronics are contained in the tent in the background. The solar panels in the background provide power to the experiment. **Figure credit:** J. Bowman (Arizona State University, left); R. Subrahmanyan (Raman Research Institute, right).

6.11 Piles and Tiles of Antennae: Mapping the Spin-Flip Background

The second strategy is to map the fluctuations in the spin-flip radiation at each frequency, in order (ideally) to build a map of variations in the hydrogen gas's density, ionization state, and temperature — and thus, indirectly, to pin-point the sources of radiation in the cosmic dawn. This requires much more sophisticated telescopes.

The first basic problem with using a single dipole antenna is that it offers almost zero angular information: regardless of the direction from which a radio wave arrives, the electrons in the rod respond in (roughly) the same way. For FM radios, this is very convenient, because we want to hear each station regardless of our orientation relative to its transmitter. But its inconvenience for astronomy is obvious, because we would usually like to know the exact location of a source on the sky. (Measurement of the *average* emission, as EDGES and CoRE are attempting, is an obvious exception.)

Fig. 6.4 The four first-generation mapping experiments. Clockwise from top left, we show one of 500 "tiles" in the Murchison Widefield Array (MWA, in Western Australia), several of the "core" stations of the Low Frequency Array (LOFAR, in the Netherlands), four of the dishes at the Giant Metrewave Radio Telescope (GMRT, in India), and one antenna for the Precision Array to Probe the Epoch of Reionization (PAPER, in West Virginia and South Africa). Note the diversity of approaches to building such a telescope. **Figure credits:** clockwise from top left: C. Lonsdale (Haystack Observatory); TopFoto (Assen); B. Premkumar (NCRA); A. Parsons (UC-Berkeley).

This is why most telescopes are dishes: the dish's surface reflects the incoming radiation to a point that depends on its precise incoming direction, allowing us to resolve individual sources. Thus one approach to mapping is with dishes; which greatly improve the ability to pin-point the radiation's source.

However, a single radio dish's angular resolution is usually still modest, because light waves with such large wavelengths tend to refract (or bend) as they reflect off the dish. To achieve even higher resolution, astronomers use a technique called *interferometry* to combine signals from multiple dishes (or antennae). In essence, light from a source on the western sky takes longer to reach a dish that sits farther east than one that is positioned in the west, while directly overhead light might strike them both at the same time. By carefully recording when each radio wave strikes each dish, the signals from multiple dishes can be combined and used to triangulate precisely the location of a radiation source. Today, astronomers can even combine signals from dishes on different continents to build maps with extraordinary angular resolution, capable of measuring the thickness of an American nickel coin at the distance of the moon!

The second difficulty posed by mapping is that one then must measure the tiny amount of radiation from a small patch of the sky (rather than the total amount from the entire sky, as EDGES and CoRE do). To compensate for the smaller intrinsic signal, one must greatly increase the collecting area of the telescope — either by building one extraordinarily large antenna, or (as in an interferometer) combining the light from many antennae into one whole.

As a result, interferometry is the method of choice for mapping the spin-flip background, combining several or even hundreds of antennae into one colossal telescope. The first generation of these telescopes will have a few *acres* of antennae; future generations may have a hundred times that. The *Murchison Wide-field Array*, or MWA, discussed in the introduction to this essay, is a prime example. Led by a consortium of US and Australian astronomers, it "tiles" 16 pairs of dipole antennae into one group and will eventually collect light from 500 such tiles — using an astonishing 16 000 individual dipoles to measure the signal!

No fewer than three other groups are also building mapping instruments all over the world. The *Low Frequency Array*, or LOFAR, is a large general-purpose radio telescope in the Netherlands, with the 21-cm background a key science goal. The *Precision Array to Probe the Epoch of Reionization*, or PAPER, is primarily a US effort but sits in South Africa. Both of these instruments are interferometers built from dipoles, although many of the details differ from the MWA.

Finally, a group led by Ue-Li Pen, of the Canadian Institute of Theoretical Astrophysics, is using an Indian telescope called the *Giant Metrewave Radio Telescope*, or GMRT, to search for the signal. This is a more traditional radio telescope (built from 30 large dishes) and is being repurposed for this effort.

6.12 Mountains to Scale: Challenges to Observing the Spin-Flip Background

Yet, despite the excitement generated by these efforts, all previous attempts to measure the 21-cm background have failed. The obstacles to observing it are formidable, and only recently has the inexorable forward march of computing technology allowed astronomers to overcome the challenges (or so we hope!).

The first hurdle, to which we have already alluded, is interference from terrestrial sources — FM radio stations, TV stations (analog broadcasts fill this range), aircraft, satellites, and countless other sources. These completely swamp the cosmological signal, so instruments require some way to work around them. The simplest approach is to go as far as possible from the contamination — unfortunately, that requires one to go very far indeed! Even in the middle of the Australian Outback — home of the MWA — some contamination remains.

So the second approach is to measure the frequency structure of the signal at very high resolution: in essence, the instrument has a huge number of "channels" separated by tiny frequency shifts. Terrestrial interference tends to produce signals that are narrow in frequency, so one can simply excise the contaminated channels and use the "clean" regions between them for the measurement.

The existing experiments use both of these approaches; MWA, PAPER, and EDGES all sit in very isolated locales (either South Africa or Western Australia) to reduce the contamination. LOFAR, on the other hand, is centered in the Netherlands, where the interference is exceptionally large. They will rely on excellent spectral resolution and fast computer software to extract the cosmological information between the local interference.

A second challenge is the *ionosphere*, the uppermost layer of our atmosphere (stretching from 50–1000 km). Here, the atoms and molecules have been ionized by solar radiation, and the free electrons scatter the low-frequency radio waves we wish to measure over most of the wavelength range of interest — this is troublesome but not fatal, because the electrons bend the waves but do not absorb them. As a result, sources appear to jitter on the sky — this occurs in a similar fashion to turbulence in the atmosphere making stars twinkle. Nevertheless, large computers are necessary to correct these ever-changing distortions to the maps.

However, at the very lowest frequencies (below about 40 MHz), the ionosphere becomes opaque to the incoming radio waves, erasing all the astronomical information. At these lowest frequencies, the only way to study the spin-flip signal is to place a telescope beyond the atmosphere — in space.

The third, and most serious, obstacle is that this spin-flip background from the cosmic dawn is far from being the only *astronomical* source in the relevant frequency range — and these "foregrounds" are *much* brighter than the cosmological signal itself.

A wide range of objects contribute to this foreground, from supermassive black holes in the distant Universe to relatively nearby clusters of galaxies to the accumulated radiation of trillions of normal galaxies. But by far the most important is the radio emission from our own Galaxy, the Milky Way. *Cosmic rays*, or subatomic particles racing through the Galaxy near the speed of light, permeate the Milky Way. The cosmic ray electrons spiral around the magnetic fields of our Galaxy and emit low-frequency radio waves. Unfortunately, over most of the sky, these cosmic rays are 10 000 times (or more) brighter than the spin-flip radiation we seek.

There is simply no way to avoid these contaminants, at least not short of traveling thousands of light years outside the Milky Way. We therefore must find ways to isolate the cosmological information embedded in the much brighter foregrounds. The key lies in the frequency structure of the signal. Recall that the hyperfine transition produces a spectral line, with a precisely defined frequency. As that radiation travels through the expanding Universe, it stretches by an amount that depends on its total path length (and hence its initial distance from us). Thus each observed wavelength corresponds to gas at a slightly different distance from us.

We therefore expect the intensity of the radiation to vary rapidly with the observed wavelength: one wavelength may correspond to gas inside an ionized bubble, while the next lies slightly farther away — outside of the bubble — so contains neutral gas, while the next sits in high-density, neutral gas so emits strongly, etc.

In contrast, the cosmic ray emission from our Galaxy varies extremely smoothly with wavelength: each electron produces a broad spectrum of radiation, and any line of sight passes trillions of individual electrons. We therefore measure the variation with wavelength along any particular direction and *subtract* the slowly-varying average component. This should "clean" the foreground contaminants but leave the rapidly-fluctuating cosmological signal.

Unfortunately, the process also removes information from the cosmological signal, because the method only works when the cosmological signal actually does vary rapidly with frequency. If one observes a small piece of the sky, this is reasonable: then the entire pixel can indeed sit inside an individual ionized bubble, followed by a rapid transition to neutral gas, etc. However, as the observed region grows larger on the plane of the sky — and in particular larger than the ionized bubbles — the contrast between the different frequencies decreases, because a single bubble no longer contains the entire pixel. In the limit of a very large region — say, the entire sky — *every* channel will contain a mixture of ionized bubbles and the neutral sea between them, and only the slow increase in the *average* number of ionized bubbles causes the signal to change. This slow increase may not vary any more rapidly than the galactic foregrounds, making their separation very difficult.

This is a particularly acute problem for the average signal measurements toward which EDGES and CoRE aim; in that case Fig. 6.2 shows rather gentle variation. Removing the foregrounds then requires precise modeling, which poses one of the major challenges for these experiments. Most likely their best opportunity is the era spanning the first stars, the first black holes, and the beginning of reionization, when the signal contains multiple reversals that foreground features are unlikely to mimic. In contrast, the range from the beginning to end of reionization has much less structure and will be more difficult to measure with confidence.

All of these problems become much worse at lower frequencies — corresponding to early cosmic times — where the Galaxy's brightness increases dramatically. Thus the most accessible part of the spin-flip background is at the highest frequencies (shortest wavelengths), or the *most recent* epochs. For this reason, all of the current mapping experiments focus on the reionization era. Conveniently, the patchwork of ionized bubbles and neutral gas also provides the clearest signals for such maps.

6.13 Sound and Fury, Signifying Statistics

Unfortunately, the sophisticated algorithms meant to clean this contamination cannot be perfect, and, in practice they leave a great deal of noise over and above the cosmological signal of interest to us. In fact, for all of the mapping experiments described above, this residual noise is larger than the signal itself, sometimes substantially so!

Thus, despite their power, these interferometers do not provide a pure view of the 21-cm radio signal; in fact the noise "static" makes its extraction extremely difficult. This means that we cannot actually use these telescopes to generate maps of the cosmic dawn. Is all the effort for naught?

Fortunately, even without detailed maps we can use *statistics* to extract information from the spin-flip signal even. Imagine that each "pixel" in our map has a small amount of real signal and a large amount of completely random noise. If we averaged over a large number of pixels, the random noise would cancel out and we would be left with a measurement of the average signal in each pixel. This is a simple example (and in fact cannot even be done in practice, because of the foreground removal algorithms), but more sophisticated statistics can be measured. We will rely on these to reveal the important physics of the cosmic dawn.

6.14 An Explosion of Telescopes

With all of these challenges, measuring the spin-flip background is certainly a daunting enterprise. Nevertheless, the impressive technical advances pioneered by the telescope teams place us today on the cusp of a truly exciting time. The teams' diverse approaches — not to mention their dedication — instill a great deal of confidence for the future.

LOFAR, the European (primarily Dutch) experiment, has the longest history, with plans stretching back about 20 years. It is the largest facility in terms of both scale and scientific range, as it is designed not only to measure the spin-flip background but to study objects in or near our Galaxy and even cosmic rays striking the Earth's atmosphere. The team faces a substantial challenge in the loud and complex radio environment of the Netherlands, but they are confident their algorithms can work between the contaminated channels and extract the signal. Their "station" approach of stacked dipoles will also require careful antenna calibration in order to ensure that the foreground removal algorithms work as hoped. However, as the largest experiment, they will come closest to producing maps. They also have a relatively fast start, with serious observations beginning in the fall of 2010.

The GMRT is also a general-purpose radio telescope. It began scientific observations in 1995 — amazingly enough, the spin-flip background was one of its initial motivations, but at that time the signal was poorly understood. Recently it has required substantial effort to redefine the purpose of the telescope for these efforts in a modern context. (For example, the team has had to contend with power lines leaking radio waves, as well as to develop novel methods to calibrate the measurements.) But the large size of the GMRT dishes, as well as the fast start enabled by having an instrument already on the ground, let the GMRT group place the first meaningful constraints on the spin-flip background during the summer of 2010. They showed that, 600 million years after the Big Bang, the hydrogen gas was either already ionized or significantly heated by black holes.

The MWA, a US/Australian collaboration situated on the Murchison sheep ranch in Western Australia (and with which this author is affiliated), was conceived and designed specifically to measure the statistics of the spin-flip background. It will be constructed of over 10 000 carefully optimized antennae grouped into 500 "tiles" (each with 16 pairs of antennae) and sits in one of the most radio-quiet areas in the world. Like LOFAR, the tile strategy increases the collecting area at the expense of complexity in the properties of the instrument. Indeed, it is so complex that the measurements must be processed in "real time", as the data is taken, so the MWA relies on powerful computers to sort the information. On the other hand, it should offer comparable statistical constraints to LOFAR, even though the telescope itself is several times smaller. As of fall 2010, the first 32 tiles are in place and taking data, with the full instrument expected to be complete in late 2011.

PAPER, led by astronomers in the USA but situated primarily in South Africa, is also designed specifically to measure the spin-flip background. It uses single antennae, rather than tiles, so is much smaller than the other telescopes; on the other hand, the data is much simpler to interpret. The PAPER team's philosophy is to start small and gradually build the instrument up, tackling obstacles as they arise. The group is therefore nimble and can react quickly as unexpected challenges arise. PAPER has already deployed two instruments (one in West Virginia and one in South Africa) and is gradually building their dataset over time.

One of the most exciting aspects of this rich field has been the four different philosophies driving the teams, leading to unexpected and complementary advances. As the biggest player, LOFAR aims to be a comprehensive low-frequency radio telescope. The MWA is optimized to measure the spin-flip background, making design choices that push the envelope to punch beyond its weight — but that will ultimately require a very careful understanding of the entire instrument. PAPER approaches the problem piece-by-piece, building bigger and better instruments as each stage is complete. Although much smaller than its competitors, PAPER is in many ways a simple instrument that can respond effectively to unanticipated challenges. Finally, the GMRT group made clever use of an existing — but suboptimal — telescope and generated the first interesting measurements. All have pushed each other forward by developing independent observing strategies, analysis methods, and calibration algorithms, and this healthy competition can only benefit the community in the long run — if only because the measurement is so difficult that multiple teams will be required to confirm any results!

6.15 Dreams for the Future

Even though these first-generation instruments are still works-in-progress, the community is already developing ambitious plans for future instruments capable of studying the spin-flip background in exquisite detail. Many of these projects call for expansion of the first-generation experiments to build bigger, better telescopes providing more sensitive statistical measurements and, eventually, even true maps of the spin-flip background. One extremely important goal is to push the observations to earlier cosmic times, or longer wavelengths. The foreground noise grows more and more intense as the wavelength increases, so measurements of the properties of the first stars and black holes require much larger instruments than even LOFAR.

A second goal is to measure the structure on ever-smaller scales: because the amount of spin-flip radiation decreases in these smaller patches, bigger and better telescopes are required to extract the signal. Ultimately, the dream is to observe the signal on the smallest possible scales — generating the 10 million billion data points mentioned previously. While this ultimate limit is far beyond anything currently contemplated, expanding the signal's range will undoubtedly enable sharper and sharper measurements of the cosmic dawn.

But other plans are even more ambitious than simply expanding the existing instruments. Recall that two of the major stumbling blocks are terrestrial interference and the Earth's ionosphere. If these prove to be substantial challenges — and they most certainly are at the lowest frequencies (corresponding to the earliest times in the cosmic dawn) — the best strategy would be to head *away* from Earth. Any satellite will sit beyond the ionosphere, but the interference problem is trickier: an Earth-orbiting satellite "sees" radio signals from the entire face of the Earth, so the interference is not necessarily any better!

Instead we must physically block the terrestrial radio waves. Fortunately, a nearly perfect block exists relatively close to the Earth: the moon. The moon's far side is shielded from terrestrial interference by thousands of kilometers of lunar rock and provides a nearly ideal location for a low-frequency radio telescope. Moreover, because the moon rotates once around its axis over exactly the same time-period that it orbits the Earth, the *same* side always faces away from the Earth: thus the "far side" is never subject to terrestrial interference and gets a permanently clean view of the radio sky.

Of course, delivering hundreds or thousands of radio antennae to the far side of the moon is itself an enormous challenge, but engineers are working on developing light, low-power antennae and electronics for these purposes. An alternative is a spacecraft that orbits the moon; during the fraction of the orbit spent on the far side, the same benefits apply. While spacecraft large enough to host mapping interferometers are all but impossible, launching a single antenna to measure the average background (just like EDGES and CoRE, but extending to earlier cosmic times) is certainly possible, and one mission concept — called the *Dark Ages Radio Explorer*, or DARE, has already been formulated by Jack Burns (of the University of Colorado) and collaborators including this author.

6.16 An Unfinished Story

We stand now at the threshold of these new instruments' first observing campaigns, leaving our story unfinished — for now. Yet, although the challenges are daunting indeed, the next few years will see the maturity of several innovative and daring experiments to study the spin-flip background from the cosmic dawn. The payoff is enormous: a brand new approach to study the epoch of the first galaxies', and one that promises unique insight into their properties and evolution. The flood of information that we anticipate from these telescopes — together with exciting new conventional telescopes ranging from the near-infrared to radio — will enable the first peeks at the complex physics governing the first stars, black holes, and galaxies. Within the next decade, our understanding of the earliest phases of galaxy formation in our Universe — and hence the progenitors of our Milky Way — will radically improve.

And yet, even such insight as this may not provide the most interesting ending to our story. Scientific exploration, whether in a laboratory, on an archaeological dig, or with a new telescope, always holds the tantalizing prospect of the completely unexpected. If we are lucky, our story will not end with answers to questions we already anticipate, but with new and unexpected puzzles that probe even deeper into the nature of the cosmos. Then the new window into the Universe's history enabled by these innovative telescopes will truly have come into its own.

CHAPTER 7

NEUTRINO MASSES FROM COSMOLOGY

OFER LAHAV AND SHAUN THOMAS
Department of Physics and Astronomy
University College London
Gower Street,
London WC1E 6BT
United Kingdom

One of the most exciting results in physics over the last decade has been the evidence that neutrinos have mass. In fact, so far neutrinos are the only form of non-baryonic dark matter that exists for sure. Particle physics experiments have confirmed that neutrinos oscillate between the three known flavors and this, in turn, implies a finite difference in their mass "eigenstates". While there is strong evidence for new physics beyond the Standard Model, this does not determine the absolute value of the masses. Cosmology not only probes this absolute mass scale, by currently determining the sum of the masses, but it is a completely independent method for which to test against. We now know that their masses are smaller than a billionth of a single hydrogen atom and that they are so elusive that they can pass uninterrupted through slab of lead with a width of a light year. Naturally, this requires understanding of exactly how the mass is measured from cosmology. Weighing the neutrinos provides perhaps one of the most beautiful interplays between physics of the largest and smallest scales and in the process highlights the synergy between particle physics and our Universe.

7.1 A Brief History of Cosmological Neutrinos and Hot Dark Matter

Cosmological measurements of Type Ia Supernovae, Large Scale Structure, the Cosmic Microwave Background and other probes are consistent with a concordance model in which the Universe is flat and contains approximately 4% baryons, 21% dark matter and 75% dark energy. However the two main ingredients, dark matter and dark energy, are still poorly understood. We do not know if they are real or whether they are the modern "epicycles", which fit the data until a new theory improves our understanding of the observations.

One may wonder if this consensus on the concordance model is a result of globalization. It is interesting to contrast the present day research in cosmology with the research in the 1970s and 1980s. This was the period of the Cold War between the former Soviet Union and the West. During the 1970s the Russian school of cosmology, led by Y. Zeldovich, advocated massive neutrinos (hot dark matter) as the prime candidate for all dark matter. As explained below, neutrinos were relativistic when they decoupled in the early Universe and so they wipe out structure on small scales. This led to the top-down scenario of structure formation: In this picture "Zeldovich pancakes" of the size of superclusters formed first, and then they fragmented into clusters and galaxies. This was in conflict with observations and cosmologists concluded that neutrinos are not massive enough to make up all of the dark matter. The downfall of this top-down hot dark matter scenario of structure formation, and the lack of evidence for neutrino masses from terrestrial experiments made the model unpopular. The Western school of cosmology, led by J. Peebles and others, advocated a bottom-up scenario, the framework that later became known as the popular cold dark matter. However, the detection of neutrino oscillations (Sec. 7.2) showed that neutrinos do indeed have mass; i.e. hot dark matter does exist, even if in small quantities. Therefore both forms, cold dark matter and hot dark matter, may exist in nature. This example illustrates that independent schools of thought can actually be beneficial for progress in cosmology and science.

7.2 Insights from Particle Physics

A tantalizing puzzle emerged in physics after observers started to detect neutrinos produced in the Sun: they observed far fewer than predicted from solar models. This became known as the solar neutrino problem. The issue was eventually alleviated by two experiments: Super-Kamiokande in 1998 (Fukuda *et al.*, 1998) and the Sudbury Neutrino Observatory (SNO) in 2001 (Ahmad *et al.*, 2001). They both indicated that neutrinos undergo flavor oscillations as they propagate. The early detectors were only sensitive to the electron neutrino and the absence of these particles was due to their transformation into other previously undetected flavors (muon and tau types). The total number of these weakly interacting neutrino flavors is known to be three. The evidence for this comes from the decay of the Z boson as more flavors would decrease its accurately measured lifetime. Additional neutrinos could exist but these would need to be sterile, meaning they do not interact via the weak force.

The most important consequence of the aforementioned oscillations is that they demonstrate that the three neutrino flavor states $(\nu_e, \nu_\mu, \nu_\tau)$ are quantum superpositions of finite mass states (m_1, m_2, m_3). This implies the oscillations can only occur if the neutrinos are massive! Mathematically, the mixing between these states is specified by a matrix, which in turn is expressed in terms of the three mixing angles: $\theta_{12}, \theta_{23}, \theta_{13}$. The Standard Model of particle physics precisely describes experimental data up to the electroweak scale. However, it provides no explanation for neutrino masses and their mixing. Accordingly, this evidence is the first indication for physics beyond the Standard Model.

The Super-Kamiokande experiment is a large underground neutrino observatory based in Japan. It consists of 50 000 tons of ultra-pure water and thousands of photomultiplier tubes. These are designed to detect Cherenkov radiation produced from the interaction of incident neutrinos with electrons or nuclei. Its predecessor even measured the burst of neutrino emission from supernova SN 1987A. Similarly SNO also observes Cherenkov radiation and is located deep underground in Canada. It was the first to directly indicate oscillations in the solar neutrinos.

Since the profound discovery more precision experiments have been set up to measure the oscillations' mixing parameters. This includes MINOS, where a muon neutrino beam is sent over 735 km from Chicago to Minnesota. The probability of the particles remaining in the muon state depends on the propagation length, energy of the particles (which are determined) and finally the splitting between mass eigenstates.

Overall there are a host of atmospheric, reactor and accelerator experiments that have given us ever-improving information on the mass splitting and mixing angles. These are summarized (Schwetz et al., 2008) below:

$$|\Delta m_{31}^2| = 2.40^{+0.12}_{-0.11} \times 10^{-3} \, \text{eV}^2, \tag{7.1}$$

$$\Delta m_{21}^2 = 7.65^{+0.23}_{-0.20} \times 10^{-5} \, \text{eV}^2, \tag{7.2}$$

$$\sin^2 \theta_{12} = 0.304^{+0.022}_{-0.016}, \tag{7.3}$$

$$\sin^2 \theta_{23} = 0.50^{+0.07}_{-0.06}, \tag{7.4}$$

$$\sin^2 \theta_{13} = 0.01^{+0.016}_{-0.011}. \tag{7.5}$$

It is interesting to note the absolute value in Eq. (7.1). It is because of this that the *hierarchy*, or mass order, of the eigenstates is undetermined. There are two possibilities: the normal hierarchy ($m_3 \gg m_2 > m_1$) and the inverted hierarchy ($m_2 > m_1 \gg m_3$). In current cosmological analyses the three masses are often assumed to be equal because the present measured limits are greater than the splittings and two hierarchies. In addition, it is clear from these expressions that all they detail is the *relative differences* and not the actual absolute scale. There are currently two methods to tackle this issue.

The first approach to measuring the absolute mass scale is with a kinematic Tritium beta-decay experiment. The idea is to analyze the ejected electron's energetic spectrum where the finite mass of the electron neutrino is expected to cause a change in shape and decrement in the end-point energy. In the decay (7.6) the emitted electron and antineutrino share a total energy of 18.6 keV.

$$^3\text{H} \rightarrow {}^3\text{He} + e^- + \bar{\nu}_e \tag{7.6}$$

If the neutrino is massless then there is no lower limit to the neutrino's share of this energy and the electron spectrum would extend up until the 18.6 keV limit. However, given that the neutrino has mass this removes some of the available energy, leaving a decrement at the end-point electron spectrum. Experimentally this is extremely challenging due to the statistical rarity of such endpoint events. The current limit from the Mainz Tritium beta-decay experiment sets a neutrino mass bound of 7 eV (Kraus *et al.*, 2005).

The KATRIN experiment hopes to make at least an order of magnitude improvement on this bound over the next few years. The electron energies are counted against a retarding threshold potential and pass through a spectrometer at just below the end-point energy. Its sensitivity on the electron neutrino mass is 0.2 eV at 90% (if massless) and will have a 5σ discovery potential for the electron neutrino mass if at least 0.35 eV. The vast gain is in part due to the large spectrometer (see Fig. 7.1) and pure Tritium source.

The second approach for an absolute mass measurement is through neutrinoless double beta decay. Normal neutrino double beta decay is a rare

Fig. 7.1 KATRIN's 200 ton spectrometer on its journey from Deggendorf to Karlsruhe. It is a colossal 23 m long and 10 m in diameter.

nuclear process whereby two neutrons in the same nuclei are spontaneously converted to two protons, electrons and antielectron neutrinos:

$$(A, Z) \to (A, Z+2) + 2e^- + 2\bar{\nu}_e. \qquad (7.7)$$

Again, the neutrinos remove some of the decay energy and a continuous emission spectrum results for the two electrons. However, if the neutrino is its own antiparticle (meaning they are Majorana particles) it is possible for the two neutrinos to annihilate and a neutrino*less* decay occurs.

$$(A, Z) \to (A, Z+2) + 2e^-. \qquad (7.8)$$

For the two electrons combined in this case one gets a peak in the energy at exactly the value of the decay limit. Furthermore, the rate of this decay is related to the square of the neutrino mass. If the neutrino is not its own antiparticle (meaning it is a Dirac particle) the process does not occur. Therefore, the observation of this phenomenon not only provides evidence of the absolute mass scale but also evidence on the neutrinos' intrinsic nature! If seen it also demonstrates the violation of lepton number conservation, which is required in some Grand Unified Theories.

One experiment looking for neutrinoless double beta decay is NEMO (1, 2, and 3), with superNEMO in the research and development stage. This next generation experiment is expecting an effective Majorana neutrino mass sensitivity of 0.05 eV by 2019. It is worth noting that if neutrinos are Dirac particles and have a small mass, then the only hope to measure the mass-scale in the near future is from cosmology. This is simply because Tritium Beta decay experiments have a finite resolution and the neutrinoless beta decay will have no signal. However, the overall importance of particle physics experiments in the search for neutrino masses cannot be overstated given they directly probe neutrino properties.

7.3 Background to Cosmology

Cosmology is a daring intellectual endeavor. It seeks to describe the origin, evolution and fate of the Universe. It also aspires to provide a complete census of its contents. These two apparently separate goals are in fact intimately related. This is because of Einstein's theory of gravity — general relativity. It imposes the concept of gravity not as a force but as a representation of the geometry, or curvature, of spacetime. Furthermore, this curvature is determined by the very presence of matter and energy. Similarly, any particle moving through such geometry will have its path affected, creating an apparent trajectory or gravitational orbit.

Applying this gravitational framework to the Universe itself therefore gives us a mathematical description where large-scale dynamics are dependent upon constituents. The resulting expression is the Friedmann equation and it is a solution to Einstein's formulae, giving

$$H^2 \equiv \left(\frac{\dot{a}}{a}\right)^2 = \frac{8\pi G}{3}(\rho_\Lambda + \rho_b + \rho_c + \rho_\nu) - \frac{k}{a^2}, \qquad (7.9)$$

where a is the scale factor describing a relative scale for expansion in the Universe; H is the Hubble parameter; ρ_i are the densities of dark energy, normal baryonic matter, cold dark matter and neutrinos, respectively; and k is the global curvature. The matter densities can have a particular critical density whereby the curvature tends to zero. In fact, there are observational and theoretical arguments explaining why this might be the case. It is therefore usual to express the physical densities above as ratios of this critical density. This renders,

$$1 = \Omega_\Lambda + \Omega_B + \Omega_C + \Omega_\nu + \Omega_K. \qquad (7.10)$$

These new matter densities can now be seen as fractions of the Universe's entire mass-energy budget, where $\Omega_K \approx 0$.

Measurements of distant Type Ia supernovae have shown that the Universe is not only expanding, as deduced by Hubble, but is actually accelerating in its expansion (i.e. the second time-derivative of a is positive). These explosive objects allow cosmologists to effectively test a form of Eq. (7.9) above to measure the rate of expansion as they look further out across the cosmos. To explain this unusual accelerated expansion it has become clear that a new exotic form of matter with negative pressure is needed. This is dubbed "dark energy".

While probes of this smooth cosmological expansion remain a powerful tool and conceptually close link to uncovering the nature of dark energy and other constituents it is actually analyzing deviations, fluctuations and inhomogeneities in large-scale cosmology that are most powerful. In particular, it is by studying perturbations in the distribution of matter that arise through the gravitational growth of structure that is most relevant to the measurement of neutrino masses.

It is first worth noting that in cosmology a theoretical model is not expected to make an exact prediction about the specific location of a galaxy or the precise temperature of a Cosmic Microwave Background (CMB) temperature fluctuation at a point in the sky. A model is, however, expected to make a prediction about the *statistical distribution* of these fluctuations. The statistic commonly used in cosmology is called the "power spectrum of fluctuations" P(k) which is the square of Fourier modes. This entity effectively describes how a field changes on varying scales in the rms sense. Specifically, if a field has fluctuations that are changing significantly over fixed distances, such that the variance of these fluctuations are large, then the power spectrum is also large at this scale. Conversely, if the field is smoother on a given scale then the power spectrum will be small.

It is from a combination of expansion and growth observations, to be described further in Sec. 7.5, that have given rise to the aforementioned *concordance model of cosmology*. As mentioned earlier, this is a Universe that is macroscopically isotropic and homogeneous, where gravity is described by general relativity and is made from dark energy ($\Omega_\Lambda \approx 0.75$), dark matter ($\Omega_C \approx 0.21$) and some ordinary baryonic matter ($\Omega_B \approx 0.04$). Finally, there is also a smaller, but unknown, contribution from neutrinos ($\Omega_\nu < 0.01$).

Despite the small contribution to the overall density the neutrinos are an integral part of the cosmological model and its interrelated parameters. It is because of this that neutrino mass estimates can be derived from experiments that have actually been designed to measure other quantities, such as dark energy or dark matter. Obtaining neutrino masses from cosmology has become both highly successful and unavoidable. Most remarkably, they provide the most unique opportunity to question the whole concordance framework by testing for consistency with particle physics experiments.

7.4 The Physics of Cosmological Neutrinos

A cosmological constraint on the sum of the neutrino masses is primarily a constraint on the relic Big Bang neutrino density Ω_ν, i.e. the energy budget consumed by the cosmic neutrino background. This background was initially in equilibrium with the very early cosmic plasma but subsequently decoupled after around 1 second as a result of its weak interaction. Despite electron-positron annihilations later heating the photon distribution it is still possible to associate the temperature of the two particle populations by equating their entropy densities. The photon distribution can be inferred directly from the CMB temperature today. From this one can relate the cosmic neutrino density to the quantity of interest — the sum of the individual mass eigenstates. This relation is given by

$$\Omega_\nu = \frac{\sum m_\nu}{93.14\, h^2\, \text{eV}} \qquad (7.11)$$

with h describing the Hubble parameter. It is this relation that helps us to probe the absolute mass scale.

The neutrinos have a large thermal velocity as a result of their low mass and therefore do not cluster or clump together like baryonic or cold dark matter. This *suppresses the growth of structure* in the Universe and consequently neutrinos smooth the natural clustering behavior of matter. While this suppression at first sight might seem subdominant it cumulatively becomes significant over the evolution of the Universe. This is how the smallest of masses is inferred from physics on the largest of scales. This effect is illustrated clearly for two different N-body simulations in Fig. 7.2. We can see that the pure cold dark matter simulation exhibits more "blobs" compared with the more "filamentary" structure when massive neutrinos are added.

In order to quantify this effect and include it in a theoretical model it is insightful to first consider the competing factors determining the growth of structure from an initial overdensity. These initial perturbations in the early Universe are the seeds of all the structure we see today and result from primordial quantum fluctuations amplified to cosmological scales by inflation.

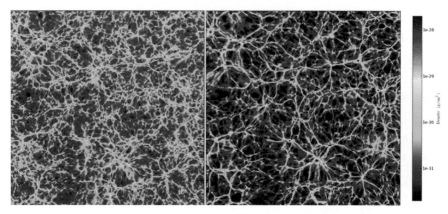

Fig. 7.2 Two simulations of the matter distribution in the Universe with massless (left panel) and massive 1.9 eV (right panel) neutrinos (Agarwal and Feldman, 2010). In the latter case the neutrinos have acted to smooth the clustering on the smallest scales.

The mechanism that provides the growth, from seed to structure, is gravitational instability. Mutual gravitational attraction will act towards the center of the initial overdensity attempting to increase it. However, during the epoch of radiation domination in the early Universe this is contested by pressure acting against gravity. Mathematically, the evolution of the fractional matter overdensity δ can be written as

$$\ddot{\delta} + 2H\dot{\delta} = \left(-\frac{c_s^2}{a^2}k^2 + 4\pi G\rho_0\right)\delta, \tag{7.12}$$

where c_s is the sound speed of the pressure-exerting fluid and the wave vector k is a consequence of writing the fractional overdensity in Fourier space. This expression is derived from perturbed energy–momentum conservation in Einstein's gravitational equations. This is equivalent to combining the Euler, continuity (conservation) and Poisson equations. The evolution of an overdensity therefore oscillates on scales smaller than the sound speed's causal influence in the plasma. Eventually matter comes to dominate over radiation and the sound speed in Eq. (7.12) drops to zero. As a result the gravitational attraction becomes uncontested and the perturbations grow on all scales to become clusters and galaxies themselves.

This subsequent growth of structure is also dependent upon the background expansion. This can be understood simply because more cumulative expansion acts to oppose collapse. It is accounted for by the second term in Eq. (7.12). Therefore measuring the clustering of matter in the Universe gives us more information on dark matter and energy, which determine the overall expansion dynamics.

As described above the neutrinos also have their part to play. They erase their own perturbations on scales smaller than the *free streaming* length k_{fs}.

$$k_{fs} = 0.82 \frac{\sqrt{\Omega_\Lambda + \Omega_m(1+z)^3}}{(1+z)^2} \left(\frac{m_\nu}{1\,\text{eV}}\right) h\,\text{Mpc}^{-1}. \qquad (7.13)$$

Here m_ν is the mass of the neutrino under consideration. Again, this is expressed as a wave vector because we are working in Fourier space. Therefore, over small scales the clustering of dark matter and galaxies is suppressed. On large scales the neutrinos cluster like normal matter and it is this scale-dependent effect that can help isolate the neutrino signal.

As stated earlier when we compare our theoretical prediction to data we do not calculate the value of a specific overdensity but instead the statistical distribution in the form of the power spectrum P(k). Therefore we expect the amplitude of this power spectrum to be suppressed in the presence of massive neutrinos over small scales (large k). This is shown in Fig. 7.3 and would describe the two different simulations shown earlier.

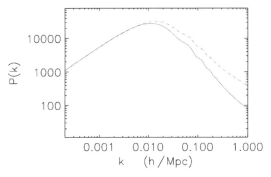

Fig. 7.3 Suppression in the statistical clustering of matter in the Universe (matter power spectrum) with massive neutrinos (solid) compared to massless neutrinos (dashed). This is seen over smaller scales (larger k).

This statistic can be measured with a variety of techniques including using galaxies as tracers of the underlying matter distribution or by analyzing the statistical distortion of images, called weak gravitational lensing. These will be described further in the next section.

7.5 Observational Methods

As described previously we can hope to measure the neutrino mass by comparing observational and theoretical models of the growth of structure in our Universe, quantified by the "matter power spectrum" $P(k)$. However, we also noted that matter is primarily dark — apparently not interacting electromagnetically. How is it then possible to maintain the observational side of the analysis?

There are a number of effects within the Universe that have become experimental procedures in cosmology. Effectively they are different ways of measuring the statistical variation of fluctuations and therefore the neutrinos' suppression in the underlying constituent fields.

It is important to stress that careful analyses of the neutrino mass in cosmology are not invalid because there is uncertainty in other related cosmological parameters. There are indeed unknowns but these are *accounted* for in the final neutrino mass limits: the other variables are marginalized over as nuisance parameters in the spirit of Bayes' theorem.

7.5.1 *The cosmic microwave background*

The tiny temperature fluctuations (at the level of 1 part in 100 000) in the CMB are the result of oscillations in the coupled photon-baryon plasma imprint in the early cosmos. The oscillations are a consequence of gravitational instability and fluid pressure — as described previously — arising from overdensities and underdensities in the very early Universe. Again, similar to the matter power spectrum, cosmologists construct an observed statistic of the resulting temperature distribution — the CMB *temperature power spectrum* — across the microwave sky.

Therefore it is natural to suppose that we can look for neutrino signals in this rms-like statistic. If the neutrinos suppress the early perturbations then the plasma's pressure is less impeded. This subsequently boosts the amplitude of the oscillation peaks and therefore the temperature power spectrum, which measures the variance of the field. However, neutrinos are relativistic in the early Universe at the time of the CMB if they are less massive than $\sim 1.3\,\text{eV}$. They therefore act like radiation and free stream uniformly over large scales. In this way the CMB is mildly insensitive to lighter neutrino masses and they become degenerate with other parameters, such as the Hubble parameter (Ichikawa et al., 2005).

Degeneracies in the CMB can be reduced, however, by considering that the CMB signal propagates to us — the observer — through the late and structural Universe. Smaller secondary effects, such as weak lensing of the CMB, which can be observed with better sensitivity (Lewis and Challinor, 2006), can therefore improve "CMB-only" neutrino mass bounds.

7.5.2 Galaxy clustering

A natural assumption is often that galaxies are likely to reside where there are large clumps and overdensities of dark matter. Therefore, we can use easily observed galaxies in a survey as tracers of the unknown underlying mass distribution; relating the subsequent clustering statistics to that of the matter power spectrum. Indeed this has been a successful approach over the last few decades and was a method that indicated indirectly the existence results of dark energy *pre*-supernovae (Efstathiou et al., 1990).

There are however two main challenges to how one interprets galaxy clustering data. The first concerns the "biasing" of galaxies in the survey: How well do the luminous galaxies trace the dark matter distribution? One prescription called "linear biasing" states that fluctuation counts in galaxies are proportional to the underlying mass fluctuations. However, it could be that this relation between the two fields is a function of scale and that it is different for different galaxy types. A scale-dependent biasing of the galaxies' power spectrum is particularly pernicious for neutrino estimates as it could act to mimic the very scale-dependent suppression we are looking for. With regards to different galaxy types one empirical approach is to use galaxy statistics for different classes of galaxies, e.g. for red and blue galaxy types, and to derive the upper limit on the neutrino mass from each of them separately (Swanson et al., 2010).

The second challenge concerns the fact that as matter collapses it gradually evolves from being a linear system to a nonlinear one. Consequently, theoretical models in the regime are challenging to solve and become increasingly uncertain over smaller scales. This is where there has been more collapse and clustering evolution; affecting the high k (small scales) in Fig. 7.3 the most severely. A common method to circumvent this has been to use N-body codes to simulate the evolution of matter clustering for various cosmological parameters. Alternatively, in order not to bias the inferred results, one can discard the data over these small and most problematic scales. This also removes data most plagued by the issue of scale-dependent biasing raised above.

7.5.3 Weak gravitational lensing

As light from distant galaxies traverses the vast distances of the Universe, towards our telescopes, it passes by the clustered and inhomogeneous distribution of baryonic and dark matter. This acts to slightly distort the path and in turn affects the apparent origin of the light ray. The consequence for a galaxy is that its image will be distorted too — becoming more elliptical, or *sheared*. In isolation this is a small and almost immeasurable 1% effect. However, by looking at the shapes of many galaxies in a survey the statistical lensing effect can be extracted. Given that deflection occurs via the intervening mass distribution the resulting statistic is sensitive to the all-important matter power spectrum. This is also what gives us more information from the CMB as mentioned previously.

Fortunately, weak lensing does not suffer the same biasing systematic as a galaxy clustering analysis. This is simply because the deflection of light is a gravitational effect and it occurs similarly whether the sourced mass is baryonic or is dark. Unfortunately, extracting this small distortion invites other technical challenges. For example, before gravitational shearing the observed galaxy already has some intrinsic, but unknown, ellipticity. Moreover, measuring the final ellipticity in a cosmologically distant object with instrumental, atmospheric and pixelization effects introduces further potential systematics.

By being a direct probe of the mass perturbations weak lensing is promising for neutrino mass estimations (e.g. Kitching *et al.*, 2008). However, in addition to the above challenges, it is crucial to accurately model nonlinear mass fluctuations in the presence of massive neutrinos. Such work is underway by several groups (e.g. Agarwal and Feldman, 2010; Hannestad, 2010).

7.5.4 *The Lyman-alpha forest*

The Lyman-alpha forest is a series of absorption lines resulting from intervening neutral hydrogen in the spectrum of distant quasars. The hydrogen resides in vast gas clouds along the observer–quasar line-of-sight and absorbs at the Lyman-alpha transition. The *forest* of absorption lines occurs simply because gas at different redshifts absorbs light at different apparent frequencies along the quasar's spectrum. The cosmological power of this observation is derived from the assumption that the gas again traces the underlying mass distribution. Therefore, statistics of the gas translate to statistics of the dark matter.

An added benefit from the Lyman-alpha probe is that much of this absorption occurs at high redshift, before the mass distribution has too much time to become nonlinear in its collapse. One can therefore strive for a more linear analysis. However, there are potentially many added systematics that can affect the gas's behavior and therefore its reliability as a dark matter tracer. Recent attempts to derive neutrino mass from Lyman-alpha include (e.g. Seljak *et al.*, 2006; Viel *et al.*, 2010).

7.6 Observational Limits as of 2010

The current limits from particle physics on the absolute scale reside at around $7\,\text{eV}$ (Kraus *et al.*, 2005), as described in Sec. 7.2. With the large influx of data over the last decade it has not taken long for cosmological bounds to supercede this. Galaxy clustering data from the 2 degree Field redshift (2dF) survey (Elgaroy *et al.*, 2002) set upper limits at

$\sum m_\nu < 1.8\,\text{eV}$ at the 95% confidence level.[1] More recently, WMAP data put the bound at $1.2\,\text{eV}$ (Komatsu et al., 2009), which is similar to the CMB-only limit described in the last section. While this is conservative compared to the values below, the CMB is a systematically well-understood observable and therefore this value can be seen as a very reliable benchmark for the current generation of experiments.

New analyses improve upon these values not just with better statistics, but by utilizing the complementarity of cosmological probes. While some probes constrain one direction of the parameter space; others will constrain another. Therefore, in combination, optimal and more aggressive results are inferred. One example of this is from the MegaZ Luminous Red Galaxy study (Thomas et al., 2010), from the largest survey to date — the Sloan Digital Sky Survey. The statistical power of nearly three quarters of a million galaxy tracers was combined with the CMB and data on the expansion history to further reduce the uncertainty on the parameter space (Fig. 7.4). This gave the limit $\sum m_\nu < 0.28\,\text{eV}$ — one of the lowest values in the literature.

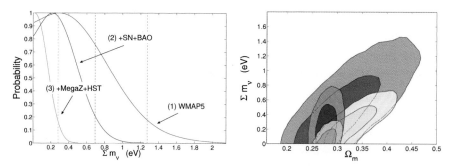

Fig. 7.4 An example of one of the current upper limits on the neutrino mass (Thomas et al., 2010). A combined analysis from the CMB (WMAP; red), plus supernovae distance probes (blue) and then SDSS galaxy clustering data (MegaZ; green) push the limit to $< 0.28\,\text{eV}$ (95% CL). Left panel: The one-dimensional probability distribution as each data set is added. The horizontal lines denoted 95% confidence limits. Right panel: 68% and 95% confidence regions for the same data combinations but with the variation in the matter density also shown. Uncertainties in other parameters are accounted for by marginalizing over them. The yellow contour represents CMB + MegaZ.

---

[1] All subsequent cosmological limits will be quoted at 95% confidence level.

In addition, certain combined analyses including Lyman-alpha data (e.g. Seljak *et al.*, 2006; Viel *et al.*, 2010) have produced equally impressive bounds $\sum m_\nu < 0.17\,\text{eV}$. However, there have been many suggestions that the Lyman-alpha data produces tension with other data sets and is systematically prone. Despite this it remains a promising tool for the future and these combined analyses represent the benchmarks for today's sub-eV limits. Most importantly, neither showed any hints for finite masses at this level and consequently cosmological bounds if correct are already suggesting that future experiments, such as KATRIN described earlier, are unlikely to make any detection.

Future Cosmological Experiments

The next benchmark in the pursuit of massive neutrinos will be governed by the approaching generation of precision surveys on the horizon and most importantly by the fundamental neutrino mass hierarchy and splittings.

By taking the square root of Eq. (7.1) it is possible to show that if the neutrinos obey a normal hierarchy the sum of the masses must be *at least* $\sim 0.05\,\text{eV}$. This will occur if the first mass eigenstate is zero. The second mass eigenstate then contributes a small additional mass as seen from (7.2). Alternatively, in the inverted hierarchy the sum of the masses can be as low as $\sim 0.1\,\text{eV}$. Therefore, the next advancements cannot continue to reduce the bound indefinitely — unless something is intrinsically wrong within the cosmological or particle physics interpretation. This means that upcoming galaxy clustering data, for example, may be able to further determine the hierarchy.

One such galaxy clustering survey on the cusp of first light is the Dark Energy Survey (DES). This will map out the statistical distribution of 300 million galaxies from the southern hemisphere and will eventually combine a variety of techniques including weak lensing, Type Ia supernovae and the analysis of large galaxy clusters. DES galaxy clustering combined with the recently launched Planck CMB surveyor is capable of setting the upper limit to $0.12\,\text{eV}$ or detecting the neutrino mass if it is significantly larger than $0.1\,\text{eV}$ (Lahav *et al.*, 2010). Future wide-field radio and spectroscopic surveys will be able to determine the neutrino mass even more accurately (e.g. Abdalla and Rawlings, 2007; Carbone *et al.*, 2010).

Further into the future the combination of weak lensing and high-precision CMB experiments may reach sensitivities down to the lower bound of $\sim 0.05\,\text{eV}$ on the sum of the neutrino masses set by the current oscillation data. It is interesting to note the great prospects for combining data from particle physics experiments and cosmological data sets (e.g. Host *et al.*, 2007; Fogli *et al.*, 2007).

Conclusions and Implications

There are very good prospects that if the neutrino mass in nature is larger than 0.1 eV then it will be *detected* by experiments such as the CMB, galaxy clustering, weak lensing, Lyman-alpha clouds and their combinations. Current upper limits have already approached 0.3 eV in a robust way. However, it is important to stress that the inference on neutrino masses from cosmology is done within the paradigm of dark matter and dark energy. Moreover, it will be increasingly crucial to control systematics in any of these probes.

So far three Nobel prizes have been awarded for breakthroughs in neutrino physics: for the first neutrino detection[2] in 1995; for the discovery of the muon neutrino[3] in 1988; and for the observation of cosmic and supernova neutrinos[4] in 2002. There is no doubt that the next neutrino milestone is the robust detection of mass. This breakthrough may come from laboratory particle experiments, but cosmology could also play an important part in confirming and complementing them. Then it will be left for theorists to explain the consequences of such an apparently "unnatural" mass scale.

Acknowledgments

It is a pleasure to thank our neutrino collaborators in recent years: Filipe Abdalla, Oystein Elgaroy, Hume Feldman, Ole Host, Angeliki Kiakotou and Molly Swanson.

---

[2] Frederick Reines.
[3] Leon Lederman, Melvin Schwartz and Jack Steinberger.
[4] Raymond Davis Jr. and Matatoshi Koshiba.

References

Abdalla, F. and Rawlings, S. (2007). *MNRAS* **381**, pp. 1313–1328.
Agarwal, S. and Feldman, H. (2010). *Mon. Not. Roy. Ast. Soc.* **410**, pp. 1647–1654.
Ahmad, Q. R. *et al.* (2001). *Phys. Rev. Lett.* **87**, p. 071301.
Bond, J. R., Efstathiou, G. and Silk, J. (1980). *Phys. Rev. Lett.* **45**, p. 1980.
Carbone, C. *et al.* (2010). arXiv:1012.2868.
Efstathiou, G., Sutherland, W. and Maddox, J. (1990). *Nature* **348**, pp. 705–707.
Elgaroy, O. *et al.* (2002). *Phys. Rev. Lett.* **89**, p. 061301.
Fogli, G. *et al.* (2007). *Phys. Rev. D* **75**, p. 053001.
Fukuda, Y. *et al.* (1998). *Phys. Rev. Lett.* **81**, pp. 1158–1162.
Hannestad, S. (2010). arXiv:1007.0658.
Host, O. *et al.* (2007). *Phys. Rev. D.* **76**, p. 113005.
Ichikawa, K., Fukugita, M. and Kawasaki, M. (2005). *Phys. Rev. D* **71**, p. 043001.
Kitching, T. *et al.* (2008). *Phys. Rev. D* **77**, p. 103008.
Komatsu, E. *et al.* (2009). *Astrophys. J. Suppl.* **180**, pp. 330–376.
Kraus, C. *et al.* (2005). *Eur. Phys. J. C.* **40**, p. 447.
Lahav, O. *et al.* (2010). *MNRAS* **405**, pp. 168–176.
Lesgourges, J. and Pastor, S. (2006). *Phys. Rept.* **429**, pp. 307–379.
Lewis, A. and Challinor, A. (2006). *Phys. Rept.* **429**, pp. 1–65.
Seljak, U., Slosar, A. and McDonald, P. (2006). *JCAP* **0610**, p. 014.
Swanson, M., Percival, W. and Lahav, O. (2010). *MNRAS* **409**, p. 1525.
Schwetz, T., Tortola, M. and Valle, J. W. F. (2008). *New Journal of Physics* **10**, p. 113011.
Thomas, S. A., Abdalla, F. B. and Lahav, O. (2010). *Phys. Rev. Lett.* **105**, p. 031301.
Viel, M., Haehnelt, M. and Springel, V. (2010). *JCAP* **06**, p. 015.

CHAPTER 8

MEASURING THE EXPANSION RATE OF THE UNIVERSE

LAURA FERRARESE

National Research Council of Canada,
Herzberg Institute of Astrophysics,
5071 West Saanich Road,
Victoria BC V9E 2E7, Canada

8.1 Introduction

Few periods in the history of astronomy have led to such revolutionary changes in our understanding of the Universe, and our place within it, as the 1920s. Today, spectacular images of distant galaxies from the Hubble Space Telescope have entered virtually every household, and it is hard to imagine that less than a century ago — a time short enough to be in the memory of many of the readers' grandparents — some of the most brilliant minds of the time believed our Galaxy, extending a staggering 300 000 light years across, *to be* the Universe. That the Universe extends far beyond the reach of the Milky Way — a discovery that can only be compared to Galileo confirmation, three centuries earlier, of Copernicus theory — was not proven until 1925. Four years later, barely enough time for people to come to grips with the fact that the Universe was far larger than previously believed (and other galaxies exist beyond our own!) Edwin Hubble produced a simple diagram that was destined to lead to one of the greatest discoveries in the history of modern science: the Universe is expanding.

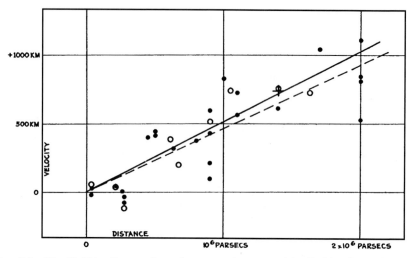

Fig. 8.1 The Hubble diagram from the paper that started it all: "A relation between distance and radial velocity among extragalactic Nebulae" (Hubble, 1929). From Hubble's original caption: "Radial velocities, corrected for solar motion (note the typo in the units), are plotted against distances estimated from involved stars and mean luminosities of nebulae in a cluster. The black discs and full line represent the solution for solar motion using the nebulae individually; the circles and broken line represent the solution combining the nebulae into groups; the cross represents the mean velocity corresponding to the mean distance of 22 nebulae whose distances could not be estimated individually." Note that the term "extragalactic" in the title of the article would have been extremely controversial only five years earlier!

Hubble's diagram, shown in Fig. 8.1, was astonishingly simple: the radial velocity of 24 galaxies plotted against their distance. His paper was modestly titled "A relation between distance and radial velocity among extra-galactic Nebulae" and contained no mention of an expanding Universe. On the contrary, Hubble suggested that the relation could be interpreted as confirmation that the Universe is *static* or, more precisely, static *and* of the kind theorized by the Dutch astronomer Willem de Sitter. Hubble had met de Sitter a year earlier, while attending a meeting of the International Astronomical Union in Leiden. De Sitter's work was mostly focused on the study of Jupiter's satellites, but in 1916, it shifted towards the astrophysical implications of Einstein's freshly published theory of general relativity. In 1917 de Sitter produced a solution of Einstein's field equations that could be interpreted as a model for an expanding

Universe.[1] This, unfortunately, failed to impress Einstein, who held a curiously strong belief that the Universe must be static, for no good reason other than to admit that otherwise would be "senseless". Such was Einstein insistence on this matter, that in a 1917 paper he added a constant term (the infamous "cosmological constant") to his field equations just so they would admit "the existence of a finite mean density in a static Universe" [quoted from (Einstein and de Sitter, 1932) the paper where the cosmological constant was finally abandoned, as a consequence of Hubble's work]. Perhaps influenced by Einstein's belief, or perhaps just for ease of comparing the two models, de Sitter rewrote his solution using new coordinates, in which the components of the metric tensor were now time-independent. The question on everybody's mind at this point was not whether the Universe was expanding, but which of Einstein's or de Sitter's *static* solution could be used to describe it.

Then Hubble entered the scene. There was a crucial difference between the two static models. De Sitter's solution predicted a singular effect: time would slow down at large distances from the observer, so that light emitted from distant galaxies would be redshifted and give the illusion that those galaxies were receding. Einstein's model predicted nothing of the kind. And indeed, Hubble's correlation left no doubt that galaxies recede away from us in all directions, with a velocity that is directly proportional to their distance, exactly as predicted by de Sitter. In the very last paragraph of his five and a half page paper, Hubble notes that "the velocity–distance relation might represent the de Sitter effect". Einstein's static model of the Universe was not viable. Unfortunately, it soon turned out that neither was de Sitter's model: beside the peculiar behavior noted above, de Sitter's universe was *empty*: his solution was a vacuum solution.[2] As the only two known static solutions seemed untenable, there was no escaping the conclusion that the Universe must be dynamic. Dynamic solutions were in fact readily

---

[1] The term "Universe" was used very differently then than it is now. Einstein's and de Sitter's Universes, at this stage, would have been equated with the Milky Way.
[2] De Sitter was of course aware of this, but argued that his solution might still describe the Universe as long as the density of matter was close to zero. This became untenable with the first measurements for the mass of the Milky Way.

available — Friedmann published a non-static model in 1922, and Lemaître published another in 1927 — but up to that point the belief in a static Universe had been so strong that non-static solutions had been overlooked, or at the very best, relegated to the role of mathematical curiosities. After Hubble's observations, Lemaître's paper was resurrected and his model of an expanding Universe held as the true solution. Hubble's diagram took on a new meaning: if the Universe is expanding uniformly in all directions, then every galaxy would be seen as moving away from any other galaxy, with a speed that is directly proportional to the distance. The constant of proportionality, which Hubble indicates by the letter "K", is now known as the Hubble constant, H_0, and is measured in $\text{km}\,\text{s}^{-1}\,\text{Mpc}^{-1}$. Its inverse, $1/H_0$, has units of time: if the Universe is expanding it must have had a beginning, and $1/H_0$ is nothing less than the age of the Universe under the assumption that the expansion has proceeded at a constant rate. Hubble's original estimate for H_0 was $\sim 500\,\text{km}\,\text{s}^{-1}\,\text{Mpc}^{-1}$, implying an age of the Universe that even at that time was in conflict with the age of the Earth inferred from radioactive decay. Something was very seriously wrong.

Since the very beginning, the difficulty of measuring the expansion rate of the Universe has always been entirely associated with the estimate of distances. Even in the 1920s, velocities could be obtained with relative ease; indeed, by 1929, radial velocities for 41 spiral nebulae had been collected, during the course of the previous 17 years, by Vesto Slipher at the Lowell Observatory in Arizona. It was distance that was elusive. The velocities listed in Table 1 of Hubble's 1929 paper,[3] are generally quite accurate; in many cases they are practically coincident with the modern values. But the distances can be off by over an order of magnitude! For a few nearby galaxies, Hubble measured distances using stars which show periodic variability in their luminosity, Cepheid variables, or "Cepheids" for short. But Cepheids could not be detected in galaxies far enough to provide good leverage for Hubble's relation. For most galaxies, Hubble then assumed that

---

[3]Except for the velocities of four galaxies, which are attributed to Humason, the others are almost certainly from Slipher, although Hubble did not cite the source in his paper.

the brightest stars, or the galaxies themselves have a constant luminosity, which he could calibrate in nearby galaxies with known Cepheid distances. In doing so, Hubble established the use of a cosmic "distance ladder": a series of distance indicators, each tailored to specific classes of galaxies and is effective over different ranges in distance, which together can be used to reach farther and farther into the expanding Universe, into the "Hubble flow". Since Hubble's time a tremendous amount of work has gone into refining this distance ladder, and for good reasons. As if measuring the expansion of the Universe was not motivation enough, distances are needed to translate virtually all *observed* properties into *physical* ones: fluxes to luminosities, angular scales into sizes. Without distances, for instance, we would be unable to calculate the masses of astrophysical objects (from planets to black holes to galaxies), to study the spatial distribution and clustering of galaxies, or to measure the rate with which stars form.[4]

When it was launched in 1990, one of the main goals of the Hubble Space Telescope, named in Edwin Hubble's honor, was the accurate measurement of the Hubble constant by establishing precise distances to nearby galaxies. As in Hubble's work, these distances would be measured using Cepheid variables and could be used to calibrate other distance indicators that would be applied to farther galaxies still. The modern value of the Hubble constant is a factor seven smaller than Hubble's original value, and the modern version of the Hubble's diagram, extending much further back in time, has revealed subtle deviations from a linear law (Riess *et al.*, 1998; Perlmutter *et al.*, 1999). This can only be interpreted as meaning that the Universe is not only expanding, but *accelerating*. This acceleration suggests that 73% of the Universe is composed of "dark energy".

The accelerating Universe is the subject of a different chapter. This chapter will focus instead on the extragalactic distance scale, the key to Hubble's diagram. A representative distance scale ladder, showing the interplay of the methods described in this chapter, is shown in Fig. 8.2.

---

[4]For the interested reader, a comprehensive collection of distances can be found online on the NASA IPAC Extragalactic Database (NED): http://nedwww.ipac.caltech.edu/Library/Distances/

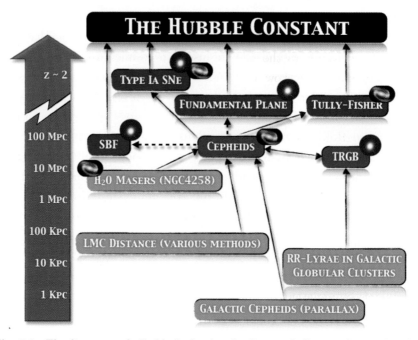

Fig. 8.2 The distance scale "ladder", showing the distance indicators discussed in this chapter. Primary distance indicators — Cepheids (Sec. 8.2) and the Tip of the Red Giant Branch (TRGB, Sec. 8.3) — are shown in red, and are calibrated based on direct distance measurements to specific objects (shown in green). Secondary distance indicators — the Surface Brightness Fluctuation method (Sec. 8.4), the Fundamental Plane (Sec. 8.5), the Tully–Fisher relation (Sec. 8.6) and Type Ia supernovae (Sec. 8.7) — are shown in blue. They are usually calibrated against Cepheids' distances (a dashed arrow indicates that the calibration is based on an indirect comparison of "group" or "cluster" distances, while a solid arrow indicates a direct comparison of distances to individual galaxies), although a calibration against TRGB distances have also been used, with consistent results. Note that distances based on the TRGB and Cepheids, being independent, can be tested against each other. A little image in the corner indicates the type of galaxy to which each distance indicator can be more easily applied: early-type or spirals. The scale to the left shows a rough estimate of the distance each method can reach given current observing facilities.

The first section of this chapter will review Cepheid variables and their role at the very foundation of the extragalactic distance scale. Subsequent sections will describe secondary distance indicators that either provide an independent test of the Cepheid distance scale (the Tip of the Red Giant Branch) or that, relying on the Cepheid distance scale, can be used to push farther than what Cepheids can: the Surface Brightness

Fluctuation Method, the Fundamental Plane of early type galaxies, the Tully–Fisher relation for spiral galaxies, and Type Ia supernovae. The final section is dedicated to the discussion of a consensus value for Hubble's famous constant and how it can help us to constrain the age of the Universe. Excellent review articles dedicated to the Hubble constant and the extragalactic distance scale are available to readers left wishing for more, in particular: Rowan-Robinson (1985), Huchra (1992), Jacoby et al. (1992), Tamman, Sandage and Reidl (2008) and Freedman and Madore (2010).

8.2 Twinkle Twinkle Little Star: Cepheid Variables

At a certain point in the lifetime of most stars, departures from simple thermo-mechanical equilibrium result in resonant oscillations of the entire star, or some significant part of its envelope. The star pulsates, causing periodic variations in its luminosity. Cepheids are but one class of variable stars, but several others exist, and indeed pulsation in stars is a very generic phenomenon, affecting most stars at one point or another (sometimes even at multiple points) of their evolution. Cepheids, however, are particularly relevant for the issue at hand as they are at the very foundation of the extragalactic distance scale, so it is on these stars that we will concentrate our attention. Detailed reviews of the Cepheids distance scale and additional references can be found in Sandage and Tammann (2006) and Freedman and Madore (2010). Cepheids are relatively massive (~ 4 to $13\,\mathcal{M}_\odot$), young stars ($< 10^8$ years) at a fairly advanced stage of evolution: having exhausted hydrogen in their cores, they are now in a relatively long-lived phase where triple-alpha reactions are taking place in the core, burning helium into carbon and oxygen. Their visual absolute magnitudes[5] can be between -2 and -7, i.e. they are, intrinsically, 500 to 50 000 times more luminous than the Sun. Moreover, their luminosity can vary, with surprising regularity, by a factor ~ 2 over a period that can range between 2 to over

---

[5]Optical astronomers, for historical reasons, like to measure luminosities in "magnitudes", defined as -2.5 times the logarithm (in base 10) of the luminosity. Larger luminosities correspond to lower magnitudes: a 0th magnitude star, for instance, is 100 times more luminous than a 5th magnitude star.

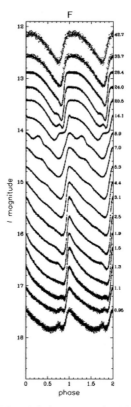

Fig. 8.3 A compilation of I-band light curves for classical Cepheids in the Small Magellanic Cloud, observed by the OGLE project. Data for these Cepheids were collected during in an eight-year period, from 2001 to 2009, with the 1.3-meter Warsaw telescope at Las Campanas Observatory, Chile. Typically, 700 individual observations were made of each Cepheid. The light curves are "phased", namely the data are folded until the best period (giving the "smoothest" light curve) is found. The period (in days) of each Cepheid is indicated by the small number to the left of each light curve. The figure was adapted from Soszyński et al. (2010).

100 days (Fig. 8.3). But the truly remarkable property is that all Cepheids of given luminosity pulsate with (very nearly) the same period. This period is longer, and *predictably* so, the more luminous the star is (see Fig. 8.4). In principle, therefore, if the *intrinsic* luminosity and period of a Cepheid are known, the distance to any other Cepheid can be estimated from its *apparent* luminosity and period. In reality, things are significantly more complicated. But let us start from the very beginning.

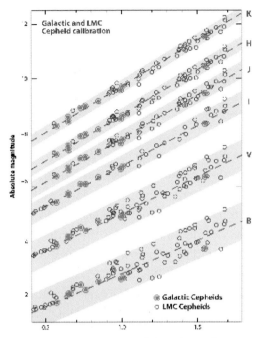

Fig. 8.4 The most recent characterization of the relation between the period (in the log) and luminosity (in magnitudes) of Cepheids, i.e. the Leavitt law, from Freedman and Madore (2010). The relation is shown in six different filters (indicated to the right), from the visual (B, V and I) to the near-infrared (J, H and K). The different symbols refer to Cepheids in our own Galaxy (in yellow), which are used to calibrate the luminosity scale of the relation, and in the Large Magellanic Cloud (in red), which are used to determine the slope of the relation.

8.2.1 *A brief historical overview*

The name itself, "Cepheids" derives from the prototype of such class, δ Cephei, one of the seven stars defining the northern constellation of Cepheus. δ Cephei is easily visible with the naked eye, and its variability — by a factor two over a ~ 5 day period — was first noted by John Goodricke in 1784.[6] But the history of Cepheids in the context of the extragalactic

---

[6]Two years earlier, working closely with Edward Piggot, Goodricke had established the variability of Algol, and correctly suggested that it could be due to eclipses from a companion star. One and a half years to the day after his discovery of d Cephei's variability, Goodricke, who had been deaf since age five, died of pneumonia contracted as a consequence of exposure while observing. He was 21 years old.

distance scale — reviewed by Fernie (1969) — started only in the early 1900s when, by comparing two photographic plates of the Small Magellanic Cloud, Henrietta Leavitt discovered an unusual number of variable stars. Working with data taken with the 24-inch refractor at Arequipa in Peru, owned by Harvard College Observatory, she started to scrutinize yet more plates and, by 1908, the number of known variables in the Small and Large Magellanic Clouds had increased to a staggering 1800 (Leavitt, 1908). A few of these were observed in enough plates that their period could be measured, and Leavitt noted that "the brighter variables have longer period". She did not elaborate any further, but a few years later (Leavitt and Pickering, 1912), with a sample of 25, Leavitt felt confident enough to state that "there is a simple relation between the brightness of the variables and their periods. The logarithm of the period increases by about 0.48 for each increase of one magnitude in brightness". She further commented that "deviations from a straight line (in the relation between magnitude and logarithmic period) might become smaller when an absolute scale of magnitude is used" and that the "periods (of the variables) are apparently associated with their actual emission of light". Although Leavitt never went as far as explicitly stating that Cepheids could be used to measure distances, the implication was clear. One year later, Hertzsprung (1913) provided the first calibration of Leavitt's period–luminosity relation[7] and used it to derive a distance to the Small Magellanic Cloud; further studies by Russell and Shapley quickly followed (Russell, 1913; Russell and Shapley, 1914; Shapley, 1918).

But it was Edwin Hubble's work in 1925 and 1929 that sealed the role of the Leavitt law as the foundation of the cosmological distance scale. By using Cepheids in M31 and M33, Hubble estimated the distance to these "nebulae" to be 285 kpc. We now know that in fact M31 is at over 700 kpc, and M33 is a little further still, but these do not matter: the relevance of Hubble's result is that it placed M31 and M33 firmly, indisputably, beyond

---

[7] In 2009, in recognition of Leavitt's fundamental contribution, the American Astronomical Society suggested that the Cepheid period–Luminosity relation should be referred to as the "Leavitt law".

the Milky Way, which was believed to be, at the very most, 100 kpc across (we now know it to be about three times smaller). This was a finding of epic proportions: it meant that "nebulae" were, in fact, galaxies in their own right, and that the Milky Way was simply one of many such nebulae.[8] Four years later, Hubble used Cepheids as standard rulers to measure, or calibrate,[9] distances to 24 galaxies with known radial velocities. The resulting Hubble diagram was to become incontrovertible proof that the Universe is expanding.

8.2.2   The theoretical perspective

As the world of observational astronomy was rocked by this remarkable string of discoveries, progress was also rapidly advancing on the theoretical front. The mechanism to explain the variability of Cepheids was developed by Sir Arthur Eddington in 1917, following earlier suggestions by Shapley.[10] The basic idea is that Cepheids (as well as many other types of periodic variables) behave as a thermodynamic heat engine: a partially ionized convective layer within the star (the astrophysical equivalent of a cylinder in a steam engine) drives radial pulsations by absorbing heat during

---

[8]This had been an item of intense contention, culminating five years earlier in one of the most fascinating (and instructive!) bits of astronomical history: the famous Great Debate between two of the greatest astronomers of the time, Harlow Shapley and Heber Curtis, regarding the nature of "spiral nebulae" [a review can be found in Trimble (1995)]. Shapley held the view that nebulae were simply clouds within our own Galaxy, which, at 100 kpc across, effectively defined the Universe. Curtis believed the Galaxy to be smaller by a factor 10, and held Kant's view that nebulae were "island universes". Ironically, to prove Curtis right (at least on the issue of island universes), Hubble used Shapley's calibration of the Leavitt's law for Cepheids. Even more ironically, Curtis did not believe Cepheids to be reliable distance indicators. In fact, pretty much the one thing Shapley and Curtis agreed upon was that dust extinction did not play a major role and could be neglected when measuring distances. This was wrong. It was later found that neglecting dust extinction was one of the main reasons why Shapley's calibration of the Leavitt law (used by Hubble) was off by a factor four (in luminosity).
[9]For a few galaxies, the distances were determined directly using Cepheids, while for the others distances were estimated by assuming that the brightest stars, or the nebulae themselves, had a constant intrinsic luminosity, which was again gauged from galaxies with Cepheid distances. So, Hubble used Cepheids both as distance indicators, and as calibrators for secondary methods, an approach that persists today.
[10]This theory was finally placed on secure physical grounds by Baker and Kippenhahn (1962, 1965) and by Cox (1963).

compression and releasing it during expansion. The variations in luminosity are therefore not driven by an *intrinsic* change in energy generation, which is stable, but by the interplay between the changes in temperature and radius of the outer envelope as the star expands and contracts. To understand why, it is sufficient to remember that a star behaves like a black body: its luminosity L, radius R and temperature T are related by Stefan–Boltzman's law: $L = 4\pi R^2 \sigma T^4$, where σ is a constant. As the star expands, its radius increases and its temperature decreases, so the two effects play against each other. However, fractional changes in temperature are larger than those in radius and this, combined with the larger exponent of the temperature in the equation above, causes the star to become fainter (and vice versa during the contraction phase). Observations indeed show that variations in luminosity are traced closely by variations in surface temperature, radius and radial velocity: as the star becomes brighter, the temperature of the outer layers increases, the radius decreases, and the radial velocity of the outer envelope is negative (i.e. the star is shrinking).[11]

To better understand Eddington's idea, let us imagine a partially ionized layer within the stellar interior. If some mechanism causes the layer to absorb the heat coming from the core, the resulting increase in pressure will drive the layer to expand, lifting the envelope above. If, during the expansion, the layer starts to radiatively loose energy, the diminished pressure support will cause it to contract, until the contraction is such that the layer starts to absorb heat again and the cycle starts anew. The mechanism that causes the layer to store heat during compression and release it during expansion is a change in opacity due to the partially ionized nature of the gas, which is really key to this process. As a rule, the opacity of a gas decreases as its temperature increases: if a blob of gas absorbs a little extra heat, it becomes more transparent, the heat escapes, equilibrium is

---

[11]To be precise, the maximum luminosity (and temperature) is not reached when the radius is smallest, but a fraction of a period later, when the star is already expanding. The reason for this is that changes in opacity (which will be described in more detail shortly) in the outer envelope cause the energy coming from the driving layer to be temporarily stored before being slowly released during expansion.

re-established, and no radial pulsations take place. But in a partially ionized medium, opacity *increases* with temperature. In this case, as the blob of gas absorbs energy, it heats up and becomes more opaque, thereby trapping even more energy. Potentially catastrophic consequences are averted only when the gas reaches a temperature and a density where the opacity again begins to fall, allowing the energy to escape. What makes a partially ionized medium so special? During contraction, only part of the work PdV (where P is the pressure and dV is the change in volume) exerted on the layer is used to raise the temperature, while the rest is used to ionize the gas. The opacity increases with density but decreases faster with temperature, but since the increase in temperature is diminished relative to the case in which no ionization is taking place, the change in opacity is driven by the change in density. In other words, the opacity increases as the gas becomes ionized, even if the temperature increases, and heat is retained. During expansion, part of the work done by the layer is used to recombine the gas, and again the opacity follows the change in density, which is now decreasing: heat is released. In most variable stars, including Cepheids, the driving layer is composed of partially ionized helium, the reason being that the relatively high ionization potential of helium is very effective in driving the changes in opacity.

Radial pulsations can only take place under very fine-tuned conditions: for instance, the partially ionized layer has to be deep enough within the star to be optically thick (otherwise radiative processes will be too efficient in dissipating energy) but not deep enough to allow the massive envelope above to dampen the instability. Effectively, this means that the star (besides having a partially ionized convective envelope) also needs to have a temperature within a very narrow range of values. A star's temperature changes dramatically during its evolution, as heavier and heavier elements are fused in the core and surrounding layers, and most stars, at one point or another, and almost independently of their luminosity, will find themselves to have just the right temperature, as well as all other conditions, to become unstable to radial pulsations. The reason why most stars are not *observed* to be variable is simply because this evolutionary phase is very short lived.

8.2.3 The Leavitt law

With a better understanding of the physics behind Cepheid pulsations, we can now return our attention to the Leavitt law, a very recent characterization of which can be seen in Fig. 8.4 (from Freedman and Madore, 2010; see also Fouqué et al., 2007). In its original form, the Leavitt law simply states that the mean Cepheid luminosity (or rather, magnitude) is directly proportional to the log of the period. But at any given period, different Cepheids can differ by as much as a factor of two in luminosity (depending on the period and the wavelength at which L is measured). In other words, the Leavitt law has significant scatter.

There are several reasons for this [a thorough review can be found in (Feast and Walker, 1987)]. The most important is that there is a third parameter linking period and luminosity, namely the temperature of the variable star, as can be easily understood from first principles. The period is roughly comparable to the time it takes for a sound wave to propagate across the star. If v_s is the sound speed, the period P is then given by $P \sim 2R/v_s$, where R is the radius of the star. The sound speed is of the same order as the thermal speed of particles in the stars, and can be estimated from the star's mass $\mathcal{M}$ and radius through the virial theorem, $v_s^2 \sim G\mathcal{M}/R$, where G is the gravitational constant. Putting two and two together, $P \sim 2(R^3/G\mathcal{M})^{1/2}$. Finally, relating the luminosity to the radius using the Stefan–Boltzman law (and removing all constants for simplicity!) $P \propto L^{0.75}T^{-3}\mathcal{M}^{-0.5}$. What this equation shows is that the period is in fact not just a function of the star's luminosity, but also of its mass and temperature. Theoretical models predict a tight relation between mass and luminosity [for instance, Valle et al. (2009) recently found $L \propto \mathcal{M}^{3.15}$] so the mass dependence can be neglected, but the dependence on temperature remains.

Although observationally temperature can be mapped into a broad band color (Sandage and Gratton, 1963; Sandage and Tammann, 1968), and therefore the scatter in the Leavitt law could be reduced by using observations in multiple filters, in practice observing Cepheids with more than a couple of filters becomes prohibitive. For this reason, a single Cepheid cannot be used to measure distances, as suggested in our introductory paragraph. Instead a galaxy distance is measured by using the Leavitt law defined by a large number of Cepheids. The best linear-fit to this Leavitt law is then matched to the Leavitt law for a galaxy of known distance: differences in temperature between different Cepheids come out in the wash, and the distance is given by the relative shift (in luminosity) between the two relations (modulo extinction by dust!).

The second complication is that the procedure above works if we are confident that the Leavitt law is the same, in slope and absolute normalization, in *every* galaxy. But, not surprisingly, the pulsation mechanism of Cepheids depends on the composition of the star: in other words, two Cepheids, with identical period but different composition, might have slightly different luminosity (and by different amounts at different wavelengths).[12] What this means is that Cepheids in two galaxies whose stellar populations differ in their average composition, might define Leavitt laws with different slopes and zero points. The magnitude of this effect is small, but surprisingly difficult to calculate, both theoretically and observationally [see Bono *et al.* (2010) for a thorough overview of the subject], and this issue has yet to be resolved to everybody's satisfaction.

The last hurdle to overcome is to find a galaxy with known distance to serve as a "Cepheid yardstick". Just like in Leavitt's time, the Large Magellanic Cloud is still the galaxy of choice when it comes to define the period–luminosity relation for Cepheids. Modern surveys, in particular the Optical Gravitational Lensing Experiment (OGLE) and the Massive Compact Halo Object survey (MACHO) have scanned the Large and Small Magellanic Clouds, as well as the bulge of our own Galaxy, in search of rare gravitational microlensing events. In doing so, MACHO collected light curves for 1709 Cepheids (Alcock *et al.*, 1999), and OGLE added another 1666 to the list, bringing the total to an astonishing 3375 (Soszyński *et al.*, 2008). Unfortunately, the distance to the LMC is not well known: the spread in published distances, based on independent methods, but sometimes also

---

[12]Indeed, the exponent in the relation between mass and luminosity depends on the composition of the star.

on different applications of the same method, is over 30% [see Gibson (2000) for a compilation, and Freedman and Madore (2010) for a recent update]. For this reason, Freedman and Madore (2010) propose to tie the zero point of the Leavitt law to a sample of ten Galactic Cepheids for which accurate geometric distances are known via parallax measurements (Benedict *et al.*, 2007). These Cepheids can be used to set the absolute scale of Leavitt's law, while the slope continues to be defined by the much larger and better defined sample of Cepheids in the Large Magellanic Cloud. Freedman and Madore estimate that this hybrid calibration of the Leavitt law allows measurement of distances with 3% accuracy, making Cepheids the most accurate distance indicator available.[13]

8.2.4 *An independent check*

A validation of the Cepheid distance scale comes from a very special galaxy, NGC 4258.[14] The circumnuclear region of NGC 4258 is host of a thin, very nearly edge-on disk of molecular clouds. The clouds produce water maser emission at 22 GHz, and therefore their motion and spatial distribution can be studied with the Very Long Baseline Array (VLBA) at spatial resolutions a factor ~ 200 higher than could be achieved with the Hubble Space Telescope in the optical. They have been shown (Greenhill *et al.*, 1995) to be orbiting in almost perfect Keplerian fashion less than one light year away from a central supermassive black hole. The proper motion of the clouds that project along our line of sight to the nucleus can be measured

---

[13] As an aside, this modern calibration of Leavitt's law is significantly different from the one used by Hubble, which was based on noisy and sparse data, neglected to correct the luminosities for absorption by interstellar dust, and failed to recognize the difference between Population I and II Cepheids. The latter are also a class of variable stars, but they are different from the classical, Population I Cepheids we have discussed so far: they are instead low mass, older stars and are about 1.5 magnitudes fainter than classical Cepheids. By Recognizing this fact, in 1952 Walter Baade announced a revision of the brightness of Cepheids by a factor four [later published in (Baade, 1956)], leading to an overnight increase in the age of the Universe by a factor two!

[14] NGC 4258 has several claims to fame. It was one of the original 12 galaxies identified by Seyfert (1943) as having peculiar nuclei. These nuclei are now recognized to be powered by black holes with mass million to billion times the mass of the Sun. The mass of the black hole in NGC 4258 is the most accurate yet measured in an external galaxy: $(3.90 \pm 0.34) \times 10^7 \mathcal{M}_\odot$ (Miyoshi *et al.*, 1995).

in projected units, and by assuming that their actual velocity equals that of the clouds 90 degrees away along the edge of the disk (which are moving along the line of sight) the distance to NGC 4258 could be determined with great precision: 7.2 ± 0.3 Mpc (Herrnstein *et al.*, 1999). One might wonder why this extraordinarily elegant method has not supplanted all others as a distance estimator. The reason is that, alas, galaxies like NGC 4258 are exceedingly rare. Nevertheless, the Cepheid distance to NGC 4258 (Mager *et al.*, 2008) agrees very well with the maser distance, thereby corroborating the Cepheid calibration described above.

8.2.5 *Hubble (the telescope) observes Cepheids*

Observing Cepheids is not an easy task. Cepheids are young stars, and as such they are found in the crowded and heavily dust-obscured disks of spiral galaxies. The period can only be determined if enough data are taken at many phases, thus requiring intensive monitoring over long stretches of time. In addition, to correct for dust absorption, data must be taken in at least two filters (the effects of dust obscuration depend on wavelength in a known way, so, by comparing measurements taken in two or more filters, it is possible to calculate how much luminosity is attenuated by dust). From the ground, these issues are compounded by blurring from the atmosphere (which makes it more difficult to disentangle the light of Cepheids from that of nearby stars) and the fact that the weather has the irritating habit of interfering with one's carefully planned observing schedule, which can have devastating consequences when observations must be taken at precise time intervals. In view of this, Leavitt's achievements appear all the more remarkable, and it should come as no surprise if she could determine the period for only little more than a handful of her 1800 Cepheids.

The Hubble Space Telescope (HST) allowed us to overcome two of the problems mentioned above: HST is unhindered by weather, and its exceptional resolving power makes it easier to account for crowding. In the 1990s, two very large programs were undertaken with HST to measure Cepheid distances to nearby galaxies: the first (The HST Key Project on the Extragalactic Distance Scale) led by Jeremy Mould, Wendy Freedman and Robert Kennicutt, targeted 18 galaxies with the goal of calibrating the Tully–Fisher relation for spiral galaxies, the Fundamental Plane for early-type galaxies, and the Surface Brightness Fluctuation method (see Secs. 8.4–8.6). The second, led by Allan Sandage, targeted eight galaxies with the goal of calibrating Type Ia supernovae (see Sec. 8.7). These programs (and others that followed) have now led to accurate Cepheid distances to 31 galaxies, and secured, for the first time since Leavitt's discovery of the law that carries her name, the extragalactic distance ladder.

8.3 The Aborted Explosion of Stars: The Tip of the Red Giant Branch

Stars are generally very predictable. Figure 8.5 shows the color–magnitude diagram for Messier 5,[15] a gravitationally bound assembly of several hundred thousand coeval stars. The band running diagonally across the bottom half of the diagram is the main sequence: it is populated by stars in the prime of their lives, burning hydrogen into helium in their cores. What differentiates stars along the main sequence is their mass: less massive stars are less luminous and redder (i.e. they have lower temperature), more massive stars are more luminous and bluer. Once hydrogen is exhausted in the core, the combustion moves from the core to a shell surrounding the core. At this stage, the entire luminosity of the star is produced within the shell, the star expands, its luminosity increases, and the temperature drops. The star then leaves the main sequence and moves upwards and to the right in the color–magnitude diagram, ascending the "red giant branch".[16]

Meanwhile, the core is doing something different altogether. Helium "ash" is deposited by the hydrogen burning shell onto the helium core, causing it contract, and its mass, density and temperature to increase. In low mass stars like the Sun, the contraction is such that the core becomes

---

[15] As mentioned earlier, color, which is plotted on the abscissa, is a good proxy for temperature.

[16] A digression: Because the time spent by a star on the main sequence is shorter for more massive stars, brighter stars "peel-off" the main sequence first. It follows that the luminosity at which the main sequence "bends" to join the red giant branch is an indication of the age of the stellar population under study, a point to which we will return in Sec. 8.8.

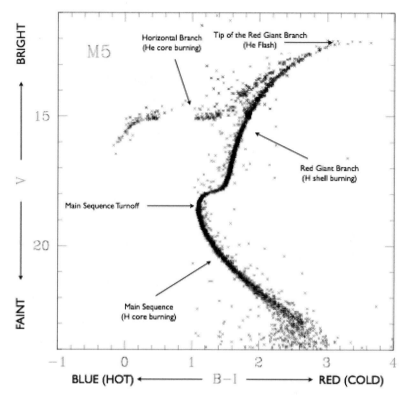

Fig. 8.5 A color–magnitude diagram for Messier 5 (M5), an old globular cluster in the Milky Way, at a distance of 7.6 Mpc (e.g. Sandquist *et al.*, 1996). The $B-I$ color on the abscissa (where B and I refer to the apparent magnitudes in, respectively, a blue and red part of the electromagnetic spectrum) is a proxy for stellar temperature, while the V magnitude in the ordinate is a measure of the luminosity in the green part of the spectrum. The various features mentioned in the text are there labeled. Note the "gap" in the horizontal branch: this is where RR-Lyrae, a class of variable stars, are found (but were not included when making the plot). The few stars falling far from the principal sequences are likely foreground field stars not associated with the cluster. The figure, which was produced using data from ground-based telescopes, was kindly provided by Peter Stetson.

degenerate, namely its equation of state, relating pressure and density, is independent of temperature: this process leads to one of the most dramatic events in stellar astrophysics, the "helium flash". When the temperature in the degenerate core reaches the point needed to burn helium (about 100 million degrees Kelvin), the star undergoes a thermonuclear runaway. The newly kindled nuclear reactions increase the temperature, but because the core is degenerate, the pressure cannot respond fast enough to allow the core to expand and cool. The temperature increases even further, which in turns accelerates the rate of nuclear reactions, and for a few seconds the star produces enough energy in the core to rival that of an entire galaxy.[17] Remarkably, none of this energy makes it out: the star does not become any brighter. Rather, all of the energy is used to lift the degeneracy, at which point the core is free to expand, its temperature drops, and the star once again settles in an equilibrium phase, quietly burning helium in its core. In the color–magnitude diagram, the star suddenly stops ascending the red giant branch and virtually instantaneously moves downwards and to the left to the "horizontal branch". This creates the sharp discontinuity known, rather unimaginatively, as the Tip of the Red Giant Branch, or TRGB. Which brings us to the reason for this long digression: The luminosity of a red giant at the time of the helium flash, i.e. the luminosity of the TRGB is, under the right circumstances, rather insensitive to the detailed nature of the star (for instance its composition and core mass). In other words, it is as close to a perfect "standard candle" as one can hope to get.

The use of the TRGB as a distance indicator began with the work of Da Costa and Armandroff (1990) and Lee *et al.* (1993); a more recent overview of the subject can be found in Bellazzini (2008). Observationally, the method is relatively straightforward, and certainly far less intensive than Cepheids: a pair of observations in two filters, and at one epoch, are sufficient. The TRGB is also relatively bright (although not as bright as Cepheids which, in the *I*-band, outshine TRGB stars by a factor of several to several tens, depending on the period). Thus it is possible, with the Hubble Space Telescope, to detect the TRGB out to distances approaching 20 Mpc[18] (Durrell *et al.*, 2007; Caldwell, 2006). Every galaxy type can be targeted, as long as its stellar population is old enough (a few billion years or so) for low-mass stars to have reached the helium flash stage. The only

---

[17]This explosive burning of helium doe not have catastrophic consequences, because the degeneracy in the core is lifted before things can get out of control. In contrast, *carbon* burning in degenerate conditions can lead to disruption of the star.

[18]The main issue here is not just one of luminosity, but also that the galaxy needs to be resolved into individual stars in order to construct a color–magnitude diagram. At large distances, this can only be done in lower density, dwarf galaxies.

slight complication comes from the fact that the magnitude of the tip does show a mild (but measurable) dependence on the composition (metallicity) and age of the stars. However, the age dependency is negligible as long as the stars are older than ~ 4 billion years, while the metallicity dependence is minimized when the observations are performed in the I-band where — as long as the stars are not unusually metal rich — it can be calibrated quite accurately based on broad band colors (Da Costa and Armandroff, 1990; Salaris and Girardi, 2005) or directly against theoretical stellar evolution models (Salaris et al., 2002). Overall, TRGB distances to individual galaxies can be measured with an accuracy of a few percent, comparable to what can be achieved using Cepheid variables.

But by far the most important virtue of the TRGB is that it is one of few distance indicators to be completely decoupled from the Cepheid distance scale, thus providing an important check on the systematics that might affect it. The absolute magnitude of the TRGB, in fact, has been calibrated using Galactic globular clusters (for which distances are known from RR Lyrae variable stars, Lee et al., 1993), Local Group dwarf galaxies [for which distances are known from the luminosity of the horizontal branch, calibrated in turn using trigonometric parallaxes measured by the Hipparcos satellite (Rizzi et al., 2007)], or directly using stars with trigonometric distances in the solar neighborhood (Tabur et al., 2009). By comparing Cepheid and TRGB distances for 15 nearby galaxies,[19] Rizzi et al. found that the two agree, on average, at better than the 0.5% level! In a parallel investigation, Mould and Sakai (2008, 2009a,b) published values of the Hubble constant based on a TRGB calibration of the Surface Brightness Fluctuation method, the Tully–Fisher relation, the Fundamental Plane, and Type Ia supernovae. By combining all methods, their final value is $H_0 = 70 \pm 6 \,\mathrm{km\,s^{-1}\,Mpc^{-1}}$. Following an identical procedure, except that secondary distance indicators are calibrated against the Cepheid distance scale instead, Freedman et al.

---

[19]Deriving TRGB and Cepheid distances for the same galaxy is possible only if the galaxy has a mix of old and young stars. This is due to the fact that Cepheids have an age of 100 million years or less (more luminous Cepheids are younger), while the TRGB is well characterized only for stars older than 4 billion years.

(2001) found $H_0 = 72 \pm 8 \, \text{km s}^{-1} \, \text{Mpc}^{-1}$. The excellent agreement is an independent validation of both the Cepheid and the TRGB distance scale.

8.4 Bumpy Galaxies: The Surface Brightness Fluctuation Method

In a paper published in 1988, John Tonry and Donald Schneider used a disarmingly simple fact to measure distances to nearby galaxies: all else being equal, when recorded on a high signal-to-noise CCD image, nearer galaxies have a more "mottled" appearance than farther ones, as shown in Fig. 8.6. This is a direct consequence of the fact that, while the (finite) number of stars recorded in each pixel increases as the square of the distance, statistical fluctuations in their number (equal to the square root of the number of stars) scale linearly with distance. If N is the number of stars per pixel, and f their average flux, what we perceive as "mottling" is simply $\sigma = \sqrt{N} \times f$. Since $N \propto d^2$ and $f \propto 1/d^2$, where d is the distance, $\sigma \propto 1/d$, giving more distant galaxies a smoother appearance. From here, measuring distances is a (conceptually) easy step away: the average stellar flux, f, is equal to the ratio of the pixel-to-pixel variance, $\sigma^2 = Nf^2$, to the mean flux per pixel, Nf. If the intrinsic luminosity, L, of the typical star is known, the distance then follows as $d^2 = L/4\pi f$. Needless to say, not all galaxies are created equal, and L will depend on

Fig. 8.6 A picture is worth a thousand words: from left to right, the panels show a globular cluster at a distance of ~ 10 kpc, M32, a companion of the Andromeda galaxy at ~ 800 kpc, and M49, the brightest galaxy in the Virgo cluster, at 16.7 Mpc. More distant objects appear smoother, which is the basis of the Surface Brightness Fluctuation method. The figure was kindly provided by John Blakeslee and John Tonry.

the mix of stellar populations in each individual galaxy. For *old* stellar populations and in redder optical bands, however, L is dominated by red giant branch stars (encountered in the previous section) and its behavior depends mostly on the stellar composition, or metallicity; the latter can be characterized empirically using images taken in (carefully chosen) multiple passbands.

The implementation of this method, aptly dubbed "Surface Brightness Fluctuations", or SBF, is complicated by several factors. Observationally, the intrinsic Poisson fluctuations in the number of stars per pixel (or, more precisely, resolution element) need to be disentangled from other effects that also cause pixel-to-pixel variations in the signal. Some fluctuations are instrument-related: noise from photon counting statistics, readout noise of the CCD, variations in the pixel-to-pixel sensitivity of the detector. Moreover, globular clusters, dust patches, H II regions and stellar associations belonging to the galaxy in question, as well as contamination from foreground stars and background galaxies, all provide unwanted modulations in the signal. SBF studies therefore require high signal-to-noise images (to minimize the impact of photo-counting statistics), an accurate instrumental calibration (such as flat-fielding and fringing corrections), and precise cataloging of all contaminants (globular clusters, stars, background galaxies, dust patches). The amplitude of the fluctuations are then determined by analyzing the spatial power spectrum of the image, from which a smooth model of the galaxy has been subtracted and contaminants have been masked. Because fluctuations in adjacent pixels are correlated through convolution with the point spread function (PSF) due to the atmosphere and telescope optics, their signature in the power spectrum is on the scale of the PSF — which must also be accurately known — making them easily separable from readout noise and photon counting statistics, both of which have a power spectrum that does not favor any particular scale. A final source of contamination to the SBF power spectrum comes from *unresolved* contaminants, for instance globular clusters or background galaxies too faint to be detectable on an individual basis. Such contribution to the power spectrum — which also shows up on the scale of the PSF — must be estimated and corrected for.

The last hurdle to overcome before SBF can be used as a distance indicator is, of course, its absolute calibration. In principle, this can be done directly using stellar evolution theory — effectively making SBF a primary distance indicator, at a par with Cepheids. However, uncertainty in the mass distribution of the stars, as well as in the details of their evolution [a particularly severe problem when SBF is measured in the near infrared (Blakeslee *et al.*, 2001; Raimondo *et al.*, 2005; Raimondo, 2009)] make it preferable to calibrate SBF against the Cepheid distance scale itself (Ferrarese *et al.*, 2000; Tonry *et al.*, 2001).

So, how has SBF fared since its first exposition in 1988? Tonry and Schneider estimated that SBF could be used to measure distances to early-type galaxies out to 20 Mpc, with an accuracy of $\sim 20\%$. This was far exceeded in the ambitious ground-based SBF survey of Tonry *et al.* (2001) — targeting 300 early-type and a few spiral galaxies with large unobstructed bulges — that reached galaxies as far as 53 Mpc with a mean error on the distance of only $\sim 10\%$. Farther horizons still were opened with the installation, in 2002, of the Advanced Camera for Surveys (ACS) on HST. The combination of large field-of-view, spatial resolution and sensitivity afforded by the ACS, has made it possible to measure SBF distances out to 100 Mpc (Biscardi *et al.*, 2008) with unprecedented precision. In the nearby (16.5 Mpc) Virgo cluster, Mei *et al.* (2007) were able to measure ACS SBF distances for 84 early-type galaxies with high enough accuracy (3% to 4% for most galaxies) that the three-dimensional structure of the cluster could, for the first time, be fully exposed: Virgo, the dominant mass concentration in the local Universe and the largest collection of galaxies within 35 Mpc, has a back-to-front depth of 2.4 Mpc, is slightly triaxial, and has yet to reach an equilibrium configuration. A small, compact cloud of galaxies, is seen falling towards the cluster from behind.

Coming back to our quest for H_0: Values[20] of the Hubble constant based on SBF measurement vary from $H_0 = 69 \pm 4 \pm 5 \, \text{km s}^{-1} \, \text{Mpc}^{-1}$ to

---

[20]Here and in the remainder of this chapter, when two errors are quoted for the Hubble constant, the first refers to the statistical (random) error, while the second (far harder to quantify!) refers to systematic uncertainties due to factors like the absolute calibration of the Leavitt law, or its metallicity dependence.

$H_0 = 77 \pm 4 \pm 7\,\mathrm{km\,s^{-1}\,Mpc^{-1}}$ (e.g. Ferrarese et al., 2000; Tonry et al., 2001; Jensen et al., 2001; Liu and Graham, 2001), the difference being due to the details of the calibration used for SBF, and the corrections applied to the sample of galaxies used to measure the rate of expansion. It is notable, however, that these values agree, within the errors, with the value of $H_0 = 76 \pm 6 \pm 5\,\mathrm{km\,s^{-1}\,Mpc^{-1}}$ derived by Biscardi et al. (2008) by calibrating SBF against theoretical stellar population models. The agreement is a confirmation of the validity of the Cepheid distance scale.

Ultimately, the inability to push deep into the Hubble flow (except for Biscardi et al., all of the works cited above are based on galaxies with a recession velocity less than $5000\,\mathrm{km\,s^{-1}}$) makes SBF more susceptible to local inhomogeneities than other distance indicators. However, the precision associated with SBF distances makes the method a very important tool indeed in mapping the distribution of galaxies in the local Universe and measuring local *departures* from a pure Hubble flow (Tonry et al., 2000). Such departures — or large-scale flows — are caused by, and therefore a tracer of, the inhomogeneous distribution of intergalactic dark matter: the main player in the evolution of cosmic structures.

8.5 The Orderly Nature of Early Type Galaxies: The Fundamental Plane

Early-type galaxies, as a whole, form a surprisingly homogeneous family. The size R, mass $\mathcal{M}$, and stellar velocity dispersion σ of a self-gravitating, stable stellar system are related by the virial theorem: $\sigma^2 \propto \mathcal{M}/R = (\mathcal{M}/L)\,\mu \times R$, where $\mu = L/\pi R^2$ is the galaxy surface brightness, and $\mathcal{M}/L$ is the mass-to-light ratio.[21] This equation describes a plane in the three-dimensional space defined by $\log(\sigma)$, $\log(\mu)$ and $\log(R)$: *if* early-type galaxies form a homologous family (i.e. their structural and kinematical properties differ only by a constant scale factor), *and* their mass-to-light ratio is constant, all early-type galaxies should lie on this plane. They very nearly do.

---

[21] Observationally, R is often defined as the galaxy "effective" radius, i.e. the radius containing half of the total light, while μ is the mean surface brightness within R. s is the central value of the velocity dispersion.

In the late 1980s, two independent teams (Djorgovski and Davis, 1987; Dressler *et al.*, 1987) discovered that, indeed, the velocity dispersion, surface brightness and size of early type galaxies define a relation that is linear in logarithmic space. Using a sample of over 200 galaxies belonging to 10 galaxy clusters within ~ 160 Mpc, Jorgensen *et al.* (1996) found that this Fundamental Plane of early type galaxies is best described as $\sigma^{1.24} \propto \mu^{0.82} \times R$: close, but not quite identical to the prediction of the virial theorem.[22] A more recent characterization of the Fundamental Plane is reproduced in Fig. 8.7: both face-on and edge-on views are shown. The tilt of the observed Fundamental Plane — relative to the predictions of the virial theorem — can be explained if the stellar mass-to-light ratio, $\mathcal{M}/L$, and/or the galaxy structural properties are not constant (as assumed in the first paragraph), but they themselves depend on mass. For instance, Jorgensen *et al.* find

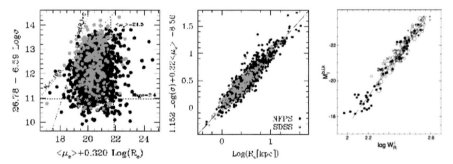

Fig. 8.7 The two left-most panels show a face-on and an edge-on representation of the Fundamental Plane for early type galaxies, from D'Onofrio *et al.* (2008). The different colors indicate data from two different surveys, representing 59 nearby clusters with redshifts between 0.04 and 0.07. The right-most panel shows the I-band Tully–Fisher relation from Tully and Pierce (2000); the quantity plotted on the abscissa is the logarithm of the inclination-corrected H I line width, in $\mathrm{km\,s^{-1}}$. The different symbols identify data from different clusters. Note that in all panels, quantities are plotted in physical units: in the Fundamental Plane plots, the radius is in kpc, while in the Tully–Fisher plot, the magnitude is in absolute units.

---

[22]The tightness of the Fundamental Plane betrays a remarkable homogeneity in the stellar population of early type galaxies. The tilt of the plane is also quite insensitive to environment: Jorgensen *et al.* find no statistical change in the slope of the Fundamental Plane from cluster to cluster; while Bernardi *et al.* (2003) find only mild differences when comparing the Fundamental Plane of field and cluster galaxies. Additional characterizations of the Fundamental Plane are given by Hudson *et al.* (1997) and Colless *et al.* (2001).

that a dependence of the type $\mathcal{M}/L \propto \mathcal{M}^{0.24}$, is sufficient to explain the observed tilt. The exact reasons as to why $\mathcal{M}/L$ should depend on $\mathcal{M}$ are not entirely clear, but a change in the stellar population is certainly implicated (as betrayed by the fact that more massive galaxies are redder), and so are structural and dynamical non-homology (Pahre et al., 1998; Graves et al., 2009; Trujillo et al., 2004).[23] The Fundamental Plane then becomes a powerful tool to study galaxy evolution, and indeed changes in its tilt and thickness as a function of redshift have been used to test hierarchical theories of galaxy formation and trace the star formation history of early-type galaxies (e.g. Treu et al., 2005; Hyde and Bernardi, 2009).

The relevance of the Fundamental Plane as a distance indicator comes from the fact that while σ and μ are distance-independent, R, as observed, is not. Early-type galaxies at different distances will therefore lie on parallel Fundamental Planes, shifted relative to each other along the R axis by an amount proportional to their relative distance. If the distance to one galaxy is known, its Fundamental Plane can then be cast in terms of the physical (as opposed to projected) radius, and the distance to all other galaxies is then given by the shift required to bring them onto the (absolute) Fundamental Plane of the first.[24] The commonly adopted absolute calibration of the Fundamental Plane was derived by Kelson et al. (2000) as part of the HST Key Project on the Extragalactic Distance Scale. Unlike the Tully–Fisher relation, to be discussed in Sec. 8.6, the Fundamental Plane cannot be calibrated directly — on a galaxy-by-galaxy basis — against the Cepheid distance scale, for the very simple reason that the young stars undergoing Cepheid pulsation are long gone in early-type galaxies, whose stellar populations are, as a rule, uniformly old. Kelson et al. (2000) therefore compared Cepheid and Fundamental Plane distances not for individual galaxies, but for galaxy *clusters*[25]: the (non-trivial) assumption here being that the spiral galaxies to which Cepheid distances are measured define the same mean cluster distance as the ellipticals used in constructing the

---

[23]Other possibilities are a change in the ratio of dark to visible matter as a function of mass (Hopkins et al., 2008), and variations in the kinematic structure.
[24]More precisely, what is derived is the "angular diameter" distance.
[25]Kelson et al. used the Fornax and the Virgo clusters, as well as the Leo I group.

Fundamental Plane. Based on statistical argument, Kelson *et al.* estimate that this assumption alone introduces a 5% error in the absolute calibration of the Fundametal Plane and accounts for a quarter of the systematic error in the value of H_0 derived from Fundamental Plane studies.

Another limitation of the method, when applied to individual galaxies, is the intrinsic (i.e. irreducible) thickness in the Fundamental Plane; for nearby galaxies this thickness (which is due to differences in stellar population form galaxy to galaxy) translates into a 17% uncertainty in distance (Jørgensen *et al.*, 1996; Blakeslee *et al.*, 2002). On the bright side, although as a distance indicator, the Fundamental Plane is not as precise as SBF, it is far less demanding from an observational point of view, and Fundamental Plane data can be obtained for large numbers of galaxies with relative ease. Additionally, it can reach to farther distances than SBF.[26] Based on 11 clusters with systemic velocity between 1 100 and $12\,000\,\mathrm{km\,s^{-1}}$ (from Jorgensen *et al.*, 1995), Kelson *et al.* derive a Hubble constant $H_0 = 78 \pm 7 \pm 8\,\mathrm{km\,s^{-1}\,Mpc^{-1}}$, where the main source of random uncertainty is associated with the slope and zero point of the Fundamental Plane, while the zero point of the distance scale for calibration, as well as uncertainties in the group calibration, are mostly responsible for the systematic error. Furthermore, Blakeslee *et al.* (2002) found good agreement between the SBF and Fundamental Plane distance scales, and derive a value of the Hubble constant of $H_0 = 73 \pm 4 \pm 8\,\mathrm{km\,s^{-1}\,Mpc^{-1}}$ from a Cepheid-based SBF calibration of the Fundamental Plane.

8.6 The (Not Quite As) Orderly Nature of Spiral Galaxies: The Tully–Fisher Relation

In the previous chapter, it was argued that the virial theorem, combined with a relative homogeneity in stellar populations and structural properties, explains why early type galaxies populate a thin Fundamental Plane. The natural question is, of course, whether the same holds for spiral

---

[26]The Fundamental Plane has been measured for galaxies up to $z \sim 1.2$, although at these distances, changes in the tilt of the plane induced by galaxy evolution interfere with its use as a distance indicator (Treu *et al.*, 2005).

galaxies. The answer is yes, however the plane for disk systems does not coincide with, and is not quite as neatly behaved as, the one defined by dynamically hot galaxies. Burstein *et al.* (1997) studied this issue in detail, and concluded that in a three-dimensional space with axes defined by carefully chosen (and physically motivated) combinations of σ, μ and R,[27] spiral galaxies do scatter on a plane, but this is offset and tilted relative to the one populated by early-type galaxies by an amount that changes continuously as one moves from early-type spirals (Sa and Sb) to later-types (Sc and Sd). Conveniently, however, when viewed edge-on, the Fundamental Plane for spiral galaxies projects into a simple relation between the galaxy total luminosity and the maximum rotational velocity of the disk[28]: this relation is known as the Tully–Fisher relation, from the two astronomers who, in 1977, first brought to public attention its tremendous potential as a distance indicator. Equally as important, although we will not dwell on it here, is the fact that the Tully–Fisher relation can be used to probe the relative spatial distribution of dark and luminous matter in spiral galaxies. As such, the relation is a strict test (and a serious challenge!) to current theories of galaxy evolution: neither analytic, semi-analytic nor hydrodynamic simulations have yet been able to recreate models of spiral galaxies that obey the relation in detail (e.g. Governato *et al.*, 2007, and references therein). A recent determination of the Tully–Fisher relation is shown in Fig. 8.7.

The application of the Tully–Fisher relation as a distance indicator is based on the fact that while the rotational velocity of the disk is distance-independent, the luminosity scales as the inverse of the square of the distance. As is the case for the Fundamental Plane, if the distance to a Tully–Fisher galaxy is known, the distance to any other galaxy can be

---

[27] These quantities are defined as in Sec. 8.5, except that in the rotationally dominated spiral galaxies s is replaced by the inclination-corrected maximum rotational velocity, V_{rot}.

[28] An equivalent relation [known as the Faber–Jackson relation; (Faber and Jackson, 1976)] exists for early-type galaxies. However, the Faber–Jackson relation is close but not exactly equal to an edge-on projection of the Fundamental Plane, and therefore has larger scatter than the plane itself. In the case of the Tully–Fisher relation, the scatter cannot be reduced by introducing a third parameter, such as size or surface brightness (Courteau and Rix, 1999).

inferred from the shift (along the luminosity axis) needed to bring it on the absolute Tully–Fisher relation of the first. Observationally, however, the Tully–Fisher relation is more difficult to characterize than the Fundamental Plane, for two reasons: the luminosity of a spiral galaxy is greatly affected by dust obscuration[29] (while early-type galaxies are, for the most part, dust-free), and the rotational velocity of the disk needs to be inferred from its component along the line of sight, the only quantity that can be measured directly. Correcting for dust obscuration is a difficult art and one that has yet to be mastered: although several empirical recipes exist, none is fully satisfactory [see Sakai *et al.* (2000) for a discussion]. Similarly, the inclination angle of the disk — needed to transform the line of sight velocity into a rotational velocity on the disk's plane — must be estimated from the observed ratio of the projected major to minor axis given some assumption on the intrinsic axial ratio, which is not known on a galaxy-to-galaxy basis. By one of those cruel jokes of Nature, inclination corrections are smallest in highly inclined galaxies, but in such galaxies, extinction corrections are fraught with uncertainties.

To make matters worse, the rotational velocity itself is not easily defined. It is generally measured by exploiting the presence, in the disks of spiral galaxies, of large quantities of hydrogen, which produces clear signatures in both the optical and the radio part of the electromagnetic spectrum. A strong emission line in the optical (Hα at 6563 Å) is produced by H II regions throughout the disk, and is commonly used to trace the disk rotation curve (e.g. Courteau, 1997). The snag is that, at the edge of the optical disk, rotation curves are not falling (as they should if all the mass were associated with luminous matter) but stay rather flat or even rise. This indicates the presence of significant amount of dark matter, whose contribution to the gravitational potential at large radii exceeds that of the luminous matter: for practical purposes, this means that optical rotation curves might not reach far enough to sample the maximum

---

[29] Absorption is also caused by dust belonging to our *own* galaxy. This is, however, known quite reliably: dust emits in the far-infrared and has been mapped in exquisite detail by Schlegel *et al.* (1998) using the 100 μm all-sky surveys from the IRAS and COBE satellites.

rotational velocity of the disk. Luckily, *neutral* hydrogen can be traced out to radial distances that typically reach twice as far as the optical disk. Spin transitions in the ground state of hydrogen atoms produce a detectable emission at 21 cm: if hydrogen is distributed throughout the disk, then lines produced at different locations within the disks are Doppler-shifted relative to one another due to differences in the rotational velocity. When a galaxy is observed within the beam of a single-dish radio telescope, the width of the 21 cm emission line is a direct measure of the maximum rotational velocity of the disk. The uncertainty here comes from the fact that the 21 cm line often has a complex and/or asymmetric profile and its wings are difficult to map as they are generally drowned in noise, making the exact definition of "width" somewhat nebulous. Catinella *et al.* (2005, 2007) have explored these issues in detail.

The last hurdle to overcome is, as usual, the calibration of the Tully–Fisher relation using galaxies for which the distance is independently known. This was accomplished by Sakai *et al.* (2000), again as part of the HST Key Project on the Extragalactic Distance Scale. The calibration, based on 21 galaxies with Cepheid distances, was applied to galaxies belonging to four distant clusters (all within $10\,000\,\mathrm{km\,s^{-1}}$) observed in four broad bands from B to H. The values of H_0 range from $74 \pm 2\,\mathrm{km\,s^{-1}\,Mpc^{-1}}$ (for the I-band sample) to $67 \pm 3\,\mathrm{km\,s^{-1}\,Mpc^{-1}}$ (for the H-band sample), giving an average of $H_0 = 71 \pm 4 \pm 10\,\mathrm{km\,s^{-1}\,Mpc^{-1}}$. The largest source of random uncertainty in the Tully–Fisher distance to an individual galaxy arises from the scatter in the relation itself. This scatter (which, as is the case for the Fundamental Plane, is also important to constrain models of galaxy formation) is known to vary along the relation, being larger for galaxies with lower rotational velocity, and is also known to depend on the wavelength chosen to measure the luminosity [as are the slope and zero point of the relation, see e.g. Sakai *et al.* (2000); Masters *et al.* (2008)]. In the optical, the scatter in the Tully–Fisher relation translates into a 15–20% error in the distance to an individual galaxy. Recent results, however, indicate that the scatter might be significantly reduced at longer wavelengths: Freedman and Madore (2010) estimate that by using $3.6\,\mu\mathrm{m}$ data from the Spitzer Space Telescope it might be possible to derive Tully–Fisher distances good to the 5% level. This is comparable to the precision that can be achieved with SBF measurements, but with the additional advantage that Tully–Fisher distances can be pushed further into the Hubble flow.[30]

8.7 Stellar Explosions: Type Ia Supernovae

Type Ia supernovae are the Ferrari of distance indicators: rare, expensive, finicky, but hard to beat when it comes to performance. At maximum brightness, they outshine Cepheids by close to six orders of magnitude, making them detectable well beyond redshift 1.0 (Riess *et al.*, 2007).[31] Sure enough, galaxies are even brighter, and the Tully–Fisher relation and the Fundamental Plane can reach at least as far, but Type Ia supernovae have two crucial advantages. First, when properly calibrated, they have far higher precision, 5% in distance, than either the Tully–Fisher relation or the Fundamental Plane (Riess *et al.*, 1996). This makes Type Ia supernovae powerful cosmological probes: the small (10%) departures observed from the linear Hubble flow due to the interplay of matter and dark energy would have been lost, had a less accurate distance indicator been used. Second, all evidence supports the fact that supernovae do not evolve with redshift (Coil *et al.*, 2000; Hook *et al.*, 2005). In contrast, galaxies do: they merge with each other and their stellar population evolves. Changes with redshift in the Tully–Fisher relation and the Fundamental Plane probe galaxy evolution, not cosmology.

But then, of course, there are disadvantages. Supernovae are rare: a volume of $30\,000\,\text{Mpc}^3$ would yield, statistically, only one Type Ia supernova per year (Neill, 2006): this is a volume large enough that if it were centered on the Milky Way, it would extend far enough to include the Virgo cluster! Supernovae are variable phenomena and unlike Cepheids, that at least have the good sense of not disappearing — if you miss one, it is gone. As will be elaborated upon below, they need to be observed

---

[30] Recently, Gnerucci *et al.* (2010) measured the Tully–Fisher relation in galaxies at $z = 3$, finding evidence that at those early epochs, galaxies are still in the process of building their stellar components.

[31] The current record holder, SN1997ff, is at a redshift of 1.755.

before, during and after (at least a month, possibly longer) maximum brightness. Observing in more than one band is necessary to establish their type and estimate dust contamination to the observed luminosity. In short, supernovae searches are observationally very expensive, requiring frequent (possibly daily) monitoring of large patches of the sky. Hundreds of nights at ground-based telescopes, and hundreds of hours of HST time have been devoted to dedicated supernovae searches (Tonry *et al.*, 2003; Barris *et al.*, 2004; Astier *et al.*, 2006; Miknaitis *et al.*, 2007; Riess *et al.*, 2007).

Finally, the fact that a detailed physical understanding of supernovae is still lacking is a cause of mild uneasiness: establishing a firm theoretical footing for Type Ia supernovae explosions would (at least that is the hope!) lift any lingering doubt that some stellar evolution conspiracy, and not cosmology, is responsible for the observed trends in their luminosity as we look further back in time. What is known with certainty is that supernovae are the consequence of a thermonuclear runaway — the same process responsible for the fact that the red giant branch has a "tip". However, while in red giant stars thermonuclear runaway is not catastrophic, for the progenitors of Type Ia supernovae it leads to the complete disruption of the star. Exactly how this happens is subject of debate (in contrast to the processes that lead to the helium flash, which rest on very solid theoretical grounds). In fact, even the progenitors of Type Ia supernovae are not known with certainty: observationally there is not a single example in which the progenitor of a Type Ia supernovae could be identified,[32] so all speculations must rest on theoretical arguments. The most credited scenario involves carbon–oxygen white dwarfs,[33] perhaps surrounded by

---

[32] In contrast, there are two cases in which massive stars were identified as the progenitors of Type II supernovae, one of which is the famed SN1987A that occurred (obviously in 1987) in the Large Magellanic Cloud.

[33] A white dwarf is the last stage in the stellar evolution of stars with mass less than $\sim 10 \mathcal{M}_\odot$ which is no longer able to support nuclear fusion. The composition of white dwarfs — whether it is helium, carbon–oxygen or oxygen–neon — depends on the mass of the original star: heavier elements are synthesized in the more massive stars. In all cases, however, a white dwarf is stable against collapse because it is supported by pressure generated by degenerate matter. Undisturbed white dwarfs are stable systems, destined to cool indefinitely, a property that can be used to estimate the age of the Universe (see Sec. 8.8).

a helium shell, although in at least some cases, it cannot be ruled out that more massive oxygen–neon white dwarfs might be involved. What causes the white dwarf to suddenly explode is also not well understood; however, it is almost certain that the white dwarf needs to have a binary companion. The reason is that, in order for a white dwarf to be knocked off its unperturbed state of quiet cooling, its mass needs to be increased above the Chandrasekhar limit the mark above which degenerate matter cannot withstand gravitational collapse. Carbon–oxygen white dwarfs have masses of order $0.6\,\mathcal{M}_\odot$, while the Chandrasekhar limit is $1.39\,\mathcal{M}_\odot$: the difference must be made up by material accreted from a companion star. The companion could be either a "normal" star, or another white dwarf — both hypotheses come with arguments in favor and against. The next step, namely the detailed physics of the explosion, is more uncertain still. Because of the complexity of the problem, progress can only be made by using three-dimensional numerical simulations that incorporate a realistic set of initial conditions and include accurate equations describing all of the physical processes involved — not an easy task. In the case where the progenitor is a carbon–oxygen white dwarf that has approached the Chandrasekhar mass, the explosion is caused by the sudden ignition of carbon and oxygen, which results in a thermonuclear runaway because of the degenerate state of the star. How exactly the explosion proceeds is unclear: at present, deflagration (i.e. a combustion that propagates subsonically by thermal conductivity) or a deflagration that degenerates into a detonation (i.e. a combustion that propagates supersonically, creating a high pressure shock wave in front of it) appear to be the leading contenders. A thorough review of the physics of Type Ia supernovae, and how they match observational data, is given in Hillebrandt and Niemeyer (2000). As the authors put it, "from a theorist's point of view, it appears to be a miracle that all the complexity seems to average out in a mysterious way to make the class so homogeneous."

And homogeneous it is, the potential of Type Ia supernovae as distance indicators was realized virtually as soon as astronomers became aware of their existence [a recent review of the use of supernovae to measure cosmological parameters is given in (Kirshner, 2009)]. Sure enough, in the early days the confusion between classical novae and supernovae muddled the waters, but this was resolved, once and for all, by Hubble's 1925 observations of Cepheids in nearby galaxies. By 1938, the famed astronomer Walter Baade observed that the magnitude reached by the supernovae at maximum had a dispersion of "only" 1.1 mag, making them one of the most precise standard rulers of the time. All of the supernovae in Baade's study happened — by sheer luck — to be of what we now call Type I, but a few years later, Minkowski (1941) discovered that supernovae could be divided in two classes, based on the absence (Type I) or presence (Type II) of hydrogen in their spectra.[34] Type II supernovae are generally fainter and have larger dispersion in their maximum magnitude than Type I's, so distinguishing the two classes was an important step in establishing the accuracy of supernovae as distance indicators. In an article published in 1968, Charles Kowal brought the dispersion in the maximum magnitude of Type I supernovae down to 0.6 mag (quite remarkable, considering that the data on which he based his study were obtained using photographic plates and that he did not account for dimming due to dust), and commented that Type I supernovae could potentially be used to measure distances to clusters of galaxies with 5% to 10% accuracy, perhaps even probe cosmic acceleration.

In the mid-80s it was further recognized that the Type I class is also not homogeneous, and a further division was made between Type Ia and Type Ib/c, the latter lacking silicon in their spectra. The progenitors of Type Ib/c are now known to be massive stars (as are the progenitors of Type IIs) while, as mentioned earlier, the progenitors of Type Ia are thought to be degenerate, compact white dwarfs. The distinction made it possible to improve the accuracy in the Type Ia distance scale — which at that point still relied on the maximum magnitude as a standard candle — by another factor of two.

In the late 80s, an increasing number of supernovae were recorded using CCDs with much higher photometric fidelity that could be achieved with

---

[34]Type II supernovae can also be used as distance indicators, although the method is based on the rate of expansion of their photosphere, rather than the magnitude of the supernova itself. A review can be found in Poznanski *et al.* (2010).

photographic plates. This made it increasingly more obvious that there is a significant variety in the light curves of Type Ia supernovae and that, in particular, the luminosity at maximum could vary by as much as a factor five from object to object (Phillips *et al.*, 1987, 1993; Hamuy *et al.*, 1996; Riess *et al.*, 1996). Luckily, there appeared to be a very good correlation between the magnitude at maximum and the shape of the light curve: supernovae with the slowest rate of decline were invariably the most luminous. By combining information from the luminosity and shape of the light curve, and using multiple bands, Riess *et al.* (1996) improved the accuracy of supernovae as distance indicator by another factor of two, to the modern 5% in distance.

But the usual last hurdle remained. In his 1968 paper, Kowal remarked that "the main problem now is one of calibration. It is not yet possible to determine the Hubble constant by means of supernovae alone, since no reliable distance moduli are available to calibrate the absolute magnitudes of the nearer supernovae." This situation was not resolved until the launch, nearly a third of a century later, of the Hubble Space Telescope. Allan Sandage, one of Kowal's colleague at Carnegie Institution of Washington, and his collaborators undertook an ambitious program to observe Cepheid variables in eight galaxies that had been hosts of Type Ia supernovae (Saha *et al.*, 2006; Sandage *et al.*, 2006), thus providing the first secure calibration of the Type Ia supernovae distance scale.[35] At the end of their 15-year long study, the group published a value of $H_0 = 62.3 \pm 1.3 \pm 5 \,\mathrm{km\,s^{-1}\,Mpc^{-1}}$ (Sandage *et al.*, 2006) Previously, based on a reanalysis of a subset of six of the galaxies observed by Sandage and collaborators and available at the time, Gibson *et al.* (2000) and Freedman *et al.* (2001) published a value 14% larger, $H_0 = 71 \pm 2 \pm 6 \,\mathrm{km\,s^{-1}\,Mpc^{-1}}$. Sadly, most of the difference between the two studies is due to differences in the Cepheid distances themselves: supernovae do not really check whether the galaxy in which

---

[35] Allan Sandage's association with Cepheids and the expansion rate of the Universe is deep-rooted. He was a graduate student at the California Institute of Technology when Walter Baade, his thesis advisor, discovered Population II Cepheids. During the same period, he was graduate student assistant to Edwin Hubble himself, continuing his work after Hubble's sudden death in 1953.

they go off is a good target for Cepheid searches, and in some galaxies only few Cepheids could be detected, resulting in very uncertain distances.

The most recent effort in this direction is by Riess *et al.* (2005, 2009) who added four new galaxies to the list of Type Ia supernova hosts with Cepheid distances [and by the time this book is in press, they will have added two more (Riess, private communication)]. They also did something different compared to previous studies: they imposed very stringent requirements on the quality of the supernovae to be used as calibrators (availability of CCD data, a light curve that must extend before maximum, low dust extinction), they worked in the near infrared (thus reducing uncertainties introduced by dust extinction and metallicity), and they tied the Cepheid scale to the maser galaxy, NGC 4258, instead of the Large Magellanic Cloud.[36] Their final calibration, based on six Type Ia supernovae host galaxies, was then used to produce a Hubble diagram for sample of 240 supernovae at $z < 1.0$ (Fig. 8.8), and derive a value of the Hubble constant of $74.2 \pm 3.6 \, \text{km s}^{-1} \, \text{Mpc}^{-1}$.

8.8 Further Reflections and Future Directions

8.8.1 *Eighty years of Hubble constants*

Figure 8.9 shows how published values of the Hubble constant, compiled by Prof. John Huchra at the Harvard Smithsonian Center for Astrophysics, have changed with time. Several striking features are apparent. First, it might be noted that Hubble's value of $500 \, \text{km s}^{-1} \, \text{Mpc}^{-1}$ was *not* the first! Two previous estimates were published by Lemaître in 1927 and Robertson in 1928, but both were based on few galaxies for which distances had been measured by Hubble. Second, there are two "transition phases" in this plot. The first, more noticeable, happened in the mid 1950s: values published before this date cluster around $\sim 500 \, \text{km s}^{-1} \, \text{Mpc}^{-1}$, while virtually no value published after 1960 is above $100 \, \text{km s}^{-1} \, \text{Mpc}^{-1}$. What

---

[36]This step bypasses uncertainties related to the still ill-determined distance to the Large Magellanic Cloud. Additionally, the same HST instrument was used to detect Cepheids in the supernova host galaxies *and* in the calibrating galaxy, NGC 4258. Therefore, the accuracy of the calibration is improved since just a relative, and not an absolute, knowledge of the Cepheids' flux is required.

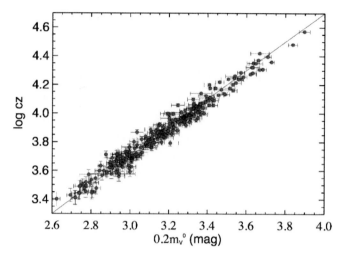

Fig. 8.8 A modern version of the Hubble diagram (compare this to Fig. 8.1!), defined by 240 Type Ia supernovae at $z \lesssim 0.1$, from (Riess *et al.*, 2009). The quantity m_v^0 in the abscissa is the peak apparent visual magnitude, corrected for extinction and accounting for the dependence of the magnitude at maximum on the shape of the light curve. In the ordinate, z is the redshift, and c is the speed of light, so that cz is the recessional velocity of the supernova host galaxy.

caused this dramatic drop was the realization, in the late 1940s, of the importance of dimming by interstellar dust, and the discovery, in the early 1950s, of population II Cepheids, with the consequent re-adjustment of the Leavitt law for Cepheids and, with it, the entire extragalactic distance scale ladder.

The second phase-transition occurred around the year 2000, and is more subtle. After 1960, there is no significant trend in the published values, rather they all scatter around a mean of $\sim 70 \,\mathrm{km\,s^{-1}\,Mpc^{-1}}$. However, the magnitude of this scatter decreases significantly after 2000; in other words, the published values become more *precise*. This was the result of two decade-long efforts to provide an accurate set of Cepheid distances to nearby galaxies, using the Hubble Space Telescope: the HST Key Project on the Extragalactic Distance Scale (Freedman *et al.*, 2001) and the Type Ia supernovae calibration project led by Allan Sandage (Sandage *et al.*, 2006). For the first time since 1929, a large number of secondary distance indicators could be calibrated in a reliable and *consistent* manner. This

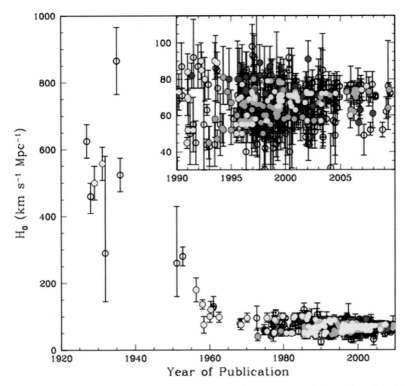

Fig. 8.9 A compilation of published values of H_0, from data carefully collected by Prof. John Huchra at the Harvard Smithsonian Center for Astrophysics. The inset at the top-right shows values published in the past two decades only (1990–2010). The symbols are color-coded according to the method used: red for the Tully–Fisher relation; blue for SBF; green for Type Ia supernovae; cyan for the Fundamental Plane; magenta for the tip of the red giant branch. Yellow points represent averages of values obtained using more than one method. Open circles represent values based on methods not discussed in this chapter, such as the Globular Cluster Luminosity Function, or the Sunyaev–Zeldovich effect.

resulted in far better agreement than ever before between the values of the Hubble constant measured using different methods.

The fact that several distance indicators, all subject to different biases and systematics, and applied to completely independent samples of galaxies, in different distance ranges, *all give a consistent answer*, is certain assurance that we must be on the right track. But could the Cepheid distance scale itself be in error? Figure 8.9 shows that accounting for systematic uncertainties is a risky business. There is no overlap in the error bars for many of the published estimates of H_0: the true error is almost always larger than the astronomers' best estimate! But even on this point, we have some reassurance: as discussed earlier, the Cepheid distance scale has been corroborated by the completely independent geometric distance to the maser galaxy, NGC 4258, *and* by distances based on the tip of the red giant branch. In addition, recent determinations based on other independent methods, such as time delay in strong gravitational lenses (e.g. Blandford and Narayan, 1986), the Sunyaev–Zel'dovich effect (Sunyaev and Zel'dovich, 1969) and measurements of anisotropies in the Cosmic Microwave Background (Peebles and Yu, 1970; Sunyaev and Zel'dovich, 1970), all converge on values of the Hubble constant between 70 and $75\,\mathrm{km\,s^{-1}\,Mpc^{-1}}$. A detailed discussion of these methods in the context of the extragalactic distance scale can be found in Freedman and Madore (2010).

Dare we conclude this chapter with a "final" value of the Hubble constant? At the end of the HST Key Project on the Extragalactic Distance Scale, Freedman *et al.* (2001) published what probably comes closest to a consistent, consensus value. The Key Project team used Cepheid distances to 31 galaxies (only 18 of which observed as part of the Key Project, but all analyzed in a consistent manner) to calibrate the Tully–Fisher relation (Sakai *et al.*, 2000), Type Ia supernovae (Gibson *et al.*, 2000), the Fundamental Plane for early type galaxies (Kelson *et al.*, 2000), and the Surface Brightness Fluctuation method (Ferrarese *et al.*, 2000). Each method, thus calibrated and applied to distant samples of galaxies, produced a value of the Hubble constant, as has been discussed in the previous sections. The weighted average of all values gave $H_0 = 72 \pm 3 \pm 7$ $\mathrm{km\,s^{-1}\,Mpc^{-1}}$. Freedman and Madore, in their recent 2010 review, revised this to $H_0 = 73 \pm 2 \pm 4\,\mathrm{km\,s^{-1}\,Mpc^{-1}}$; the smaller systematic uncertainty results from the improved calibration of the Leavitt law discussed in Sec. 8.2.3.

8.8.2 *The age of the Universe*

In 1929, Hubble's estimate of the Hubble constant implied an age of the Universe of 1.3 Gyr, embarrassingly lower than the age of the Earth estimated from radioactive decay. Since then, estimating the age of the oldest objects has been a litmus test for any value of the Hubble constant aspiring to public acceptance. The Earth does not actually provide a stringent test at all since, at a mere age of 4.5 Gyr, our Solar System is a relative newcomer. Stricter tests are set by the oldest stars in the Universe: very metal-poor stars in the halo of the Milky Way, and stars in globular clusters. The chemical composition of a star is a good tracer of its age: older stars are formed from gas that has not had the time yet to become polluted by the heavy elements synthesized in the core of massive stars, and have released into interstellar space during supernovae explosions. Globular clusters are gravitationally bound swarms of 100 000 to a million stars (an example is shown in Fig. 8.6), and are believed to have been between the very first structures to coalesce from small-scale density fluctuations in the early Universe. Ages of individual stars can be estimated from radioactive dating of Thorium and Uranium, while ages of globular clusters can be estimated by identifying, in their color–magnitude diagrams, specific features that are particularly sensitive to age. One is the "main sequence turnoff", the sharp bend where the main sequence joins the red giant branch (see Fig. 8.5). The other is the "pile-up" of stars at the faint end of the cooling sequence defined by white dwarfs.

Radioactive dating is based on a comparison between the present abundance ratio of radioactive isotopes to stable elements, and the initial value of the same ratio, i.e. the ratio of radioactive isotopes to stable elements at the time the star formed. The age, t, simply follows from the fact that radioactive isotopes decay exponentially: $r = r_0 \exp(-\ln 2 \times t/t_{1/2})$, where r and r_0 are the present and initial abundance ratios, respectively, and $t_{1/2}$ is the time necessary for the isotope to be depleted by a factor two (i.e. the half-life), which is generally very well known. As one might imagine, there are serious difficulties when applying this method to date a star. The first is of an observational nature: the absorption lines produced by radioactive isotopes in stellar spectra are exceedingly weak, and can only be detected in stars that are over-abundant in such elements. Such stars (which also need to be old, i.e. metal-poor) are very rare: it is estimated that only one star in one hundred thousand to a million is a suitable candidate for radioactive dating (Hayek *et al.*, 2009). The other difficulty comes from uncertainties in the theoretical nucleosynthesis models which are needed to predict the initial abundance ratios. As a result, to-date, abundances of radioactive isotopes have been measured in only six stars (see Hayek *et al.*, 2009 and references therein) and constraints on their ages are very loose: for instance, Cayrel (2006) estimate an age of 12.5 Gyrs with an error of 3 Gyr from the decay of ^{238}U ($t_{1/2} = 4.468$ Gyr) in CS 31082-001, while using the decay of ^{232}Th ($t_{1/2} = 14.05$ Gyr), Hayek *et al.* could only constrain the age of CS 29491-069 to be between 9.6 and 17.6 Gyr.

Observing the white dwarf cooling sequence is no walk-in-the-park either. Although our only mention of white dwarfs so far has been in the context of supernovae explosions, the overwhelming majority of white dwarfs leads a very boring life indeed. Their degenerate carbon–oxygen core is not hot enough to ignite the next sequence of nuclear reactions. With any hope of a temperature increase shattered by the degenerate nature of the core (which resists contraction), a white dwarf has no other option but to disperse its remaining latent heat into the surrounding space. A white dwarf becomes progressively fainter and cooler the older it gets, so that in the color–magnitude diagram of an old globular clusters' (where low mass stars have had enough time to get to the white dwarfs stage) white dwarfs define a "cooling" sequence roughly parallel to the main sequence, but shifted to lower luminosities and higher temperatures. Since the cooling rate slows down the fainter and cooler the white dwarf gets, the stars "pile up" at the faint end of the sequence, which is abruptly truncated. The luminosity at which this truncation occurs is an indication of the age of the cluster, since it simply reflects the fact that there are no white dwarfs old enough to have cooled and faded any further. Once again, however, how this cooling sequence and its low luminosity cutoff translate into an age requires an exact theoretical understanding of the white dwarf cooling time, and this is still subject to significant uncertainties (e.g. Salaris *et al.*, 2009). Furthermore, white dwarfs are extremely faint (the cooling sequence can extend to luminosities that are intrinsically $\sim 40\,000$ times fainter than the solar luminosity) and observing the cooling sequence is beyond the reach of ground-based telescopes. Globular cluster ages based on the white dwarf cooling sequence have been estimated, using Hubble Space Telescope data, in two galactic globular clusters: M4, which appears to be 11.6 ± 0.3 Gyr old (Bedin et al., 2009), and NGC 6397, essentially coeval at 11.4 ± 0.5 Gyr (Hansen et al., 2007). The quoted errors, however, are just those associated with the observational data: uncertainties arising from the theoretical models are of the order of ± 2 Gyr.

Last, but not least, the most precise and observationally affordable method of estimating the ages of globular clusters involves fitting stellar evolutionary models to the observed main sequence, and in particular to the "turn-off" that marks the transition between the core hydrogen burning phase, the ignition of hydrogen in a shell and the beginning of the red giant phase. As the cluster (which is assumed to be an ensemble of coeval stars[37]) ages, the turnoff moves to lower luminosities (i.e. less massive stars) since the time spent in the core hydrogen burning phase is inversely proportional to the stellar mass (elevated to the 2.5 power). As in the case of the white dwarf cooling sequence, the application of the method does rely on the stellar population models but, on the upside, the main sequence is perhaps the best understood of all phases of stellar evolution. The main source of uncertainty is (somewhat ironically) the distance to the cluster: this must be known quite precisely to be able to translate the observed turn-off luminosity into an absolute luminosity before fitting the models. In their very illustrative review, Krauss and Chaboyer (2003) estimate that uncertainties in the distance lead to a 13% error in the ages of the cluster. In the early 80s, ages for globular clusters measured in this fashion were believed to be in the 16 to 20 Gyr range, and indeed this was the main reason why "large" values of the Hubble constant were viewed with some suspicion. In a flat matter dominated Universe (no dark energy), the age of the Universe is given by $2/3\,H_0^{-1}$: anything above $40\,\mathrm{km\,s^{-1}Mpc^{-1}}$ would therefore have been seriously inconsistent with the age of the oldest stars! With updated stellar population models, and more accurate distances, the globular cluster ages have been revised downwards. In the most recent

---

[37] In some clusters, this long-held belief has actually been shattered by recent observations (see e.g. Piotto, 2009).

contribution to the subject, Krauss and Chaboyer (2003) estimate a firm (95% confidence) lower limit to the ages of globular cluster of 10.4 Gyr. Once accounting for the fact that it would take some time for the cluster to form after the Big Bang, this translates into a lower limit for the age of the Universe of 11.2 Gyr.

All three methods, therefore, point (some more forcefully than others) to a lower limit of the age of the Universe of about 11 Gyr. How does this fare with the value of $H_0 = 73 \pm 2 \pm 4 \,\mathrm{km\,s^{-1}\,Mpc^{-1}}$? The Hubble time is $1/H_0 = 13.4 \pm 1\,\mathrm{Gyr}$, but this is derived under the assumption that the Universe has expanded uniformly since the Big Bang. In a flat, matter dominated Universe with no cosmological constant, matter slows down the expansion, and the age of the Universe is reduced by a factor 2/3 relative to the Hubble time: $8.9 \pm 0.7\,\mathrm{Gyr}$. This is significantly smaller than the age of the Universe's building blocks: stars and globular clusters. The solution of this conundrum is the subject of the next chapter: observations of Type Ia supernovae have shown that the Universe is dominated by a dark energy component. A Universe where dark energy overcomes the gravitational pull of matter and accelerates the expansion is older than a matter dominated Universe. For the current concordance cosmology, which points to a flat Universe composed of 73% dark energy, the age of the Universe for $H_0 = 73 \pm 2 \pm 4 \,\mathrm{km\,s^{-1}\,Mpc^{-1}}$ is $13.3 \pm 1.1\,\mathrm{Gyr}$.[38]

8.8.3 *Room for improvement*

With the discovery of the accelerating Universe, cosmology has moved into a new regime. Tremendous efforts are directed towards measuring the equation of state of the Universe (essentially, the ratio between pressure and energy density) and its time evolution. Since the equation of state, referred to as "w", is directly related to the evolution of the energy density, its knowledge teaches us about the modalities by which the Universe expands and helps us discriminate between different scenarios for the nature of

---

[38] Readers interested in playing with different cosmological parameters will like New Wright's cosmology calculator: http://www.astro.ucla.edu/~wright/CosmoCalc.htmlhttp://www.astro.ucla.edu/~wright/CosmoCalc.html

dark energy. In the recently released and widely publicized 2010 Decadal Survey for Astronomy and Astrophysics of the US National Research Council (Blandford et al., 2010) the two top priority programs for space and ground-based astronomy are missions designed largely with the goal of measuring w.[39]

Where does the Hubble constant figure in all this? Uncertainties in H_0 translate directly into uncertainties in all physical scales: the sizes, luminosities and masses of galaxies and galaxy clusters, the masses and accretion rates of black holes, the formation rates of stars, all depend on precise distances and estimates of H_0. But an accurate knowledge of H_0 is also needed to realize the promise of "precision cosmology". Freedman and Madore (2010) show that by combining the results from the new generation of experiments — measurements of the Cosmic Microwave Background by the Planck satellite, new Type Ia supernovae searches, as well as constraints from the clustering of galaxies — a precise value of w cannot be nailed down unless H_0 is known with 1% accuracy or better. For instance, in a flat Universe and for constant w, the current uncertainties in H_0 would allow 5% excursions of w around a value of -1, including solutions that would lead to vacuum instabilities and propagation of energy outside the light cone. The quest for the Hubble constant, and the measurement of precise distances, is just as hot a field of research today as it was when Hubble started it all, eighty years ago.

References

Alcock, C. et al. (1999). *ApJ* **511**, p. 185.
Astier, P. et al. (2006). *A&A* **447**, p. 31.
Baade, W. (1938). *ApJ* **88**, p. 285.
Baade, W. (1956). *PASP* **68**, p. 5.
Barris, B. J. et al. (2004). *ApJ* **602**, p. 571.
Baker, N. and Kippenhahn, R. (1962). *Zeitschrift für Astrophysik* **54**, p. 114.
Baker, N. and Kippenhahn, R. (1965). *ApJ* **142**, p. 868.
Bedin, L. R. et al. (2009). *ApJ* **697**, p. 965.
Bellazzini, M. (2008). *MmSAI* **79**, p. 440.
Benedict, G. F. et al. (2007). *ApJ* **79**, p. 453.

---

[39]It is likely not a coincidence that the acronym for the top priority space-based mission, the Wide Field InfraRed Space Telecope, is "WFIRST".

Bernardi, M. et al. (2003). *AJ* **125**, p. 1866.
Biscardi, I. et al. (2008). *ApJ* **678**, p. 168.
Blakeslee, J. P., Vazdekis, A. and Ajhar, E. A. (2001). *MNRAS* **320**, p. 193.
Blakeslee, J. P. et al. (2002). *MNRAS* **330**, p. 443.
Blandford, R. and Narayan, R. (1986). *ApJ* **310**, p. 568.
Blandford, R. and the Committee for a Decadal Survey of Astronomy and Astrophysics. (2010). *New Worlds, New Horizons in Astronomy & Astrophysics*, National Research Council, National Academy Press.
Bono, G. et al. (2010). *ApJ* **715**, p. 277.
Burstein, D. et al. (1997). *AJ* **114**, p. 1365.
Cayrel, R. (2006). *Rep. Prog. Phys.* **69**, p. 2823.
Caldwell, N. (2006). *ApJ* **651**, p. 822.
Catinella, B., Haynes, M. P. and Giovanelli, R. (2005). *AJ* **130**, p. 1037.
Catinella, B., Haynes, M. P. and Giovanelli, R. (2007). *AJ* **134**, p. 334.
Coil, A. L. et al. (2000). *ApJ* **544**, p. L111.
Colless, M. et al. (2001). *MNRAS* **321**, p. 277.
Courteau, S. (1997). *AJ* **114**, p. 2402.
Courteau, S. and Rix, H. W. (1999). *ApJ* **513**, p. 561.
Cox, J. P. (1963). *ApJ* **138**, p. 487.
Da Costa, G. S. and Armandroff, T. E. (1990). *AJ* **100**, p. 162.
De Sitter, W. (1917). *MNRAS* **78**, p. 3.
Djorgovski, S. and Davis M. (1987). *ApJ* **313**, p. 59.
D'Onofrio, M. et al. (2008). *ApJ* **685**, p. 875.
Dressler, A. et al. (1987). *ApJ* **313**, p. 42.
Durrell, P. R. et al. (2007). *ApJ* **656**, p. 746.
Eddington, A. S. (1917). *The Observatory* **40**, p. 290.
Einstein, A. (1917). *Sitzungsberichte der Königlich Preußischen Akademie der Wissenschaften (Berlin)*, Seite 142.
Einstein, A. and de Sitter, W. (1932). *PNAS* **18**, p. 213.
Faber, S. M. and Jackson, R. E. (1976). *ApJ* **204**, p. 668.
Fernie, J. D. (1969). *PASP* **81**, p. 707.
Feast, M. W. and Walker, A. R. (1987). *ARA&A* **25**, p. 345.
Ferrarese, L. et al. (2000). *ApJ* **529**, p. 745.
Fouqué, P. et al. (2007). *A&A* **476**, p. 73
Freedman, W. L. and Madore, B. F. (2010). *ARA&A* **48**, in press
Freedman, W. L. et al. (2001). *ApJ* **553**, p. 47.
Friedman, A. (1922). Über die Krümmung des Raumes, *Zeitschrift für Physik* **10**, p. 377.
Gibson, B. (2000). *MmSAI* **71**, p. 693.
Gibson, B. K. et al. (2000). *ApJ* **529**, p. 723.
Gnerucci, A. et al. (2010). arXiv:1007.4180.
Governato, F. et al. (2007). *MNRAS* **374**, p. 1479.
Graves, G. J., Faber, S. M. and Schiavon, R. P. (2009). *ApJ* **698**, p. 1590.
Greenhill, L. J. et al. (1995). *ApJ* **440**, p. 619.
Hamuy, M. et al. (1996). *AJ* **112**, p. 2398.
Hansen, B. M. S. et al. (2007). *ApJ* **671**, p. 380.

Hayek, W. et al. (2009). A&A **504**, p. 511.
Herrnstein, J. R. et al. (1999). Nature **400**, p. 539.
Hertzsprung, E. (1913). AN **196**, p. 201.
Hillebrandt, W. and Niemeyer, J. C. (2000). ARA&A **38**, p. 191.
Hook, I. M. et al. (2005). AJ **130**, p. 2788.
Hopkins, P. F., Cox, T. J. and Hernquist, L. (2008). ApJ **689**, p. 17.
Hyde, J. B. and Bernardi, M. (2009). MNRAS **396**, p. 1171
Hubble, E. (1925). Obs. **48**, p. 139.
Hubble, E. (1929). PNAS **15**, p. 168.
Huchra, J. P. (1992). Science **256**, p. 321.
Hudson, M. J. et al. (1997). MNRAS **291**, p. 488.
Jacoby, G. H. et al. (1992). PASP **104**, p. 599.
Jensen, J. B. et al. (2001). ApJ **550**, p. 503.
Jørgensen, I., Franx, M. and Kjærgaard P. (1996). MNRAS **280**, p. 167.
Kelson, D. D. et al. (2000). ApJ **529**, p. 768.
Kirshner, R. (2009). in *Dark Energy — Observational and Theoretical Approaches*, ed. Ruiz-Lapuente, P., Cambridge, Cambridge University Press.
Kowal, C. T. (1968). AJ **73**, p. 1021.
Krauss, L. M. and Chaboyer, B. (2003). Science **299**, p. 65.
Leavitt, H. S. (1908). Harvard Obs. Ann. **60**, p. 87.
Leavitt, H. S. and Pickering, E. C. (1912). Harvard Coll. Obs. Circ. **173**, p. 1.
Lee, M. G., Freedman, W. L. and Madore, B. F. (1993). ApJ **417**, p. 553.
Lemaître, G. (1927). Annales de la Societe Scientifique de Bruxelles **47**, p. 49.
Liu, M. C. and Graham, J. R. (2001). ApJ **557**, p. L31.
Mager, V., Madore, B. F. and Freedman, W. F. (2008). ApJ **689**, p. 721.
Masters, K. L., Springob, C. M. and Huchra, J. P. (2008). AJ **135**, p. 1738.
Mei, S. et al. (2007). ApJ **655**, p. 144.
Miknaitis, G. et al. (2007). ApJ **666**, p. 674.
Minkowski, R. (1941). PASP **53**, p. 224.
Miyoshi, M. et al. (1995). Nature **373**, p. 127.
Mould, J. and Sakai, S. (2008). ApJ **686**, p. L75.
Mould, J. and Sakai, S. (2009a). ApJ **694**, p. 1331.
Mould, J. and Sakai, S. (2009b). ApJ **697**, p. 996.
Neill, J. D. (2006). AJ **132**, p. 1126.
Pahre, M. A., Djorgovski, S. G. and de Carvalho, R. R. (1998). AJ **116**, p. 1591.
Peebles, P. J. E. and Yu, J. T. (1970). ApJ **162**, p. 815.
Perlmutter, S. et al. (1999). ApJ **517**, p. 565.
Phillips, M. M. (1993). ApJ **413**, p. L10.
Phillips, M. M. et al. (1987). PASP **99**, p. 592.
Piotto, G. (2009). IAU Symposium **258**, p. 233.
Poznanski, D., Nugent, P. E. and Filippenko, A. V. (2010). ApJ **721**, p. 956.
Raimondo, G. (2009). ApJ **700**, p. 1246
Raimondo, G. et al. (2005). AJ **130**, p. 2625.
Rowan-Robinson, M. (1985). *The Cosmological Distance Ladder: Distance and Time in the Universe*. W. H. Freeman & Co., New York.
Riess, A. G., Private communication.

Riess, A. G. et al. (1998). *AJ* **116**, p. 1009.
Riess, A. G. et al. (2005). *ApJ* **627**, p. 579.
Riess, A. G., Press, W. H. and Kirshner, R. P. (1996). *ApJ* **473**, p. 88.
Riess, A. G. et al. (2007). *ApJ* **659**, p. 98.
Riess, A. G. et al. (2009). *ApJ* **699**, p. 539.
Rizzi, L. et al. (2007). *ApJ* **661**, p. 815.
Robertson, H. (1928). *Phil. Mag.* **5**, p. 835.
Russell, H. N. (1913). *Science* **37**, p. 651.
Russell, H. N. and Shapley, H. (1914). *ApJ* **40**, p. 417.
Saha, A. et al. (1997). *ApJ* **486**, p. 1.
Saha, A. et al. (2006). *ApJS* **165**, p. 108.
Sakai, S. et al. (2000). *ApJ* **529**, p. 698.
Salaris, M. et al. (2009). *ApJ* **692**, p. 1013.
Salaris, M. and Girardi, L. (2005). *MNRAS* **357**, p. 669.
Salaris, M., Cassisi, S. and Weiss, A. (2002). *PASP* **114**, p. 375.
Sandage, A. R. and Gratton, L. (1963). *Star Evolution*, p. 11.
Sandage, A. R. and Tammann, G. A. (1968). *ApJ* **151**, p. 531.
Sandage, A. R. and Tammann G. A. (2006). *ARA&A* **44**, p. 93.
Sandage, A. et al. (2006). *ApJ* **653**, p. 843.
Sandquist, E. L. et al. (1996). *ApJ* **470**, p. 910.
Schlegel, D. J., Finkbeiner, D. P. and Davis, M. (1998). *AJ* **500**, p. 525.
Seyfert, C. K. (1943). *ApJ* **97**, p. 28.
Shapley, H. (1918). *ApJ* **48**, p. 89.
Soszyński, I. et al. (2008). *AcA* **58**, p. 163.
Soszyński, I. et al. (2010). *AcA* **60**, p. 17.
Sunyaev, R. and Zel'dovich, Y. (1969). *Astrophys. Space Sci.* **4**, p. 301.
Sunyaev, R. and Zel'dovich, Y. (1970). *Astrophys. Space Sci.* **7**, p. 3.
Tabur, V., Kiss, L. L. and Bedding, T. R. (2009). *ApJ* **703**, p. L72.
Tammann, G. A., Sandage, A. R. and Reindl, B. (2008). *A&A Rev.* **15**, p. 289.
Tonry, J. L. and Schneider, D. P. (1988). *AJ* **96**, p. 807.
Tonry, J. L. et al. (2000). *ApJ* **530**, p. 625.
Tonry, J. L. et al. (2001). *ApJ* **546**, p. 681.
Tonry, J. L. et al. (2003). *ApJ* **594**, p. 1.
Treu, T. et al. (2005). *ApJ* **633**, p. 174.
Trimble, V. (1995). *PASP* **107**, p. 1133.
Trujillo, I., Burkert, A. and Bell, E. F. (2004). *ApJ* **600**, p. L39.
Tully, R. B. and Fisher, J. R. (1977). *A&A* **54**, p. 661.
Tully, R. B. and Pierce, M. J. (2000). *ApJ* **553**, p. 744.
Valle, G. et al. (2009). *A&A* **507**, p. 1541.

CHAPTER 9

PARTICLES AS DARK MATTER

DAN HOOPER
Theoretical Astrophysics Group,
Fermi National Accelerator Laboratory,
Department of Astronomy and Astrophysics,
The University of Chicago
dhooper@fnal.gov

Although the nature of our Universe's dark matter remains a mystery, it is likely to consist of a new species of particles, perhaps associated with the electroweak scale. Such weakly interacting massive particles (WIMPs) are theoretically attractive in part because they can be thermally produced in the early Universe with a density consistent with the measured dark matter abundance. In this chapter, I will describe some of the most popular particle candidates for dark matter and summarize the current status of the quest to discover dark matter's particle identity, including direct and indirect searches for WIMPs, and efforts being pursued at particle accelerators.

9.1 The Evidence for Dark Matter

Over the past several decades, a great deal of evidence has accumulated in support of dark matter's existence. At galactic and sub-galactic scales, this evidence includes galactic rotation curves (Borriello and Salucci, 2001), the weak gravitational lensing of distant galaxies by foreground structure (Hoekstra, Yee and Gladders, 2002), and the weak modulation of strong lensing around individual massive elliptical galaxies (Metcalf *et al.*, 2003; Moustakas and Metcalf, 2003). Furthermore, velocity dispersions of stars in some dwarf galaxies imply that they contain as much as $\sim 10^3$ times more mass than can be attributed to their luminosity. On the scale of galaxy clusters, observations (of radial velocities, weak lensing, and X-ray emission) indicate a total cosmological matter density of $\Omega_M \approx 0.2$–0.3 (Bahcall and Fan, 1998; Kashlinsky, 1998; Carlberg et al., 1999; Tyson, Kochanski and Dell'Antonio, 1998; Dahle, 2007), which is much larger than the corresponding density in baryons. In fact, they were measurements of velocity dispersions in the Coma cluster that led Fritz Zwicky to claim for the first time in 1933 that large quantities of non-luminous matter are required to be present (Zwicky, 1933). On cosmological scales, observations of the anisotropies in the cosmic microwave background have led to a determination of the total matter density of $\Omega_M h^2 = 0.1326 \pm 0.0063$, where h is the Hubble parameter in units of $100 \,\text{km/s}$ per Mpc (this improves to $\Omega_M h^2 = 0.1358^{+0.0037}_{-0.0036}$ if distance measurements from baryon acoustic oscillations and Type Ia supernovae are included) (Komatsu et al., 2008). In contrast, this information combined with measurements of the light chemical element abundances leads to an estimate of the baryonic density given by $\Omega_B h^2 = 0.02273 \pm 0.00062$ ($\Omega_B h^2 = 0.02267^{+0.00058}_{-0.00059}$ if baryon acoustic oscillations and supernovae are included) (Komatsu et al., 2008; Olive, Steigman and Walker, 2000). Taken together, these observations strongly lead us to the conclusion that 80–85% of the matter in the Universe (by mass) consists of non-luminous and non-baryonic material.

The process of the formation of large scale structure through the gravitational clustering of collisionless dark matter particles can be studied using N-body simulations. When the observed structure in our Universe (Tegmark et al., 2004) is compared to the results of cold (non-relativistic at time of structure formation) dark matter simulations, good agreement has been found. The large scale structure predicted for hot dark matter, in contrast, is in strong disagreement with observations.

Although there are many pieces of evidence in favor of dark matter's existence, it is worth noting that they each infer dark matter's presence uniquely through its gravitational influence. In other words, we currently have no conclusive evidence for dark matter's electroweak or other non-gravitational interactions. Given this fact, it is natural to contemplate whether, rather than being indications of dark matter's existence, these observations might instead be revealing departures from the laws of gravity as described by general relativity.

Since first proposed by Milgrom (1983), efforts have been made to explain the observed galactic rotation curves without dark matter within the context of a phenomenological model known as modified Newtonian dynamics, or MOND. The basic idea of MOND is that Newton's second law, $F = ma$, is modified to $F = ma \times \mu(a)$, where μ is very closely approximated by unity except in the case of very small accelerations, for which μ behaves as $\mu = a/a_0$ (where a_0 is simply a constant). Applying the modified form of Newton's second law to the gravitational force acting on a star outside of a galaxy of mass M leads us to

$$F = \frac{GMm}{r^2} = ma\mu, \tag{9.1}$$

which in the low acceleration limit (large r, $a \ll a_0$) yields

$$a = \frac{\sqrt{GMa_0}}{r}. \tag{9.2}$$

Equating this with the centrifugal acceleration associated with a circular orbit, we arrive at

$$\frac{\sqrt{GMa_0}}{r} = \frac{v^2}{r} \implies v = (GMa_0)^{1/4}. \tag{9.3}$$

In other words, MOND yields the prediction that galactic rotation curves should become flat (independent of r) for sufficiently large orbits. This result is in good agreement with galaxy-scale observations for a value of $a_0 \sim 1.2 \times 10^{-10}$ m/s^2, even without the introduction of dark matter. For this value of a_0, the effects of MOND are imperceptible in laboratory or Solar System scale experiments.

MOND is not as successful in explaining the other evidence for dark matter, however. In particular, MOND fails to successfully describe the observed features of galaxy clusters. Other evidence, such as the cosmic microwave background anisotropies and large scale structure, are not generally able to be addressed by MOND, as MOND represents a phenomenological modification of Newtonian dynamics and thus is not applicable to questions addressed by general relativity, such as the expansion history of the Universe. Efforts to develop a viable, relativistically covariant theory which yields the behavior of MOND in the non-relativistic, weak-field limit have mostly been unsuccessful. A notable exception to this is Tensor–Vector–Scalar gravity, or TeVeS (Bekenstein, 2004a,b). TeVeS, however, fails to explain cluster-scale observations without the introduction of dark matter (Skordis *et al.*, 2006). This problem has been further exacerbated by recent observations of two merging clusters, known collectively as the bullet cluster. In the bullet cluster, the locations of the baryonic material and gravitational potential (as determined using X-ray observations and weak lensing, respectively) are clearly spatially separated, strongly favoring the dark matter hypothesis over modifications of general relativity (Clowe, 2006).

9.2 The Production of Dark Matter in the Early Universe

Many of the protons, neutrons, electrons and neutrinos that inhabit our Universe can trace their origin back to the first fraction of a second following the Big Bang. Although we do not know for certain how the dark matter came to be formed, a sizable relic abundance of weakly interacting massive particles (WIMPs) is generally expected to be produced as a by-product of our Universe's hot youth. In this section, I discuss this process and the determination of the relic abundance of a WIMP (Srednicki, Watkins and Olive, 1988; Gondolo and Gelmini, 1991; Kolb and Turner, 1994).

Consider a stable particle, X, which interacts with Standard Model particles, Y, through some process $X\bar{X} \leftrightarrow Y\bar{Y}$ (or $XX \leftrightarrow Y\bar{Y}$ if X is its own antiparticle). In the very early Universe, when the temperature was much higher than the particle's mass, m_X, the processes of $X\bar{X}$ creation and annihilation were equally efficient, leading X to be present in large quantities alongside the various particle species of the Standard Model. As the temperature of the Universe dropped below m_X, however, the process of $X\bar{X}$ creation became exponentially suppressed, while $X\bar{X}$ annihilation continued unabated. In thermal equilibrium, the number density of such particles is given by

$$n_{X,\text{eq}} = g_X \left(\frac{m_X T}{2\pi}\right)^{3/2} e^{-m_X/T}, \qquad (9.4)$$

where g_X is the number of internal degrees of freedom of X.

If these particles were to remain in thermal equilibrium indefinitely, their number density would become increasingly suppressed as the Universe cooled, quickly becoming cosmologically irrelevant. There are ways that a particle species might hope to avoid this fate, however. For example, baryons are present in the Universe today because of a small asymmetry which initially existed between the number of baryons and antibaryons; when all of the antibaryons had annihilated with baryons, a small residual of baryons remained. The baryon-antibaryon asymmetry prevented the complete annihilation of these particles from taking place.

While it is possible that a particle-antiparticle asymmetry is also behind the existence of dark matter, there is an even simpler mechanism which can lead to the survival of a sizable relic density of weakly interacting particles. In particular, the self-annihilation of weakly interacting species can be contained by the competing effect of Hubble expansion. As the expansion and corresponding dilution of WIMPs increasingly dominate over the annihilation rate, the number density of X particles becomes sufficiently small that they cease to interact with each other, and thus survive to the present day. Quantitatively, the competing effects of expansion and annihilation are described by the Boltzmann equation:

$$\frac{dn_X}{dt} + 3Hn_X = -\langle \sigma_{X\bar{X}}|v|\rangle(n_X^2 - n_{X,\text{eq}}^2), \qquad (9.5)$$

where n_X is the number density of WIMPs, $H = (8\pi^3 \rho/3 M_{\text{Pl}})^{1/2}$ is the expansion rate of the Universe, and $\langle \sigma_{X\bar{X}}|v|\rangle$ is the thermally averaged $X\bar{X}$ annihilation cross section (multiplied by their relative velocity).

From Eq. (9.5), we can identify two clear limits. As I said before, at very high temperatures ($T \gg m_X$) the density of WIMPs is given by the equilibrium value, $n_{X,\text{eq}}$. In the opposite limit ($T \ll m_X$), the equilibrium density is very small, leaving the terms $3Hn_X$ and $\langle \sigma_{X\bar{X}}|v|\rangle n_X^2$ to each further deplete the number density. For sufficiently small values of n_X, the annihilation term becomes insignificant compared to the dilution due to Hubble expansion. When this takes place, the comoving number density of WIMPs becomes fixed — thermal freeze-out has occurred.

The temperature at which the number density of the species X departs from equilibrium and freezes out is found by numerically solving the Boltzmann equation. Introducing the variable $x \equiv m_X/T$, the temperature at which freeze-out occurs is approximately given by

$$x_{\rm FO} \equiv \frac{m_X}{T_{\rm FO}} \approx \ln\left[c(c+2)\sqrt{\frac{45}{8}\frac{g_X}{2\pi^3}}\frac{m_X M_{\rm Pl}(a+6b/x_{\rm FO})}{g_\star^{1/2} x_{\rm FO}^{1/2}}\right]. \tag{9.6}$$

Here, $c \sim 0.5$ is a numerically determined quantity, $g_\star$ is the number of external degrees of freedom available (in the Standard Model, $g_\star \sim 120$ at $T \sim 1\,{\rm TeV}$ and $g_\star \sim 65$ at $T \sim 1\,{\rm GeV}$), and a and b are terms in the non-relativistic expansion, $\langle \sigma_{X\bar{X}}|v|\rangle = a + b\langle v^2\rangle + \mathcal{O}(v^4)$. The resulting density of WIMPs remaining in the Universe today is approximately given by

$$\Omega_X h^2 \approx \frac{1.04 \times 10^9\,{\rm GeV}^{-1}}{M_{\rm Pl}}\frac{x_{\rm FO}}{g_\star^{1/2}(a+3b/x_{\rm FO})}. \tag{9.7}$$

If X has a GeV–TeV scale mass and a roughly weak-scale annihilation cross section, freeze-out occurs at $x_{\rm FO} \approx 20$–30, resulting in a relic abundance of

$$\Omega_X h^2 \approx 0.1 \left(\frac{x_{\rm FO}}{20}\right)\left(\frac{g_\star}{80}\right)^{-1/2}\left(\frac{a+3b/x_{\rm FO}}{3\times 10^{-26}{\rm cm}^3/{\rm s}}\right)^{-1}. \tag{9.8}$$

In other words, if a GeV–TeV scale particle is to be thermally produced with an abundance similar to the measured density of dark matter, it must have a thermally averaged annihilation cross section on the order of $3\times 10^{-26}\,{\rm cm}^3/{\rm s}$. Remarkably, this is very similar to the numerical value arrived at for a generic weak-scale interaction. In particular, $\alpha^2/(100\,{\rm GeV})^2 \sim {\rm pb}$, which in our choice of units (and including a factor of velocity) is $\sim 3\times 10^{-26}\,{\rm cm}^3/{\rm s}$. The similarity between this result and the value required to generate the observed quantity of dark matter has been dubbed the "WIMP miracle". While far from constituting a proof, this argument has led many to conclude that dark matter is likely to consist of particles with weak-scale masses and interactions and certainly provides us with motivation for exploring an electroweak origin of our Universe's missing matter.

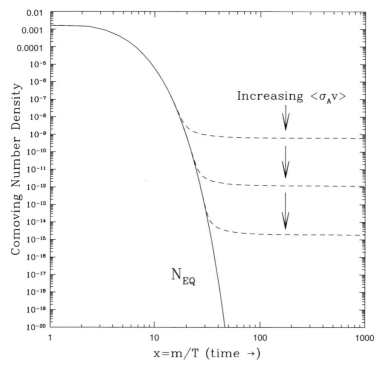

Fig. 9.1 A schematic of the comoving number density of a stable species as it evolves through the process of thermal freeze-out.

9.2.1 Case example — the thermal abundance of a light or heavy neutrino

At first glance, the Standard Model itself appears to contain a plausible candidate for dark matter in the form of neutrinos. Being stable and weakly interacting, neutrinos are a natural place to start in our hunt for dark matter's identity.

In the case of a Standard Model neutrino species, the relatively small annihilation cross section ($\langle \sigma |v| \rangle \sim 10^{-32}\,\text{cm}^3/\text{s}$) and light mass lead to an overall freeze-out temperature on the order of $T_{\rm FO} \sim$ MeV, and a relic density of

$$\Omega_{\nu+\bar{\nu}} h^2 \approx 0.1 \left(\frac{m_\nu}{9\,\text{eV}} \right). \tag{9.9}$$

As constraints on Standard Model neutrino masses require m_ν to be well-below 9 eV, we are forced to conclude that only a small fraction of the dark matter could possibly have consisted of Standard Model neutrinos. Furthermore, even if these constraints did not exist, such light neutrinos would be highly relativistic at the time of freeze-out ($T_{\rm FO}/m_\nu \sim$ MeV$/m_\nu \gg 1$) and thus would constitute hot dark matter, in conflict with observations of large scale structure.

Moving beyond the Standard Model, we could instead consider a heavy 4th generation Dirac neutrino. In this case, the annihilation cross section can be much larger, growing with the square of the neutrino's mass up to the Z pole, $m_\nu \sim m_Z/2$, and declining with m_ν^{-2} above $m_\nu \sim m_Z/2$. For a GeV–TeV mass neutrino, the process of freeze-out yields a cold relic ($T_{\rm FO}/m_\nu \sim \mathcal{O}(0.1)$), with an abundance approximately given by

$$\Omega_{\nu+\bar{\nu}} h^2 \approx 0.1 \left(\frac{5.5\,\text{GeV}}{m_\nu}\right)^2, \quad \text{MeV} \ll m_\nu \ll m_Z/2, \qquad (9.10)$$

$$\Omega_{\nu+\bar{\nu}} h^2 \sim 0.1 \left(\frac{m_\nu}{400\,\text{GeV}}\right)^2, \quad m_Z/2 \ll m_\nu. \qquad (9.11)$$

Thus, if the relic density of a heavy neutrino species is to constitute the bulk of the observed dark matter abundance, we find that it must have a mass of approximately 5 GeV (Lee and Weinberg, 1977), or several hundred GeV. The former case is excluded by LEP's measurement of the invisible width of the Z, however, ruling out the possibility of a 4th active neutrino species lighter than half of the Z mass (Amsler et al., 2008). The latter case, although consistent with the bounds of LEP and other accelerator experiments, is excluded by the limits placed by direct dark matter detection experiments (which I will discuss later in this chapter).

9.3 Beyond the Standard Model Candidates for Dark Matter

There has been no shortage of dark matter candidates proposed over the years. A huge variety of physics models that are beyond the Standard Model and that include a stable, electrically neutral and colorless particle have been constructed; many of these models could serve as a phenomenologically viable candidate for dark matter. I could not possibly list, much less review, all of the proposed candidates here. Finding the "WIMP miracle" to be fairly compelling (along with the hierarchy problem, which strongly suggests the existence of new particles at or around the electroweak scale), I choose to focus my attention on dark matter in the form of weak-scale particles. So although the dark matter of our Universe could plausibly consist of particles ranging from 10^{-6} eV axions to 10^{16} GeV WIMPzillas, I will ignore everything but those particle physics frameworks which predict the existence of a stable, weakly interacting particle with a mass in the few GeV to few TeV range.

9.3.1 Supersymmetry

For a number of reasons, supersymmetry is considered by many to be among the most attractive extensions of the Standard Model. In particular, weak-scale supersymmetry provides us with an elegant solution to the hierarchy problem (for a review of supersymmetry phenomenology, see, Martin, 1997), and enables grand unification by causing the gauge couplings of the Standard Model to evolve to a common scale (Ellis, Kelley and Nanopoulos, 1991). From the standpoint of dark matter, the lightest superpartner is naturally stable in models that conserve R-parity, which is defined by $R = (-1)^{3B+L+2S}$ (B, L and S denoting baryon number, lepton number and spin), and is thus assigned with $R = +1$ for all Standard Model particles and $R = -1$ for all superpartners. R-parity conservation, therefore, requires superpartners to be created or destroyed in pairs, leading the lightest supersymmetric particle (LSP) to be stable, even over cosmological timescales.

The identity of the LSP depends on the hierarchy of the supersymmetric spectrum, which in turn is determined by the details of how supersymmetry is broken. The list of potential LSPs which could constitute a plausible dark matter candidate is somewhat short, however. The only electrically neutral and colorless superpartners in the minimal supersymmetric standard model (MSSM) are the four neutralinos (superpartners of the neutral gauge and Higgs bosons), three sneutrinos (superpartners of the neutrinos), and the gravitino. The lightest neutralino, in particular, is a very attractive and thoroughly studied candidate for dark matter (Goldberg, 1983; Ellis et al., 1984).

Before discussing the possibility of neutralino dark matter in more detail, let us briefly contemplate the possibility that sneutrinos might make up the dark matter of our Universe. In many respects, sneutrino dark matter would behave very similarly to a heavy 4th generation neutrino, as discussed in Sec. 9.2.1. In particular, like neutrinos, sneutrinos are predicted to annihilate to Standard Model fermions efficiently through the s-channel exchange of a Z boson (as well as through other diagrams). As a result, sneutrinos lighter than about 500–1000 GeV would be under-produced in the early Universe (a ~ 10 GeV sneutrino would also be produced with approximately the measured dark matter abundance, but is ruled out by LEP's invisible Z measurement).

The Feynman diagram corresponding to sneutrino annihilation into quarks through an s-channel Z exchange can be turned on its side to produce an elastic scattering diagram with quarks in nuclei. When the elastic scattering cross section of a ~ 100–1000 GeV sneutrino is calculated, we find that it is several orders of magnitude larger than current experimental constraints (Falk, Olive and Srednicki, 1994). We are thus forced to abandon MSSM sneutrinos as candidates for dark matter.

In the MSSM, the superpartners of the four Standard Model neutral bosons (the bino, wino and two neutral higgsinos) mix into four physical states known as neutralinos. Often times, the lightest of these four states is simply referred to as "the neutralino". This particle can annihilate through a wide variety of Feynman diagrams. Which of these diagrams dominate the process of thermal freeze-out in the early Universe depends on the composition of the lightest neutralino, and on the masses and mixings of the exchanged particles (Jungman, Kamionkowski and Griest, 1996). Since so many different diagrams can potentially contribute to neutralino annihilation, the resulting relic density depends on a large number of supersymmetric parameters and is not trivial to calculate accurately. Publicly available tools such as DarkSUSY (Gondolo et al., 2004) and MicroOmegas (Belanger et al., 2007) are often used for this purpose.

Over much of the supersymmetric parameter space, the relic abundance of neutralinos is predicted to exceed that of the observed dark matter density. To avoid this conclusion, we are forced to consider the regions of parameter space which lead to especially efficient neutralino annihilation in the early Universe. Some of these possibilities include the supersymmetric parameter combinations in which the lightest neutralino has fairly large couplings, that can annihilate through a resonance, or can efficiently "co-annihilate" with another supersymmetric state in the early Universe. It is also possible (if not likely) that the MSSM does not adequately describe the supersymmetric particle spectrum found in our Universe, in which case entirely new regions of viable parameter space may be possible.

9.3.2 *Kaluza–Klein dark matter in models with universal extra dimensions*

Supersymmetry is not by any means the only particle physics framework from which a viable dark matter candidate can arise. As an alternative, I will also discuss the possibility of Kaluza–Klein dark matter in models with a universal extra dimension.

In recent years, interest in theories with extra spatial dimensions has surged. In particular, a colossal amount of attention has been given to two classes of extra dimensional theories over the past decade: scenarios featuring one or more large (millimeter-scale), flat extra dimensions (ADD) (Arkani-Hamed, Dimopoulos and Dvali, 1998, 1999), and the Randall–Sundrum scenario, which introduces an additional small dimension with a large degree of spatial curvature (Randall and Sundrum, 1999). A somewhat less studied but interesting class of extra-dimensional models, which goes by the name of universal extra dimensions (UED), postulates the existence of a flat extra dimension (or dimensions) through which all of the Standard Model (SM) fields are free to propagate (rather than being confined to a brane as some or all of them are in the ADD and Randall–Sundrum models) (Appelquist, Cheng and Dobrescu, 2001).

For a number of reasons, extra dimensions of size $R \sim \text{TeV}^{-1}$ are particularly well motivated within the context of UED (Hooper and

Profumo, 2007) (much smaller than those found in the ADD model). Among these reasons is the fact that a TeV-scale Kaluza–Klein (KK) state, if stable, colorless, and electrically neutral, could potentially serve as a viable candidate for dark matter (Servant and Tait, 2003; Cheng, Feng and Matchev, 2002).

Standard Model fields with momentum in an extra dimension appear as heavy particles, called KK states. This leads to a tower of KK states for each Standard Model field, with tree-level masses given by:

$$m^2_{X^{(n)}} = \frac{n^2}{R^2} + m^2_{X^{(0)}}, \qquad (9.12)$$

where $X^{(n)}$ is the nth Kaluza–Klein excitation of the Standard Model field, X, and $R \sim \text{TeV}^{-1}$ is the size of the extra dimension. $X^{(0)}$ denotes the ordinary Standard Model particle (known as the zero mode).

If the extra dimensions were simply wrapped (compactified) around a circle or torus, then extra dimensional momentum conservation would ensure the conservation of KK number (n) and make the lightest first level KK state stable. Realistic models, however, require an orbifold to be introduced, which leads to the violation of KK number conservation. A remnant of KK number conservation called KK-parity, however, can remain and lead to the stability of the lightest KK particle (LKP) in much the same way that R-parity conservation prohibits the decay of the lightest supersymmetric particle.

In order for the LKP to be a viable dark matter candidate, it must be electrically neutral and colorless. Possibilities for such a state include the first KK excitation of the photon, Z, neutrinos, Higgs boson, or graviton. Assuming that R^{-1} is considerably larger than any of the Standard Model zero mode masses, Eq. (9.12) leads us to expect a highly degenerate spectrum of Kaluza–Klein states at each level (although this picture is somewhat modified when radiative corrections and boundary terms are included). Of our possible choices for the LKP, the relatively large zero-mode mass of the Higgs makes its first-level KK excitation an unlikely candidate. Furthermore, KK neutrinos are excluded by direct detection experiments, just as sneutrinos or heavy fourth generation Dirac neutrinos are. For these reasons, we focus on the mixtures of the KK photon and KK

Z as our dark matter candidate. (Note that, unlike higgsino and gauginos, the KK Higgs has a different spin than the KK photon and KK Z, and thus does not mix with these states.)

The mass eigenstates of the KK photon and KK Z are very nearly identical to their gauge eigenstates, $B^{(n)}$ and $W^{3(n)}$. The reason for this can be seen from their mass matrix:

$$\begin{pmatrix} \frac{n^2}{R^2} + \delta m^2_{B^{(n)}} + \frac{1}{4}g_1^2 v^2 & \frac{1}{4}g_1 g_2 v^2 \\ \frac{1}{4}g_1 g_2 v^2 & \frac{n^2}{R^2} + \delta m^2_{W^{(n)}} + \frac{1}{4}g_2^2 v^2 \end{pmatrix}. \qquad (9.13)$$

Here $v \approx 174\,\text{GeV}$ is the Higgs vacuum expectation value. In the well known zero-mode case ($n = 0$), there is significant mixing between $B^{(0)}$ and $W^{3(0)}$ ($\sin^2 \theta_W \approx 0.23$). In the absence of radiative corrections ($\delta m^2_{B^{(n)}} = \delta m^2_{W^{(n)}} = 0$), the same mixing angle is found at the first KK level as well. If the difference between $\delta m^2_{B^{(n)}}$ and $\delta m^2_{W^{(n)}}$ is larger than the (rather small) off-diagonal terms, however, the mixing angle between these two KK states is driven toward zero. Using typical estimates of these radiative corrections, the effective first KK level Weinberg angle is found to be approximately $\sin^2 \theta_{W,1} \sim 10^{-3}$. Thus the mass eigenstate often called the "KK photon" is not particularly photon-like, but instead is nearly identical to the state $B^{(1)}$.

The KK state $B^{(1)}$ annihilates largely to Standard Model (zero-mode) fermions through the t-channel exchange of KK fermions, with a cross section given by

$$\sigma v(B^{(1)} B^{(1)} \to f\bar{f}) = \frac{95}{32\,256} \sum_f \frac{N_c (Y_{f_L}^4 + Y_{f_R}^4) g_1^4}{\pi\, m^2_{B^{(1)}}}. \qquad (9.14)$$

As the $B^{(1)}$ couples to the fermion hypercharge, the cross section scales with $Y_{f_L}^4 + Y_{f_R}^4$ and most of its annihilations proceed to charged lepton pairs.

Because the first-level KK spectrum is expected to be quasi-degenerate, co-annihilations are likely to play an important role in the thermal freeze-out of Kaluza–Klein dark matter. Depending on the details of the KK spectrum, this leads to a thermal relic abundance equal to the measured dark matter density for LKP masses in the approximate range of 500 GeV to a few TeV (Burnell and Kribs, 2006; Kong and Matchev, 2006).

9.3.3 *A note on other possibilities for TeV-scale dark matter*

At this time, I would like to make a general comment about some of the many other possibilities for the particle identity of dark matter. It is interesting to note that a wide range of solutions to the gauge hierarchy problem also introduce a candidate for dark matter. In particular, in order to be consistent with electroweak precision measurements, new physics at the TeV scale typically must possess a discrete symmetry. Supersymmetry, for example, accomplishes this through R-parity conservation. Similarly, Little Higgs models can avoid problems with electroweak precision tests by introducing a discrete symmetry called T-parity. And just as R-parity stabilizes the lightest superpartner, T-parity can enable the possibility of dark matter by stabilizing the lightest T-parity odd state in the theory (Cheng and Low, 2003). So, thinking outside of any specific particle physics framework, we can imagine a much larger class of TeV-scale physics scenarios which address the hierarchy problem without violating electroweak precision measurements by introducing a discrete symmetry, which in turn leads to a stable dark matter candidate. In this way, many of the motivations and much of the phenomenology discussed within the context of supersymmetry can actually be applied to a more general collection of particle physics models associated with the electroweak scale.

9.4 Direct Detection

The most straightforward of the strategies used to detect particle matter is to observe the energy that is deposited when dark matter particles elastically scatter with nuclei in a detector. Experiments designed with this purpose in mind include CDMS (Ahmed *et al.*, 2009), XENON (Angle *et al.*, 2008; Aprile *et al.*, 2010), ZEPLIN (Alner *et al.*, 2005, 2007), EDELWEISS (Sanglard *et al.*, 2005), CRESST (Angloher *et al.*, 2005),

CoGeNT (Aalseth et al., 2010), DAMA/LIBRA (Bernabei et al., 2010), COUPP (Behnke et al., 2008), WARP (Benetti et al., 2007; Brunetti et al., 2005), and KIMS (Lee et al., 2007). This class of techniques is collectively known as direct detection, in contrast to indirect detection efforts which attempt to observe the annihilation products of dark matter particles.

A WIMP striking a nucleus will induce a recoil of energy given by

$$E_{\text{recoil}} = \frac{|\vec{q}|^2}{2M_{\text{nucleus}}} = \frac{2\mu^2 v^2 (1-\cos\theta)}{2M_{\text{nucleus}}} = \frac{m_X^2 M_{\text{nucleus}} v^2 (1-\cos\theta)}{(m_X + M_{\text{nucleus}})^2}, \tag{9.15}$$

where m_X is the mass of WIMP, $\vec{q}$ is its momentum, v is its velocity, and μ is the reduced mass. For $m_X \gg M_{\text{nucleus}}$ and a velocity of $\sim 300\,\text{km/s}$, we expect typical recoil energies of $E_{\text{recoil}} \sim M_{\text{nucleus}} v^2 \sim 1\text{--}100\,\text{keV}$.

WIMPs scatter with nuclei in a target at a rate given by

$$R \approx \int_{E_{\min}}^{E_{\max}} \int_{v_{\min}}^{v_{\max}} \frac{2\rho}{m_X} \frac{d\sigma}{d|\vec{q}|} \, v\, f(v) \, dv \, dE_{\text{recoil}}, \tag{9.16}$$

where ρ is the dark matter density, σ is the WIMP–nuclei elastic scattering cross section, and $f(v)$ is the velocity distribution of WIMPs. The limits of integration are set by the galactic escape velocity, $v_{\max} \approx 650\,\text{km/s}$, and kinematically by $v_{\min} = (E_{\text{recoil}} M_{\text{nucleus}}/2\mu^2)^{1/2}$. The minimum energy is set by the energy threshold of the detector, which is typically in the range of several keV to several tens of keV.

WIMPs can potentially scatter with nuclei through both coherent interactions with the entire target nucleus (spin-independent interactions) or through couplings to the spin of the nucleus (spin-dependent interactions). The experimental sensitivity to spin-independent couplings benefits from coherence, which leads to cross sections (and rates) proportional to the square of the atomic mass of the target nuclei. The cross sections for spin-dependent scattering, in contrast, are proportional to $J(J+1)$, where J is the spin of the target nucleus, and thus do not benefit from large target nuclei. As a result, the current experimental sensitivity to spin-dependent scattering is far below that of spin-independent interactions. For this reason, we focus on the case of spin-independent scattering of WIMPs with nuclei.

The spin-independent WIMP–nucleus elastic scattering cross section is given by

$$\sigma \approx \frac{4m_{\chi^0}^2 m_{\text{nucleus}}^2}{\pi(m_{\chi^0} + m_{\text{nucleus}})^2}[Zf_p + (A-Z)f_n]^2, \qquad (9.17)$$

where Z and A are the atomic number and atomic mass of the nucleus, f_p and f_n are the WIMP's couplings to protons and neutrons, given by (Jungman, Kamionkowski and Griest, 1996)

$$f_{p,n} = \sum_{q=u,d,s} f_{T_q}^{(p,n)} a_q \frac{m_{p,n}}{m_q} + \frac{2}{27} f_{TG}^{(p,n)} \sum_{q=c,b,t} a_q \frac{m_{p,n}}{m_q}, \qquad (9.18)$$

where a_q are the WIMP-quark couplings and the f's are quantities measured in nuclear physics experiments (Bottino et al., 2002; Ellis et al., 2005). The first term in Eq. (9.18) corresponds to interactions with the quarks in the target nuclei, while the second term accounts for scattering with gluons through a heavy quark loop.

9.4.1 *Direct detection of neutralino dark matter*

Neutralinos can elastically scatter with quarks through either t-channel CP-even Higgs exchange, or s-channel squark exchange. The amplitudes of these processes depend on masses and couplings of the lightest neutralino, the scalar Higgs bosons and the squarks. The largest cross sections occur in cases in which the heavy Higgs boson (H) is fairly light and reasonably well coupled to the lightest neutralino. In such cases, neutralino cross sections with nucleons as large as 10^{-41} to 10^{-44} cm^2 are possible. If the exchange of the lightest Higgs boson (h) or squarks dominate the elastic scattering, considerably smaller cross sections are generally predicted, in many models within the range of 10^{-44} to 10^{-46} cm^2.

Using Eq. (9.16), we can crudely estimate the minimum target mass required to potentially detect neutralino dark matter. A detector made up of Germanium targets (such as CDMS or Edelweiss, for example) would expect a WIMP with a nucleon-level cross section of 10^{-42} cm^2 to yield approximately 1 elastic scattering event per kilogram-day of exposure. Such a target mass could thus be potentially sensitive to the most optimistic supersymmetric models. The strongest current limits on spin-independent scattering have been obtained using $\sim 10^3$ kilogram-days of exposure. In contrast, reaching sensitivities near the $10^{-46}\,\mathrm{cm}^2$ level will require ton-scale detectors capable of operating for weeks, months or longer with very low backgrounds.

Currently, the strongest direct detection constraints come from the CDMS (Ahmed et al., 2009) and XENON100 (Aprile, 2010) experiments, each of which restrict the dark matter's elastic scattering cross section with nucleons to be less than a few times $10^{-44}\,\mathrm{cm}^2$ (for WIMPs with masses between a few tens and a few hundred GeV). Even stronger constraints are expected from XENON100 in the relatively near future. Although the more distant future is more difficult to predict, it is generally anticipated that experiments approaching the ton-scale will reach sensitivities near 10^{-45} or $10^{-46}\,\mathrm{cm}^2$ within in the next several years, either discovering or severely constraining the possibility of neutralino dark matter.

It is also worth noting that the DAMA/LIBRA collaboration has reported evidence for an annual modulation in its rate of nuclear recoil events, and has been claimed that this signal is the result WIMP interactions (Bernabei et al., 2010). Although this claim has been controversial, the DAMA/LIBRA collaboration has not been able to identify any other systematic effects capable of producing this signal. Very recently, results from the CoGeNT collaboration appear to provide some support for a dark matter interpretation of the DAMA/LIBRA result (Aalseth et al., 2010), although it is not clear at this time whether CoGeNT is observing dark matter, or a not-yet-understood background.

9.4.2 *Direct detection of Kaluza–Klein dark matter*

In the case of Kaluza–Klein dark matter (as described in Sec. 9.3.2), the WIMP–quark coupling a_q receives contributions from the s-channel exchange of KK quarks and the t-channel Higgs boson exchange (Cheng, Feng and Matchev, 2002; Servant and Tait, 2002). This leads to a contribution to the cross section from Higgs exchange which is proportional to $1/(m_{B^{(1)}}^2\, m_h^4)$ and a contribution from KK–quark exchange which is approximately proportional to $1/(m_{B^{(1)}}^6 \Delta^4)$, where $\Delta = (m_{q^{(1)}} - m_{B^{(1)}})/m_{B^{(1)}}$ is the fractional mass splitting of the KK quarks and the $B^{(1)}$. Numerically, the $B^{(1)}$-nucleon cross section is approximately given by

$$\sigma_{B^{(1)}n,\mathrm{SI}} \approx 1.2 \times 10^{-46}\,\mathrm{cm}^2 \left(\frac{1\,\mathrm{TeV}}{m_{B^{(1)}}}\right)^2$$
$$\times \left[\left(\frac{100\,\mathrm{GeV}}{m_h}\right)^2 + 0.09 \left(\frac{1\,\mathrm{TeV}}{m_{B^{(1)}}}\right)^2 \left(\frac{0.1}{\Delta}\right)^2\right]^2. \qquad (9.19)$$

With such small cross sections, we will most likely have to wait for ton-scale detectors before this model will be tested by direct detection experiments.

9.4.3 Some model-independent comments regarding direct detection

In the cases of the two dark matter candidates discussed thus far in this section, we are led to expect a rather small elastic scattering cross sections between WIMPs and nuclei — typically below or well below current experimental constraints. This is *not* a universal prediction for a generic WIMP, however. In fact, both neutralinos and Kaluza–Klein dark matter represent somewhat special cases in which direct detection rates are found to be particularly low.

To illustrate this point, consider a Dirac fermion or a scalar WIMP which annihilates in the early Universe to fermions with roughly equal couplings to each species — a heavy fourth generation neutrino or sneutrino, for example. We can take the Feynman diagram for the process of this WIMP annihilating to quarks and turn it on its side, and then calculate the resulting elastic scattering cross section. What we find is that, if the interaction is of scalar or vector form, such a WIMP will scatter with nuclei several orders of magnitude more often than is allowed by the limits of CDMS, XENON and other direct detection experiments. Similar conclusions are reached for many otherwise acceptable WIMP candidates (Beltran et al., 2009). A warning well worth keeping in mind for any WIMP model builder is, "Beware the crossing symmetry!"

So what is it about neutralinos or Kaluza–Klein dark matter that enable them to evade these constraints? In the case of neutralinos, the single most important feature is the suppression of its couplings to light fermions. Being a Majorana fermion, a neutralino's annihilation cross section to fermion pairs (at low velocity) scales with $\sigma v \propto m_f^2/m_{\chi^0}^2$. As a result, neutralinos annihilate preferentially to heavy fermions (top quarks, bottom quarks, and taus) or gauge/Higgs bosons. As heavy fermions (and gauge/Higgs bosons) are largely absent from nuclei, the potentially dangerous crossing symmetry does not apply. More generally speaking, current direct detection constraints can be fairly easily evaded for any WIMP which interacts with quarks through Higgs exchange, as the Yukawa couplings scale with the fermion's mass.

Alternatively, if the WIMP's couplings are simply very small, direct detection constraints can also be evaded. Small couplings, however, leave us in need of a mechanism for efficiently depleting the WIMP in the early Universe. But even with very small couplings, a WIMP might efficiently co-annihilate in the early Universe, or annihilate through a resonance, leading to an acceptable relic abundance. In this way, co-annihilations and resonances can considerably suppress the rates expected in direct detection experiments.

9.5 Indirect Detection

Direct detection experiments are by no means the only techniques being pursued in the hope of identifying the particle nature of dark matter. Another major class of dark matter searches are those which attempt to detect the products of WIMP annihilations, including gamma rays, neutrinos, positrons, electrons, and antiprotons. These methods are collectively known as indirect detection and are the topic of this section.

Dark matter particles annihilate at a rate proportional to the square of their density, and as a consequence the most dense regions are often the most promising for indirect detection. This is particularly true in the case of dark matter searches using photons. Unlike charged particles, gamma rays are not deflected by magnetic fields, and thus can potentially provide valuable angular information. For example, point-like sources of dark matter annihilation radiation might appear from high density regions such as the Galactic Center (Bergstrom, Ullio and Buckley, 1998a; Zaharijas and Hooper, 2006; Dodelson, Hooper and Serpico, 2008) or dwarf spheroidal galaxies (Evans, Ferrer and Sarkar, 2004; Bergstrom and Hooper, 2006; Strigari *et al.*, 2007). Furthermore, at GeV-scale energies, gamma rays are not significantly attenuated, and thus retain their spectral information. This makes it possible to detect dark matter annihilation products over cosmological distances (Ullio *et al.*, 2002).

The gamma ray flux from dark matter annihilations is simply given by

$$\Phi_\gamma(E_\gamma, \psi) = \frac{1}{2} \langle \sigma_{XX} |v| \rangle \frac{dN_\gamma}{dE_\gamma} \frac{1}{4\pi m_X^2} \int_{\text{los}} \rho^2 dl d\psi. \qquad (9.20)$$

Here, $\langle \sigma_{XX}|v|\rangle$ is the WIMP's annihilation cross section (times relative velocity), ψ is the angle observed, ρ is the dark matter density, the integral is performed over the line-of-sight, and dN_γ/dE_γ is the gamma ray spectrum generated per WIMP annihilation.

Telescopes potentially capable of detecting gamma rays from dark matter annihilations include the satellite-based Fermi Gamma Ray Space Telescope (FGST), and a number of ground-based Atmospheric Cerenkov Telescopes, including HESS, MAGIC and VERITAS. These two classes of experiments play complementary roles in the search for dark matter. On one hand, Fermi is able to continuously observe a large fraction of the sky, but with an effective area far smaller than possessed by ground-based telescopes. Ground-based telescopes, in contrast, study the emission from a small angular field, but with far greater exposure. Furthermore, while ground based telescopes are limited to studying gamma rays with energy greater than $\sim 100\,\text{GeV}$, Fermi is able to directly study gamma rays with energies over the range of 100 MeV to 300 GeV.

Data from the FGST have recently been used to place limits on dark matter annihilations taking place in dwarf spheroidal galaxies (Abdo *et al.*, 2010a; Scott *et al.*, 2010), in dark matter substructures (Buckley and Hooper, 2010), in the Galactic Center (Goodenough and Hooper, 2009), in the galactic halo, and cosmologically (Abdo *et al.* 2010b). While the details of each of these analyses vary, they each find constraints which are within an order of magnitude of the flux predicted from the annihilations of a simple, electroweak-scale, thermal relic. In the relatively near future, simple thermal WIMPs will likely become within the reach of this experiment.

Although charged particles produced in WIMP annihilations are deflected by the Galactic Magnetic Field, they may also be used to constrain the annihilation rate of dark matter in the Milky Way. Furthermore, by studying the spectrum of these particles (in particular, the positrons and antiprotons present in the cosmic ray spectrum) it may be possible to identify signatures of dark matter annihilations. This has been a topic of great interest in recent years.

The PAMELA experiment, which began its three-year satellite mission in June of 2006, has reported an anomalous rise in the cosmic ray positron fraction (the ratio of positron to positron-plus-electron) above 10 GeV (Adriani et al., 2009), confirming earlier indications from HEAT (Barwick et al., 1997; Coutu et al., 2001) and AMS-01 (Olzem, 2006). These measurements, compared to the standard astrophysical predictions, suggest the presence of a relatively local (within ~ 1 kpc) source or sources of energetic cosmic ray electrons and positrons. Furthermore, the WMAP experiment has revealed an excess of microwave emission from the central region of the Milky Way that has been interpreted as synchrotron emission from a population of electrons/positrons with a hard spectral index (Dobler and Finkbeiner, 2008; Finkbeiner, 2004; Hooper, Finkbeiner and Dobler, 2007). Taken together, these observations suggest that energetic electrons and positrons are surprisingly ubiquitous throughout our galaxy.

Although the origin of these electrons and positrons is not currently known, interpretations of the observations have focused on two possibilities: emission from pulsars (Hooper, Blasi and Serpico, 2008; Profumo, 2008; Aharonian, Atoyan and Volk, 1995; Zhang and Cheng, 2001; Buesching et al., 2008), and dark matter annihilations (Cholis et al., 2009; Cirelli and Strumia, 2008; Arkani-Hamed et al., 2009; Fox and Poppitz, 2009). At this point in time, it is not possible to empirically discriminate between these different possibilities. With the deployment of AMS-02 on-board the

International Space Station in 2010, however, it is likely that this situation will soon be clarified.

As an alternative to observing the products of WIMP annihilations taking place in the Galactic Halo, efforts have been made to detect neutrinos that are produced by dark matter particles annihilating in the core of the Sun. As the Solar System moves through the halo of the Milky Way, WIMPs become swept up by the Sun. Although dark matter particles interact only weakly, they occasionally scatter elastically with nuclei in the Sun and lose enough momentum to become gravitationally bound. Over the lifetime of the Sun, a sufficient density of WIMPs can accumulate in its center so that an equilibrium is established between their capture and annihilation rates. The annihilation products of these WIMPs include neutrinos, which escape the Sun with minimal absorption, and thus potentially constitute an indirect signature of dark matter.

Beginning with a simple estimate, we expect WIMPs to be captured in the Sun at a rate approximately given by:

$$C^\odot \sim \phi_X (M_\odot/m_p)\, \sigma_{Xp}, \tag{9.21}$$

where ϕ_X is the flux of WIMPs in the Solar System, $M_\odot$ is the mass of the Sun, and σ_{Xp} is the WIMP–proton elastic scattering cross section. Reasonable estimates of the local distribution of WIMPs lead to a capture rate of $C^\odot \sim 10^{20}\,\mathrm{sec}^{-1} \times (100\,\mathrm{GeV}/m_X)(\sigma_{Xp}/10^{-42}\,\mathrm{cm}^2)$. This neglects, however, a number of potentially important factors, including the gravitational focusing of the WIMP flux toward the Sun, and the fact that not every scattered WIMP will ultimately be captured. Taking these and other effects into account leads us to a solar capture rate of (Gould, 1991):

$$C^\odot \approx 3.35 \times 10^{20}\,\mathrm{sec}^{-1} \left(\frac{\rho_\mathrm{local}}{0.3\,\mathrm{GeV/cm^3}}\right)\left(\frac{270\,\mathrm{km/s}}{\bar{v}_\mathrm{local}}\right)^3 \left(\frac{100\,\mathrm{GeV}}{m_X}\right)^2$$
$$\times \left(\frac{\sigma_{X\mathrm{H,SD}} + \sigma_{X\mathrm{H,SI}} + 0.07\,\sigma_{X\mathrm{He,SI}}}{10^{-42}\,\mathrm{cm}^2}\right), \tag{9.22}$$

where ρ_local and $\bar{v}_\mathrm{local}$ are the density and rms velocity of dark matter particles in the vicinity of the Solar System, and the cross sections denote spin-independent and spin-dependent scattering off of hydrogen and helium nuclei in the Sun. For heavier WIMPs, scattering off of oxygen nuclei can also be significant.

WIMPs can potentially generate neutrinos through a variety of annihilation channels. Annihilations to heavy quarks, tau leptons, gauge bosons and/or Higgs bosons can each generate energetic neutrinos in their subsequent decays. In some models, WIMPs can also annihilate directly to neutrino-antineutrino pairs. Annihilations to light quarks or muons, however, do not contribute to the high energy neutrino spectrum, as these particles come to rest in the solar medium before decaying.

Once they reach Earth, neutrinos can potentially be detected in large volume neutrino telescopes. Muon neutrinos can produce muons through charged current interactions with ice or water nuclei inside or near the detector volume of a high energy neutrino telescope. When completed, the IceCube experiment will possess a full square kilometer of effective area and kilometer depth, and will be sensitive to muons above approximately 50 GeV (DeYoung, 2005; Ahrens et al., 2003). The Deep Core extension of Icecube will be sensitive down to 10 GeV. The Super-Kamiokande detector, in contrast, has 10^{-3} times the effective area of IceCube and a depth of only 36.2 meters (Desai et al. 2004a,b). For low mass WIMPs, however, Super-Kamiokande benefits over large volume detector such as IceCube by being sensitive to muons with as little energy as ~ 1 GeV.

Because of the relatively stringent limits from CDMS (Ahmed et al., 2009) and XENON100 (Aprile et al., 2010) on the dark matter's spin-independent elastic scattering cross section with nuclei, neutrino telescopes are primarily sensitive to WIMPs that are captured in the Sun through spin-dependent scattering. The strongest bound on the WIMP-proton spin-dependent cross section has been placed by the COUPP (Behnke et al., 2008) collaboration. A WIMP with a mass of a few hundred GeV and a spin-dependent elastic scattering cross section near COUPP's upper limit ($\sigma_{XH,SD} \sim 10^{-37}$ cm^2) would be expected to yield up to 10^6 events per year in a kilometer-scale neutrino telescope. WIMPs with spin-dependent cross sections as small as $\sim 10^{-41}$ cm^2 could potentially be probed by such an experiment.

9.6 Dark Matter at Particle Colliders

Among other new states, particles with TeV scale masses and strong interactions are generic features of models of electroweak symmetry breaking. These particles appear as counterparts to the quarks to provide new physics associated with the generation of the large top quark mass. In many scenarios, including supersymmetry, electroweak symmetry breaking arises as a result of radiative corrections due to these particles, enhanced by the large coupling of the Higgs boson to the top quark.

Any particle with these properties will be pair-produced at the Large Hadron Collider (LHC) with a cross section of tens of picobarns (Dawson, Eichten and Quigg, 1985) (and somewhat smaller in the current 7 TeV center-of-mass energy operating mode). That particle (or particles) will then decay into particles including quark or gluon jets and the lightest particle in the new sector (i.e. the dark matter candidate) which proceeds to exit the detector unseen. For any such model, the LHC experiments are, therefore, expected to observe large numbers of events with many hadronic jets and an imbalance of measured momentum. These "missing energy" events are signatures of a wide range of models that contain an electroweak scale candidate for dark matter.

If TeV-scale supersymmetry exists in nature, it will very likely be within the discovery reach of the Large Hadron Collider (LHC). The rate of missing energy events depends strongly on the mass of the colored particles that are produced and only weakly on other properties of the model. If squarks or gluinos have masses below 1 TeV, the missing energy events can be discovered with an integrated luminosity of roughly $100\,\mathrm{pb}^{-1}$ (or tens of pb at 7 TeV), a small fraction of the LHC's design luminosity. Thus, we may very well know early in the LHC program that a WIMP candidate is being produced.

By studying the decays of squarks and/or gluinos it will also be possible to discover other superpartners at the LHC. For example, in many models, decays of the variety, $\tilde{q} \to \chi_2^0 q \to \tilde{l}^{\pm} l^{\mp} q \to \chi_1^0 l^+ l^- q$, provide a clean signal of supersymmetry in the form of $l^+ l^-$ + jets + missing E_T. By studying the kinematics of these decays, the quantities $m_{\tilde{q}}$, $m_{\chi_2^0}$, $m_{\tilde{l}}$ and $m_{\chi_1^0}$ could each be potentially reconstructed (Bachacou, Hinchliffe and Paige, 2000; Drees et al., 2001; Lafaye, Plehn and Zerwas, 2004; Bechtle et al., 2006). More generally speaking, the LHC is, in most models, likely to measure the mass of the lightest neutralino to roughly 10% accuracy, and may also be able to determine the masses of one or more of the other neutralinos, and any light sleptons (Bachacou, Hinchliffe and Paige, 2000; Allanach et al., 2000; Baer et al., 1995, 1996; Abdullin and Charles, 1999). Charginos are more difficult to study at the LHC.

Measurements of particle masses and other properties at the LHC can provide an essential cross-check for direct and indirect detection channels. In particular, neither direct nor indirect detection experiments provide information capable of identifying the overall cosmological abundance of a WIMP, but instead infer only combinations of density and interaction cross section, leaving open the possibility that an observation may be generated by a sub-dominant component of the cosmological dark matter with a somewhat larger elastic scattering or annihilation cross section. Collider measurements can help to clarify this situation.

9.7 Conclusions

Although the conclusion that most of the mass in our Universe is made up of dark matter is virtually inescapable, the particle nature of this substance remains a mystery. In this chapter, I have summarized the evidence for dark matter and discussed some of the phenomenology of weakly interacting massive particles (WIMPs). A wide range of experimental efforts are currently underway to detect such particles, including direct and indirect astrophysical searches, and programs at particle accelerators, such as the Large Hadron Collider. In the coming several years, these experimental programs are expected to collectively reach the sensitivity required to definitively detect dark matter particles. If this does not occur, we will have little choice but to reconsider the viability of the WIMP-paradigm.

References

Aalseth, C. E. et al. [CoGeNT collaboration]. (2010). arXiv:1002.4703 [astro-ph.CO].
Abdo, A. A. et al. (2010a). Astrophys. J. **712**, p. 147.
Abdo, A. A. et al. [Fermi-LAT Collaboration]. (2010b). JCAP **1004**, p. 014.

Abdullin, S. and Charles, F. (1999). *Nucl. Phys. B* **547**, p. 60.
Adriani, O. *et al.* [PAMELA Collaboration]. (2009). *Nature* **458**, p. 607.
Aharonian, F. A., Atoyan, A. M. and Volk, H. J. (1995). *Astron. Astrophys.* **294**, pp. L41–L44.
Ahmed, Z. *et al.* [The CDMS-II Collaboration]. (2009). arXiv:0912.3592 [astro-ph.CO].
Ahrens, J. *et al.* [The IceCube Collaboration]. (2003). *Nucl. Phys. Proc. Suppl.* **118**, p. 388.
Allanach, B. C. *et al.* (2000). *JHEP* **0009**, p. 004.
Alner, G. J. *et al.* [UK Dark Matter Collaboration]. (2005). *Astropart. Phys.* **23**, p. 444.
Alner, G. J. *et al.* (2007). *Astropart. Phys.* **28**, p. 287.
Amsler, C. *et al.* (Particle Data Group). (2008).*Phys. Lett. B* **667**, p. 1.
Angle, J. *et al.* [XENON Collaboration]. (2008). *Phys. Rev. Lett.* **100**, p. 021303.
Angloher, G. *et al.* (2005). *Astropart. Phys.* **23**, p. 325.
Appelquist, T., Cheng, H. C. and Dobrescu, B. A. (2001). *Phys. Rev. D* **64**, p. 035002.
Aprile, E. *et al.* [XENON100 Collaboration]. (2010). arXiv:1005.0380 [astro-ph.CO].
Arkani-Hamed, N., Dimopoulos, S. and Dvali, G. R. (1998). *Phys. Lett. B* **429**, p. 263.
Arkani-Hamed, N., Dimopoulos, S. and Dvali, G. R. (1999). *Phys. Rev. D* **59**, p. 086004.
Arkani-Hamed, N. *et al.* (2009). *Phys. Rev. D* **79**, p. 015014.
Bachacou, H., Hinchliffe, I. and Paige, F. E. (2000). *Phys. Rev. D* **62**, p. 015009.
Baer, H. *et al.* (1995). Phys. Rev. D **52**, p. 2746.
Baer, H. *et al.* (1996). Phys. Rev. D **53**, p. 6241.
Bahcall N. and Fan, X. (1998). *Astrophys. J.* **504**, p. 1.
Barger, V. D. *et al.* (2002). *Phys. Rev. D* **65**, p. 075022.
Barwick, S. W. *et al.* [HEAT Collaboration]. (1997). *Astrophys. J.* **482**, p. L191.
Bechtle, P., Desch, K. and Wienemann, P. (2006). *Comput. Phys. Commun.* **174**, p. 47.
Behnke, E. *et al.* [COUPP Collaboration]. (2008). *Science* **319**, p. 933.
Bekenstein, J. D. (2004a). *Phys. Rev. D* **70**, p. 083509.
Bekenstein, J. D. (2004b). *Phys. Rev. D* **71**, p. 069901.
Belanger, G. *et al.* (2007). *Comput. Phys. Commun.* **177**, p. 894.
Beltran, M. *et al.* (2009). *Phys. Rev. D* **80**, p. 043509.
Benetti, P. *et al.* (2007). arXiv:astro-ph/0701286.
Bergstrom, L., Edsjo, J. and Gondolo, P. (1997). *Phys. Rev. D* **55**, p. 1765.
Bergstrom, L., Ullio, P. and Buckley, J. H. (1998a). *Astropart. Phys.* **9**, p. 137.
Bergstrom, L., Edsjo, J. and Gondolo, P. (1998b). *Phys. Rev. D* **58**, p. 103519.
Bergstrom, L. and Hooper, D. (2006). *Phys. Rev. D* **73**, p. 063510.
Bernabei, R. *et al.* (2010). *Eur. Phys. J. C* **67**, p. 39.
Borriello, A. and Salucci, P. (2001). *Mon. Not. Roy. Astron. Soc.* **323**, p. 285.
Bottino, A. *et al.* (2000). *Astropart. Phys.* **13**, p. 215.
Bottino, A. *et al.* (2002). *Astropart. Phys.* **18**, p. 205.

Brunetti, R. et al. (2005). *New Astron. Rev.* **49**, p. 265.
Buckley, M. R. and Hooper, D. (2010). arXiv:1004.1644 [hep-ph].
Buesching, I. et al. (2008). arXiv:0804.0220 [astro-ph].
Burnell, F. and Kribs, G. D. (2006). *Phys. Rev. D* **73**, p. 015001.
Carlberg, R. G. et al. (1999). *Astrophys. J.* **516**, p. 552.
Cheng, H. C., Feng, J. L. and Matchev, K. T. (2002). *Phys. Rev. Lett.* **89**, p. 211301.
Cheng, H. C. and Low, I. (2003). *JHEP* **0309**, p. 051.
Cholis, I. et al. (2009). *Phys. Rev. D* **80**, p. 123511.
Cirelli, M. and Strumia, A. (2008). *PoS* **IDM2008**, p. 089.
Clowe, D. et al. (2006). *Astrophys. J.* **648**, p. L109.
Coutu, S. et al. [HEAT-pbar Collaboration]. (2001). In Proceedings of 27th ICRC.
Dahle, H. (2007). arXiv:astro-ph/0701598.
Dawson, S., Eichten, E. and Quigg, C. (1985). *Phys. Rev. D* **31**, p. 1581.
Desai, S. et al. [Super-Kamiokande Collaboration]. (2004a). *Phys. Rev. D* **70**, p. 083523.
Desai, S. et al. (2004b). *Phys. Rev. D* **70**, p. 109901.
DeYoung, T. [IceCube Collaboration]. (2005). *Int. J. Mod. Phys. A* **20**, p. 3160.
Dobler, G. and Finkbeiner, D. P. (2008). *Astrophys. J.* **680**, p. 1222.
Dodelson, S., Hooper, D. and Serpico, P. D. (2008). *Phys. Rev. D* **77**, p. 063512.
Drees, M. et al. (2001). *Phys. Rev. D* **63**, p. 035008.
Ellis, J. R. et al. (1984). *Nucl. Phys. B* **238**, p. 453.
Ellis, J. R., Kelley, S. and Nanopoulos, D. V. (1991). *Phys. Lett. B* **260**, p. 131.
Ellis, J. R. et al. (2005). *Phys. Rev. D* **71**, p. 095007.
Evans, N. W., Ferrer, F. and Sarkar, S. (2004). *Phys. Rev. D* **69**, p. 123501.
Falk, T., Olive, K. A. and Srednicki, M. (1994). *Phys. Lett. B* **339**, p. 248.
Finkbeiner, D. P. (2004). arXiv:astro-ph/0409027.
Fox, P. J. and Poppitz, E. (2009). *Phys. Rev. D* **79**, p. 083528.
Goldberg, H. (1983). *Phys. Rev. Lett.* **50**, p. 1419.
Gondolo, P. and Gelmini, G. (1991). *Nucl. Phys. B* **360**, p. 145.
Gondolo, P. et al. (2004). *JCAP* **0407**, p. 008.
Goodenough, L. and Hooper, D. (2009). arXiv:0910.2998 [hep-ph].
Gould, A. (1991). *Astrophys. J.* **388**, p. 338.
Halzen, F. and Hooper, D. (2006). *Phys. Rev. D* **73**, p. 123507.
Hoekstra, H., Yee, H. and Gladders, M. (2002). *New Astron. Rev.* **46**, p. 767.
Hooper, D. and Profumo, S. (2007). *Phys. Rept.* **453**, p. 29.
Hooper, D., Finkbeiner, D. P. and Dobler, G. (2007). *Phys. Rev. D* **76**, p. 083012.
Hooper, D., Blasi, P. and Serpico, P. D. (2008). arXiv:0810.1527 [astro-ph];
Jungman, G., Kamionkowski, M. and Griest, K. (1996). *Phys. Rept.* **267**, p. 195.
Kashlinsky, A. (1998). *Phys. Rep.* **307**, p. 67.
Kolb, E. W. and Turner, M. S. (1994). *The Early Universe.*
Komatsu, E. et al. [WMAP Collaboration] (2008). arXiv:0803.0547 [astro-ph].
Kong, K. and Matchev, K. T. (2006). *JHEP* **0601**, p. 038.
Lafaye, R., Plehn, T. and Zerwas, D. (2004). arXiv:hep-ph/0404282.
Lee, B. W. and Weinberg, S. (1977). *Phys. Rev. Lett.* **39**, p. 165.
Lee, H. S. et al. [KIMS Collaboration]. (2007). *Phys. Rev. Lett.* **99**, p. 091301.

Martin, S. P. (1997). arXiv:hep-ph/9709356.
Metcalf, R. B. et al. (2003). arXiv:astro-ph/0309738.
Milgrom, M. (1983). *Astrophys. J.* **270**, p. 365.
Moustakas, L. A. and Metcalf, R. B. (2003). *Mon. Not. Roy. Astron. Soc.* **339**, p. 607.
Olive, K. A., Steigman, G. and Walker, T. P. (2000). *Phys. Rept.* **333**, p. 389.
Olzem, J. [AMS Collaboration]. (2006). Talk given at the 7th UCLA Symposium on Sources and Detection of DarkMatter and Dark Energy in the Universe, Marina del Ray, CA, Feb 22–24.
Profumo, S. (2008). arXiv:0812.4457 [astro-ph].
Randall, L. and Sundrum, R. (1999). *Phys. Rev. Lett.* **83**, p. 3370.
Sanglard, V. et al. [The EDELWEISS Collaboration]. (2005). *Phys. Rev. D* **71**, p. 122002.
Scott, P. et al. (2010). *JCAP* **1001**, p. 031.
Servant, G. and Tait, T. M. (2002). *New J. Phys.* **4**, p. 99.
Servant, G. and Tait, T. M. (2003). *Nucl. Phys. B* **650**, p. 391.
Skordis, C. et al. (2006). *Phys. Rev. Lett.* **96**, p. 011301.
Srednicki, M., Watkins, R. and Olive, K. A. (1988). *Nucl. Phys. B* **310**, p. 693.
Strigari, L. E. et al. (2007). arXiv:0709.1510 [astro-ph].
Tegmark, M. et al. [SDSS Collaboration]. (2004).*Astrophys. J.* **606**, p. 702.
Tyson, J. A., Kochanski, G. P. and Dell'Antonio, I. P. (1998). *Astrophys. J.* **498**, p. L107.
Ullio, P. et al. (2002). *Phys. Rev. D* **66**, p. 123502.
Zaharijas, G. and Hooper, D. (2006). *Phys. Rev. D* **73**, p. 103501.
Zhang, L. and Cheng, K. S. (2001) *Astron. Astrophys.* **368**, pp. 1063–1070.
Zwicky, F. (1933). *Helv. Phys. Acta* **6**, p. 110.

CHAPTER 10

DETECTION OF WIMP DARK MATTER

SUNIL GOLWALA

*Division of Physics, Mathematics and Astronomy,
California Institute of Technology,
Pasadena, CA 91125, USA*

DAN MCKINSEY

*Department of Physics, Yale University,
New Haven, CT 06520, USA*

10.1 Introduction

The chapter in this volume by Hooper has described the motivation for Weakly Interacting Massive Particle dark matter, some specific WIMP candidates from supersymmetry, universal extra dimensions, and generic TeV-scale new physics, and some aspects of their direct and indirect detection. In this chapter, we delve into direct detection in more detail, describing the generic experimental signatures and the experimental techniques used.

10.2 Direct Detection of WIMPs via WIMP–Nucleon Elastic Scattering

"Direct detection" consists of observing WIMP–nucleus scattering events. The plausibility of direct detection was first demonstrated by Goodman and

Witten (1985) following the proposal of a neutral-current neutrino-detection method by Drukier and Stodolsky (1984). In the 25 years since, both the theoretical framework for calculating interaction rates and sufficiently sensitive experimental techniques have been developed. Once a WIMP–nucleus cross section has been calculated along the lines described in the chapter in this volume by Hooper, it is straightforward to calculate the expected spectrum of recoil energies due to WIMP–nucleus scattering events. Lewin and Smith (1996) is the standard reference. Particularly critical for the experimentalist is understanding the dependence of the event rate on target nucleus and recoil energy. Numerical results are provided at the end of the section.

The WIMP velocity distribution is given by the WIMP phase-space distribution. For simplicity, it is standard to assume the WIMPs occupy an isothermal, isotropic phase-space distribution, appropriate for a fully gravitationally relaxed population of WIMPs. It is known that this is not completely accurate. However, these uncertainties are small compared to the theoretical uncertainty in the WIMP–nucleon cross section. Moreover, a simple standard model ensures that results presented by direct-detection experiments are easily interpreted. The full phase-space distribution function for an isothermal halo composed of particles of mass M_δ is

$$f(\vec{x}, \vec{v}) \, d^3x \, d^3v \propto \exp\left(-\frac{M_\delta v^2/2 + M_\delta \phi(\vec{x})}{k_B T}\right) \qquad (10.1)$$

which is just a Boltzmann distribution with the gravitational potential $\phi(\vec{x})$ included. M_δ is the mass of the WIMP δ. At a given point, the position-dependent factor is fixed and can be included in the normalization, leaving a simple Maxwellian velocity distribution:

$$f(\vec{v}, \vec{v}_E) = \exp\left(-\frac{(\vec{v} + \vec{v}_E)^2}{v_0^2}\right) \qquad (10.2)$$

where $\vec{v}$ is the WIMP velocity relative to the Earth and $\vec{v}_E$ is the Earth's velocity relative to the nonrotating halo of the galaxy. Note that the definition of $\vec{v}$ has been changed relative to Eq. (10.1). The quantity v_0^2 is characteristic of the WIMP kinetic energy, $k_B T_\delta = M_\delta v_0^2/2$ and has value $v_0 \approx 220 \,\mathrm{km\,s^{-1}}$. It corresponds to the most probable velocity and is known as the *velocity dispersion*. It is related to the root-mean-square WIMP velocity $v_{\rm rms}$ by $v_{\rm rms} = v_0\sqrt{3/2}$, as holds for any Maxwellian distribution. The Maxwellian distribution is cut off at $|\vec{v} + \vec{v}_E| = v_{\rm esc}$ by the halo escape velocity. Note that the cutoff is isotropic in the galactocentric WIMP velocity $\vec{v} + \vec{v}_E$, not in the Earth-centric $\vec{v}$. In practice, the escape velocity $v_{\rm esc} \approx 550$ km s^{-1} (Smith, *et al.*, 2007) is so large compared to v_0 that it has negligible effect at most WIMP masses of interest. It must be included, though, to obtain accurate recoil spectra for low-mass WIMPs.

With the above velocity distribution, we can calculate the rate of scattering of this incoming "beam" of particles on nuclei in a detector, obtaining a recoil energy spectrum in the lab frame. The WIMP "beam" flux per unit WIMP energy has the form $E\exp(-E/E_0)$ from phase-space and the Boltzmann distribution. The differential cross section goes like $v^{-2} \propto 1/E$ and is independent of momentum transfer $q^2 \propto E_R$ when details of nuclear structure are ignored. Therefore, the product of the beam flux and differential cross section is an exponential in E. This product is integrated from the minimum possible E that can yield E_R to infinity. This lower limit is proportional to E_R, leaving an exponential in E_R with normalization factors set by the WIMP velocity (energy) distribution. With $\vec{v}_E = 0$ and $v_{\rm esc} \to \infty$ to simplify this integral, one finds that the spectrum (in keV^{-1} kg^{-1} d^{-1}) has the form

$$\frac{dR}{dE_R}(E_R)\bigg|_{\substack{\vec{v}_E=0 \\ v_{\rm esc}\to\infty}} = \frac{R_0}{E_0 r}e^{-E_R/E_0 r}F^2(q^2 = 2M_N E_R) \qquad (10.3)$$

with

$$E_0 = \frac{1}{2}M_\delta v_0^2, \quad r = 4\frac{M_\delta M_N}{(M_\delta + M_N)^2},$$

$$R_0 = \frac{2}{\sqrt{\pi}}\frac{N_0}{A}n_0\,\sigma_0\,v_0, \quad \sigma_0 = \sigma_{\delta n}\frac{m_{r,\delta N}^2}{m_{r,\delta n}^2}A^2, \qquad (10.4)$$

being the mean energy, a unitless mass ratio, the integral rate assuming $F^2(q^2) = 1$, and the cross section on the nucleus. $F^2(q^2)$ is a "form factor" that parameterizes the breakdown of coherent scattering with the nucleus as the momentum transfer increases and the WIMP sees the nucleus's internal structure. M_δ is the WIMP mass, M_N is the nucleus mass, $m_{r,\delta n}$ and $m_{r,\delta N}$ are the WIMP–nucleon and WIMP–nucleus reduced masses.

In the general case, with nonzero $\vec{v}_E$ and finite $v_{\rm esc}$, the energy dependence becomes more complex. We quote the full formula, Eq. (3.13) of (Lewin and Smith, 1996):

$$\left.\frac{dR}{dE_R}(E_R)\right|_{\vec{v}_E, v_{\rm esc}}$$
$$= \frac{k_0}{k_1}\frac{R_0}{E_0 r}\left(\frac{\sqrt{\pi}}{4}\frac{v_0}{v_E}\left[\mathrm{erf}\left(\frac{v_{\min}+v_E}{v_0}\right) - \mathrm{erf}\left(\frac{v_{\min}-v_E}{v_0}\right)\right] - e^{-v_{\rm esc}^2/v_0^2}\right) \quad (10.5)$$

where $v_{\min} = q/2m_r = v_0\sqrt{E_R/E_0 r}$ and $k_0 = (\pi v_0^2)^{3/2}$ are as defined before and

$$k_1 = k_0\left[\mathrm{erf}\left(\frac{v_{\rm esc}}{v_0}\right) - \frac{2}{\sqrt{\pi}}\frac{v_{\rm esc}}{v_0}e^{-v_{\rm esc}^2/v_0^2}\right]. \quad (10.6)$$

This form immediately leads us to annual modulation. $\vec{v}_E$ is a function of the time of year with, to a good approximation for these purposes,

$$v_E = 232 + 15\cos\left(2\pi\frac{t - 152.5}{365.25}\right)\,\mathrm{km\,s}^{-1}, \quad (10.7)$$

where t is in days (00:00 January 1 corresponds to $t = 0$). Because of the sinusoidal variation in v_E, there is a small modulation of the recoil-energy spectrum on an annual basis, which is discussed in some detail in the context of the DAMA experiment, below. The mean spectrum can be calculated by setting $v_E = \langle v_E \rangle = 232\,\mathrm{km\,s}^{-1}$.

We show the dependence of the differential and cumulative recoil-energy spectra on A in Fig. 10.1 for a 100 GeV c^{-2} WIMP. The change in the energy dependence is especially critical because detectors have thresholds in the few to tens keV region. A germanium target offers a clear advantage over a silicon target, and iodine yields a benefit over germanium if the recoil-energy threshold is below 10 keV.

The above formulae can be investigated numerically to get a feel for the energies and event rates involved. Lewin and Smith calculate a useful

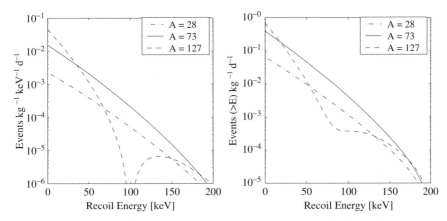

Fig. 10.1 Variation of recoil-energy spectrum with A. Left: Differential spectrum. Right: Integral of spectrum from E_R to 200 keV ($\sim\infty$), indicating the integral rate above a given threshold. Spectra are shown for silicon ($A = 28$), germanium ($A = 73$), and iodine ($A = 127$). Note that xenon ($A = 131$) has a response very similar to iodine.

form for R_0,

$$R_0 = \frac{361}{M_\delta M_N}\left(\frac{\sigma_0}{10^{-36}\,\text{cm}^2}\right)\left(\frac{\rho_\delta}{0.3\,\text{GeV}\,\text{c}^{-2}\,\text{cm}^{-3}}\right)\left(\frac{v_0}{220\,\text{km}\,\text{s}^{-1}}\right)\text{kg}^{-1}\text{d}^{-1} \quad (10.8)$$

where M_δ and M_N are in $\text{GeV}\,\text{c}^{-2}$. Recall that R_0 is the integral rate with no form-factor correction and so is only indicative of the detectable event rate. A typical WIMP mass of interest today is $M_\delta = 100\,\text{GeV}\,\text{c}^{-2}$. Halo parameters typically used are $\rho_\delta = 0.3\,\text{GeV}\,\text{c}^{-2}\,\text{cm}^{-3}$ and $v_0 = 220\,\text{km}\,\text{s}^{-1}$. This v_0 corresponds to an rms WIMP velocity of $v_\text{rms} = 270\,\text{km}\,\text{s}^{-1}$. An event rate detectable by present experiments is $0.1\,\text{kg}^{-1}\,\text{d}^{-1}$. For germanium ($A \approx 73$), this corresponds to $\sigma_0 \approx 20 \times 10^{-37}\,\text{cm}^2$, or $\sigma_{\delta n} \approx 2 \times 10^{-43}\,\text{cm}^2$.

The average energy deposition is given by

$$E_0 r = \left(\frac{M_\delta}{100\,\text{GeV}\,\text{c}^{-2}}\right)\left(\frac{v_0}{220\,\text{km}\,\text{s}^{-1}}\right)^2 \frac{4 M_\delta M_N}{(M_\delta + M_N)^2} \times 26.9\,\text{keV}. \quad (10.9)$$

For the standard halo parameters given above, $M_\delta = 100\,\text{GeV}\,\text{c}^{-2}$, and for a germanium target, this gives $E_0 r \approx 26$ keV. For reference, R_0 and $E_0 r$ are calculated for a number of common targets and WIMP masses of 10, 100, and 1000 $\text{GeV}\,\text{c}^{-2}$ in Table 10.1.

Table 10.1 R_0 and $E_0 r$ for hydrogen, silicon, germanium, and iodine and for WIMP masses of 10, 100, and 1000 $\text{GeV}\,c^{-2}$.

Target	R_0 [kg^{-1} d^{-1}] M_δ [GeV c^{-2}]			$E_0 r$ [keV] M_δ [GeV c^{-2}]		
	10	100	1000	10	100	1000
H ($A=1$)	3.9×10^{-6}	3.9×10^{-5}	3.9×10^{-7}	0.8	1.0	1.0
Si ($A=28$)	7.8×10^{-2}	5.5×10^{-2}	8.1×10^{-3}	2.2	17.6	26.6
Ge ($A=73$)	3.0×10^{-1}	5.4×10^{-1}	1.3×10^{-1}	1.2	25.9	64.1
I ($A=127$)	5.8×10^{-1}	1.7×10^{0}	6.4×10^{-1}	0.8	26.7	101.7

10.3 Inelastic Scattering

Instead of elastic scattering, some models of dark matter allow inelastic coherent scattering off nuclei, where the dark matter particle can be excited preferentially into a slightly heavier state, with energy splitting δ between the ground and excited states (Smith and Weiner, 2001). This inelastic dark matter (IDM) can arrive from sneutrinos models with small L violation, warped fermion seesaw, warped scalars, supersymmetric doublets, hidden sector U(1)x IDM (Arkani-Hamed et al., 2009), and magnetic inelastic dark matter (Chang et al., 2010). This minimum velocity to scatter with a deposited energy E_R is

$$v_{\min} = \frac{1}{\sqrt{2m_N E_R}} \left(\frac{m_N E_R}{\mu_N} + \delta \right). \tag{10.10}$$

If δ is of order 100 keV, this can lead to modified kinematics in direct detection experiments. The overall signal rate is suppressed (especially for light targets), low energy events are suppressed, and modulation signals can be enhanced. As a result, energy spectra are modified to give a peak in energy deposition instead of the exponentially falling spectrum expected of elastic scattering models. Typical modifications of energy spectra and modulation spectra are shown in Fig. 10.2.

10.4 Background Sources

Equal in importance to the expected WIMP interaction rate is the interaction rates of non-WIMP particles, termed "background" particles. Excellent discussions are given in (Heusser, 1995; Da Silva, 1996; Formaggio

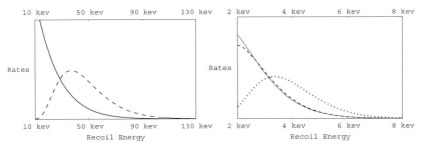

Fig. 10.2 Left: Normalized spectrum of events on germanium for ordinary WIMP (solid) and inelastic WIMP (dashed) with $\delta = 100\,\text{keV}$, both with $m = 50\,\text{GeV}\,\text{C}^{-2}$. Right: Normalized modulation as a function of energy for ordinary WIMP scenario (solid), inelastic WIMP scenario with $\delta = 100\,\text{keV}$ (dashed), and inelastic WIMP scenario with $\delta = 150\,\text{keV}$ (dotted), all with $m = 60\,\text{GeV}\,\text{C}^{-2}$. Figure taken from (Smith and Weiner, 2001).

and Martoff, 2004) among many others. The background sources are numerous and complex.

- Natural radioactive decays emit a number of high-energy photons and electrons reaching up to 2.6 MeV. Particular isotopes of interest, because they are primordially created and naturally occurring, include the ^{238}U and ^{232}Th decay chains (which include ^{222}Rn and ^{210}Pb) and ^{40}K. It is important to note that the highest energy particle emitted by these chains is the 2.6 MeV photon from ^{208}Th, yielding an electromagnetic background energy spectrum that terminates at this energy. As these photons and electrons penetrate material, they can create additional particles via pair-production and bremsstrahlung.

In a process termed *cosmogenic activation*, the hadronic component of cosmic-ray air-shower particles create a number of isotopes via interactions with naturally abundant nuclei like N, O, and Ar. Some of these isotopes of special interest to dark matter experiments, because of their prevalence in typical target materials or their likelihood of contaminating such experiments, are ^{3}H, ^{14}C, ^{39}Ar, ^{81}Kr, and ^{85}Kr. Such production rates are much higher at the surface and are substantially reduced when underground, but they may be important even deep underground. Similar cosmogenic activation occurs in important experimental construction materials such as copper, steel, and germanium, creating radioactive isotopes of manganese, iron, cobalt, zinc, gallium, and germanium among others.

Nuclear testing and the Chernobyl meltdown released so-called *anthropogenic* isotopes, including ^{3}H and ^{14}C, into the atmosphere. More exhaustive lists can be found in (Heusser, 1995; Da Silva, 1996) and references therein.

^{3}H, ^{14}C, and ^{210}Pb are of special interest because they all β-decay with low-energy endpoints of 18.6 keV, 156 keV, and 17.1 and 63.6 keV, respectively — energies of interest to WIMP searches. They are dangerous if present on the detector material or on nearby surfaces because of their short penetration depth (few to tens of microns). ^{39}Ar, ^{81}Kr, and ^{85}Kr are present in noble liquids such as argon and xenon. Cosmogenic production of ^{68}Ge occurs in germanium.

It is standard to use radiopure materials to shield against these electromagnetic activities. Lead and copper are excellent shielding materials because they are naturally low in U and Th due to the way they are produced. In general, lead contains ^{210}Pb, an isotope with a 22-year half-life. Its daughter ^{210}Bi decays via a 1.2 MeV β^-, which is energetic enough to yield continuum bremsstrahlung. Therefore, any lead shield must be lined on its inside with so-called "ancient lead,"[1] copper, or some other material. Iron and steel are typically contaminated by ^{60}Co used in furnaces to monitor lining thickness. Recently, due to the increasing cost of high-purity lead and copper, it has become clear that water is an excellent option. While water is less effective as a shield against electromagnetic activity, it is of course inexpensive and so a greater thickness can be used as a compensation. It simultaneously serves as a neutron moderator (see below). A number of experiments have now demonstrated purification of water to the necessary levels. If the experiment can be submerged in the water directly, then there is no inner tank wall whose radiopurity must be considered.

- There are many alpha-emitting isotopes in the U and Th decay chains because of their high atomic mass. The alpha is usually quite energetic,

---

[1] Lead much older than ^{210}Pb's half-life, which can be found in the ballast of sunken ships and in archaeological Roman plumbing, among other sources.

with energy of MeV, and so does not obviously pose a problem. Additionally, it penetrates only a very short distance, so only alpha activity in or very near the detector is a problem. But the alpha particle's energy deposition can be substantially "quenched" (quenching is discussed below), yielding an observed energy that is in the region of interest for dark-matter nuclear recoils. This can be particularly problematic in noble liquids. An associated problem is the recoiling nucleus, which typically has approximately 100 keV of energy (e.g. for the 5.3 MeV α from ^{210}Po) and *is* a nuclear recoil. If, for some reason, the associated alpha is not detected, the recoil mimics a WIMP interaction.

- Alpha particles (primarily from the U and Th chains due to their natural presence) can produce neutrons via (α, n) reactions. Because alphas have such a short penetration depth, the reaction must be with the material in which the emitter is embedded. (α, n) reactions with copper, silicon, and oxygen (in rock), etc. produce low-energy neutrons with an exponentially falling spectrum whose average energy is typically about 1 MeV. Tens of cm of a neutron moderator — any material with a lot of hydrogen (e.g. water, polyethylene), to which neutrons can efficiently lose energy via elastic scattering — exponentially attenuates this flux.

- Uranium and thorium isotopes also can decay via spontaneous fission, emitting neutrons with an energy spectrum similar to those from (α, n) reactions. Uranium has a much shorter half-life and so dominates the fission neutron production. These neutrons are also easily moderated.

- Cosmic-ray air showers contain a surfeit of energetic hadrons and electromagnetic particles that can interact in detectors, requiring experiments to be sited underground so that such particles are absorbed in the overburden. The conventional unit of measurement for overburden is "meters of water equivalent" (mwe), which is simply the number of meters of water that would provide equivalent stopping power. A few mwe is sufficient to stop the hadronic component and most of the electromagnetic component of cosmic rays, leaving only muons to penetrate to deep sites.

- The above penetrating muons generate their own large variety of backgrounds. Energetic muons penetrate even to very deep sites (> 1000 mwe), with mean energies of hundreds of GeV. Such energetic muons produce energetic secondaries via the interactions of the virtual photons in their EM field. These photons produce very high-energy neutrons and pions via direct interactions with nucleons, yielding appreciable flux above 100 MeV. These interactions yield hadronic showers, also. The nucleus with which the muon interacts can emit lower-energy "evaporative" neutrons as it recovers, with an exponentially decaying spectrum similar to the (α, n) spectrum. Electromagnetic showers can also be initiated via bremsstrahlung and pair production.

Most of this activity can be tagged and rejected by surrounding the experiment with a muon detector, usually made of plastic scintillator, or, more recently, PMT-instrumented water. Some experiments are planning to use liquid scintillator shields to ensure maximum detection efficiency for these particles. However, muons can interact in the rock surrounding the underground cavern housing the experiment, leaving the muon undetected.

Most of this activity produced by muon interactions in the rock is stopped by the experiment's electromagnetic and neutron shielding. However, neutrons with tens of MeV are particularly pernicious and can present an effectively irreducible background. Such neutrons can penetrate the moderator. If they reach high atomic mass material, they have dramatic interactions in which many secondary neutrons are released, some of which have enough energy to create a WIMP-like nuclear recoil in a detector. While these very high-energy neutrons are a small fraction of the flux, they effectively increase the low-energy neutron flux in a nonlinear, substantial fashion via this multiplication effect. Neutron multiple scattering (multiple scatters of the same secondary neutron or simultaneous detection of multiple secondaries) and/or high-efficiency neutron detectors under development (borated liquid scintillators, gadolinium-doped water) may enable efficient tagging and rejection of high-energy neutrons. The ultimate solution is greater depth: the high-energy neutron flux decreases by about one order of magnitude for every 1500 meters water equivalent of overburden (Mei and Hime, 2006).

10.5 Backgrounds and WIMP Sensitivity

Consider first the case of a detector that cannot distinguish WIMPs from background particles. Any such detector observes some spectrum of background-particle interactions. Such a detector can detect a WIMP by observing the expected exponential recoil-energy spectrum on top of its background spectrum while conclusively showing that such a spectrum could not arise from interactions of any other particle. Conversely, such a detector can exclude a WIMP model if the model predicts a spectrum with a higher event rate than the observed background. This is depicted naively in Fig. 10.3. In such an experiment, if no counts are observed, such a detector's sensitivity improves as MT where M is detector mass and T is exposure time — the 90% CL upper limit on 0 observed counts is 2.3 events, so the event rate excluded at 90% CL in some energy bin is $2.3/MT$.[2] This mode is termed "zero-background" operation. However, once the experiment observes background events, they begin to present a limit to the experiment's sensitivity. Initially, one must account for both the above "Poisson" fluctuation as well as the statistical uncertainty on the measured background spectrum. Once the fractional statistical uncertainty on the background becomes small, then the WIMP sensitivity will also only improve by fractionally small amounts, eventually asymptoting to the value set by the allowed Poisson fluctuations of WIMP interactions above the background spectrum. Increasing detector mass or the integration time yields no further improvement. Such an experiment is termed "background-limited". Modifying the experiment to have lower background rates is the only avenue for improvement. With electromagnetic backgrounds, this becomes increasingly more difficult because the requirements on the residual radioactivity of shielding materials and the detectors themselves become more stringent. Even in perfectly radiopure materials, cosmogenic

---

[2] The analysis is usually done in a somewhat more sophisticated fashion, but the result is usually not very different.

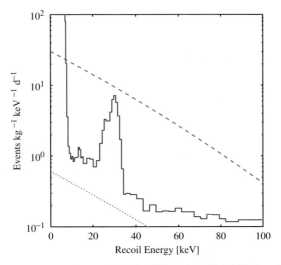

Fig. 10.3 Energy spectrum observed by the UCSB/LBNL/UCB Oroville experiment, which used Ge ionization-mediated detectors (Caldwell *et al.*, 1988). The two lines indicate recoil spectra for a 100 GeV c^{-2} WIMP with WIMP–nucleon cross sections of $\sigma_{\delta n} = 2 \times 10^{-39}$ cm^2 and $\sigma_{\delta n} = 4 \times 10^{-41}$ cm^2; the dashed spectrum is excluded, the dotted spectrum is allowed by these data. Note that QF ≈ 0.3 for this data, so the line at about 30 keV is the 10.4 keV gallium X-ray due to cosmogenic activation.

activation presents a natural limit, requiring development of underground material-refining and detector-production facilities. Such facilities are currently being planned for the Majorana neutrinoless double-beta decay experiment.

The natural next step is background subtraction. If one has independent knowledge that predicts the background spectrum, then it is possible to subtract the background spectrum and look for a WIMP spectrum in the residuals. For example, in germanium ionization-mediated detectors, one searches for radioisotope-identifying photon lines in the high-energy (> few hundred keV) spectrum, uses these to set contamination levels for the associated radioisotopes, and then simulates the continuum background of below few hundred keV expected from these radioisotopes and the associated decay chains. If the background model fits the observed spectrum perfectly, then the fractional residuals between the observed spectrum and the background model decrease as $1/\sqrt{MT}$. The shrinking fractional residuals can reveal a WIMP spectrum with sensitivity that improves as

$\sqrt{MT}$ (limiting $\sigma \propto 1/\sqrt{MT}$). The limit of such an analysis is that the fractional systematic error on the background model must also improve as $1/\sqrt{MT}$ in order for statistical errors to continue to dominate and for WIMP sensitivity to improve. Most experiments of this type eventually reach a point at which it is impossible to further improve the background model, leading the sensitivity to saturate.

Background rejection is the best method for quickly improving sensitivity. This takes many forms, but always is based on finding some signature that uniquely distinguishes WIMP events from background events. The most popular, and so far most effective, method is nuclear recoil discrimination. Most electromagnetic-background particles interact with electrons, producing electron recoils. Only WIMPs and neutrons interact with nuclei, producing nuclear recoils. Moderator and depth reduce neutron backgrounds without introducing new neutron sources, while shielding against electromagnetic backgrounds tends to introduce such new sources of such backgrounds. Therefore, rejecting the electron-recoil background enables the experiment to operate in zero-background mode, with sensitivity improving linearly with MT, as long as the neutron background is negligible.

However, every electron-recoil event has some finite probability of being mistaken for a nuclear recoil event. When such leakage results in nuclear-recoil candidates, the experiment can enter the background-subtraction phase: the misidentification rate is measured using a calibration data set (usually a gamma-ray source producing electron recoils in the detector) and this misidentification rate and the electron-recoil background are used to predict the number of misidentified electron recoils. This prediction is subtracted from the observed nuclear-recoil rate. As discussed above, this works until systematics dominate.

It should be noted that experiments using electromagnetic background rejection may eventually become subject to cosmogenic neutron backgrounds, depending on their depth, and must either move to a deeper site or find a way to detect and reject high-energy neutron backgrounds. Another possible worry is from photonuclear reactions, wherein the ambient photons from natural radioactivity or untaggable cosmogenic muons interact with nuclei to generate neutrons, protons, and/or alpha particles along with a daughter nucleus. Fortunately, only deuterium and ^{9}Be have photonuclear thresholds below the 2.6 MeV electromagnetic endpoint of natural radioactivity, while depth attenuates muon-induced electromagnetic activity.

A separate class of background-subtraction techniques utilizes the unique kinematics of WIMP interactions. One such signature is diurnal variation. Because the Earth and Sun move at approximately $232 \, \text{km s}^{-1}$ through an isotropic WIMP halo (the $220 \, \text{km s}^{-1}$ disk velocity plus the additional velocity of $12 \, \text{km s}^{-1}$ relative to the disk), WIMPs primarily come from the direction into which the solar system is moving. Because of the Earth's rotation about its axis, this direction completes a circle in the lab frame once per day. A detector that is sensitive to the direction of the recoiling nucleus will see a diurnal variation in the angular distribution of nuclear recoils with a specific phase.

A similar technique is annual modulation. As discussed earlier, the Earth's velocity with respect to the halo is sinusoidally modulated due to the Earth's orbit around the Sun. The lab-frame velocity of the WIMPs thus changes with the time of the year, yielding a small variation in the recoil-energy spectrum. It is not necessary to measure the recoil direction, only the spectrum. The modulation amplitude is typically a few percent of the total WIMP interaction rate. There is one experiment that claims a detection of annual modulation, DAMA/LIBRA, though their claim is not consistent with most other experimental limits under typical assumptions of how WIMP interactions depend on atomic mass and spin. DAMA/LIBRA will be discussed further below.

Diurnal- and annual-modulation searches are particularly intriguing because the known expected phase of the signal provides an additional handle to identify the signal in the presence of background events and instrumental effects. On the other hand, such experiments require incredible stability because the signal is a small modulation on a large time-independent background signal, and so such stability must be explicitly demonstrated to validate a signal. Additionally, sensitivity only improves as $\sqrt{MT}$ because the modulated signal must be visible above the Poisson fluctuations in the time-independent count rate.

10.6 Direct Detection Techniques

Because of the numerous ways to obtain sensitivity to WIMPs — high A, high target mass, low threshold, low background, background-rejection and -subtraction techniques — a variety of different technologies are presently in use for WIMP searches. In this section, we discuss many of these methods.

We discuss the phenomenon called "quenching." Most detectors are calibrated and provide a linear response to electron recoils. However, WIMPs produce nuclear recoils, which tend to produce less ionization, scintillation, etc. than electron recoils. The quenching factor, or QF, is the ratio of observed energy for nuclear and electron recoils of the same recoil energy. For example, in germanium ionization-mediated detectors described below, a nuclear recoil produces approximately 1/3 the ionization of an electron recoil of the same recoil energy. This effect is quite important, as it raises the threshold for WIMP recoils by 1/QF. Because detectors are usually calibrated using photon sources that yield electron recoils, energies and thresholds for such detectors are usually quoted in "keV electron-equivalent" or "keVee", which must be multiplied by 1/QF to obtain the total recoil energy (essentially, a Jacobian transformation). An additional point is that the QF may be energy-dependent. For example, Ge ionization-mediated detectors follow $QF = 0.14 E_R^{0.19}$ (Baudis et al., 1999). There has been recent controversy about the appropriate QF to use at low energy in liquid xenon (Collar and McKinsey, 2010a,b). Where a distinction between nuclear recoil and electron-equivalent energy is required, we use keVnr and keVee, respectively.

Because exclusion limits on the WIMP-nucleon cross section presented here will be out of date by the time of publication, we encourage the reader to find up-to-date results at http://dmtools.brown.edu.

10.6.1 Detectors without nuclear-recoil discrimination

As noted above, detectors incapable of nuclear-recoil discrimination operate in background-free mode until they see background, and then in background-subtraction mode if possible. Beyond this point, reducing the background is necessary for further improvement.

10.6.1.1 Germanium spectrometers

The first technique used for direct WIMP searches was the detection of ionization produced by the recoiling particle (Ahlen et al., 1987). A recoiling nucleus travels through the target material before stopping, ionizing atoms as it passes. In a semiconducting crystal, electron-hole pairs are created. An electric field applied to the crystal drifts these charges to collection electrodes, where they are sensed with a transimpedance amplifier.

The applicability of such a detector to dark matter lies in its good energy resolution and concomitant low threshold. Germanium is unique in that it is possible to drift charges through a crystal of substantial size (many cm) without any appreciable trapping. This is possible in part because the zone-refining and crystal-growth processes can yield impurity levels of order $10^{10}\,\text{cm}^{-3}$, yielding a very low density of traps for drifting charges. Additionally, germanium crystals of this purity can hold a field of hundreds to thousands of V cm^{-1} without breakdown, sufficient to deplete the crystal of thermally generated charges at 77 K. Thus, the energy resolution of such a detector is limited by readout noise rather than variations in trapping with event position. The readout noise is limited by the JFET[3] used in the amplifier and the detector capacitance and possibly microphonics. Large detectors have large capacitances, so there is a tradeoff between individual detector size and energy resolution/threshold.

The first WIMP dark matter searches employed detectors using high-purity germanium crystals with masses of order 1 kg. These detectors were initially used to look for the spectral-line signature of neutrinoless double-beta decay of an isotope in the detector itself. Search for this unique signature requires very good energy resolution, so such detectors also had very low thresholds, 5 to 10 keVee. Since QF $\sim 1/3$ for ionization in Ge, the recoil-energy threshold for WIMP interactions was approximately 15 to 30 keV. This is interesting for WIMP searches, as the WIMP spectra shown in Fig. 10.1 demonstrate. The first group to use this technology was the PNL-USC Collaboration (Ahlen et al., 1987). The UCSB/LBNL/UCB

---

[3] Junction field-effect transistor.

Collaboration also used such detectors (Caldwell et al., 1988). The final background spectrum from the latter experiment is shown in Fig. 10.3. The most recent experiments of this type were the IGEX and Heidelberg–Moscow experiments, which obtained 90% CL WIMP–nuclear cross-section limits of approximately 1×10^{-41} cm^2 at 75 GeV c^{-2} (Baudis et al., 1999; Morales et al., 2000). A variant on the latter experiment, the Heidelberg Dark Matter Search (HDMS), used two concentric Ge spectrometers in anticoincidence. They obtained better sensitivity at low masses, reaching 2×10^{-41} cm^2 at 30 GeV c^{-2} (Baudis et al., 2001).

More recently, the CoGeNT Collaboration has applied a detector of this style optimized for very low-energy depositions, below 10 keVee, the p-type point-contact detector. This detector has better noise performance and thus lower threshold than the earlier experiments thanks to its small size and thus low capacitance. Additionally, the contact geometry and electric field enables rejection of surface events at the n$^+$ contact, which covers the bulk of the detector surface, thereby reducing the overall background level. They obtained sensitivity to low-mass WIMPs of approximately 10^{-40} cm^2 at 7 GeV c^{-2} WIMP mass (Aalseth et al., 2011). Their sensitivity at higher masses is not competitive due to the low mass of the detector. Aside from their special application to low-mass WIMPs, these germanium spectrometer detectors have largely given way to germanium detectors employing electon-recoil discrimination (see below).

10.6.1.2 *Scintillation-mediated detectors*

A long-standing puzzle in the field of dark matter is the results from the DAMA/LIBRA experiment (Bernabei et al., 2010). DAMA/LIBRA contains 250 kg of sodium iodide crystals, which scintillate brightly from ionizing radiation. The crystals are viewed by low-radioactivity photomultipliers, and the crystals are themselves very low in radioactivity, built by Saint-Gobain using a proprietary method.

With a total exposure exceeding 1.17 ton-years over 13 years, a highly statistically significant (8.9 σ) variation in rate is seen, with peak rate of 0.0116 events kg^{-1} keVee^{-1} d^{-1} seen over a background of about 1 per

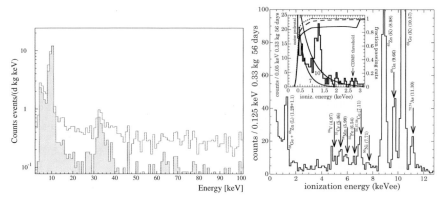

Fig. 10.4 Spectra obtained in germanium diode experiments, HDMS (left) and CoGeNT (right). The yellow (shaded) spectrum on the left is after cutting events that were coincident with a surrounding shield detector. On the right, the inset shows overlays of low-mass WIMP recoil spectra for WIMP masses of 7 and 10 GeV c^{-2} and $\sigma_{\delta n} = 10^{-40}$ cm^{-2}. Figures are taken from (Aalseth et al., 2011; Baudis et al., 2001).

day per kg per keV. Note that keVee refers to keV electron-equivalent. Due to quenching, the background rates are much lower when converted to kg^{-1} keVnr^{-1} d^{-1}. If interpreted through the standard elastic scattering paradigm, this large rate is largely inconsistent with experiments that have electron recoil/nuclear recoil discrimination, such as XENON and CDMS, though for very low WIMP masses these constraints may be evaded (Hooper et al., 2010). However, analyses by these experiments focused on low mass WIMPS disfavor even this region of possible compatability (Akerib et al., 2010, Ahmed et al., 2011, Angle et al., 2011).

At the time of this writing, the DAMA/LIBRA signal has not been independently verified. An effort underway to test the DAMA signal with other sodium iodide detectors is ANAIS. ANAIS is a large mass sodium iodide experiment intended to look for annual modulation in the dark matter signal. It will be carried out at the Canfranc Underground Laboratory and will consist of ten 10.7 kg NaI(Tl) crystals selected from a set of 14 which have been stored underground at the Canfranc laboratory since 1988 (Amare et al., 2006). In addition, in winter 2010, two sodium iodide detectors were sunk into the Antarctic ice by the IceCube Collaboration as a prototype experiment designed to test whether the ice

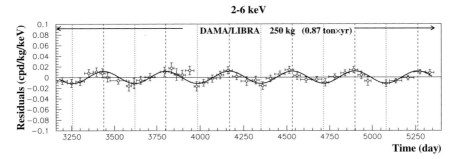

Fig. 10.5 Background residuals in the DAMA/LIBRA experiment, clearly showing an annual modulation.

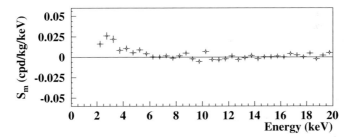

Fig. 10.6 Energy dependence of the modulated rate as observed in the DAMA/LIBRA experiment. Here keV refers to keVee here, so the data should be divided by a factor of approximately 11 to convert to cpd/kg/keVnr where keVnr refers to nuclear recoil energy. This factor is derived from QF = 0.09 in NaI as justified by a summary of measurements given in Bernabei et al. (2003). **Figure credit:** Bernabei et al. (2010).

shielding, stable environment and muon veto features of IceCube might be put to good use in testing the DAMA/LIBRA result.

The KIMS collaboration uses CsI crystals to search for dark matter. CsI crystals can be grown relatively large (8.7 kg/crystal), and pulse-shape can be used to decrease electromagnetic bakground. KIMS is installed in the YangYang underground laboratory in South Korea and is shielded by copper, polyethylene, lead, and an organic scintillator active veto.

The most significant backgrounds in KIMS are ^{137}Cs and ^{87}Rb in the crystals, with a total of 6 events per keV per kg per day at low energies (Kim et al., 2008). ^{137}Cs can be introduced through the water used to extract the Cs from ore prior to crystal production, and commercial CsI powder samples can have ^{137}Cs levels of 9 to 80 mBq per kg. Using ultra-pure water to extract the Cs can greatly reduce the level of ^{137}Cs to 2 mBq per kg. Rb is a contaminant in the CsI because Rb is similar in chemical properties to Cs. Rb levels can be reduced through a repeated crystallization technique, and the KIMS Collaboration has used this to lower ^{87}Rb levels to less than 1 ppb, or 1.1 mBq/kg.

In 2007, the KIMS Collaboration reported the results of a dedicated WIMP search, with low-background CsI crystals, and a total exposure of 3409 kg d. KIMS is the first experiment to exclude the DAMA spin-independent signal region using a detector with an ^{127}I target, which is the dominant nucleus for the spin-independent interaction of the NaI crystal (Lee et al., 2007). KIMS also gives a stringent limit on proton spin-dependent WIMP interactions and excludes the DAMA spin-dependent signal region for scattering with I for $M > 20\,\text{GeV}\,c^{-2}$.

The XMASS detectors (Suzuki, 2000) are based on liquid xenon and rely entirely on scintillation light detection. Liquid xenon is a very bright scintillator (see Sec. 10.6.2.3) and is also very effective for shielding from gamma rays and low-energy neutrons. Currently under construction in the Kamioka mine in Japan, XMASS will contain about 860 kg of active liquid xenon. Expected to be operational in late 2010, it is predicted to have a sensitivity of $2 \times 10^{-45}\,\text{cm}^2$ after 1 year of operation.

While liquid xenon operated in scintillation-only mode has little discrimination power, the XMASS strategy is to overcome this with substantial self-shielding and by taking great care in minimizing radioactive backgrounds. Event positions will be determined by measuring the photomultiplier hit patterns and comparing these data to Monte Carlo-generated hit pattern predictions. A signal strength of 4.4 photoelectrons per keVee is predicted. The 5-cm in diameter photomultipliers used in XMASS have extremely low background levels: about 0.7 mBq in uranium, 1.5 mBq in thorium, and 2.9 mBq in ^{60}Co. The detector vessel and photomultiplier mounting hardware is built from low-radioactivity copper. The xenon is distilled to lower krypton contamination from ppm to ppt levels, so as to reduce background from ^{85}Kr beta decay. XMASS will be surrounded by a water shield 8 m in diameter to reduce gamma ray and neutron flux from the cavern rock. A 24-ton (10 ton fiducial mass) detector is planned for the future and will not only be sensitive to dark matter particles but will also be able to measure the solar pp neutrino flux.

10.6.2 *Detectors with nuclear-recoil discrimination*

In the last decade, detectors with nuclear-recoil discrimination have taken the lead in sensitivity for WIMP searches. In all cases, they discriminate nuclear recoils from electromagnetic backgrounds via the differing yield of various measurable excitations — scintillation, ionization, destruction of metastable states — for nuclear and electron recoils. This difference arises due to the very different density of energy deposition for nuclear and electron recoils. A typical WIMP-induced nuclear recoil with ~ 30 keV kinetic energy deposits all its kinetic energy in a few tens of angstroms in a solid or liquid target. An electron recoil of the same energy travels many microns in the same target. The much higher density of energy deposition of nuclear recoils suppresses ionization and scintillation (the quenching phenomenon described above) because more energy goes into "invisible" channels such as heat/acoustic vibrations and atomic displacement. For metastable detectors, the tuning of the detector state ensures that it phase transitions only in response to the dense energy deposition of nuclear recoils. We will see below that directional detectors are in general also able to discriminate based on density of deposition, though of course their greatest advantage is their potential sensitivity to the recoiling particle's direction.

10.6.2.1 *Metastable or threshold detectors*

Metastable, or threshold, detectors operate by monitoring a medium held in a metastable state for which the phase transition to a distinct state can be nucleated by a nuclear recoil. There is a wide range of such phase transitions available — bubble formation, cloud formation, superconducting phase transition, etc. These techniques offer the possibility of tuning out any sensitivity to electron recoils because they do not provide a high enough energy density to nucleate the phase transition. They also tend to be quite inexpensive because the shielding requirements are minimal. They have the disadvantage of no energy information beyond the threshold energy of the phase transition: an energy spectrum would only be recoverable via a scan of the detector threshold. If the event rate is too high, the lifetime fraction can be low, which can make it challenging to acquire sufficient amounts of neutron calibration data. We make no attempt to survey all such techniques, but rather focus on bubble chambers because they have progressed to the point of yielding interesting sensitivities.

Following the invention of the bubble chamber by Glaser (1953), Seitz (1958) developed the basic theory of its operation. This theory relates threshold for bubble formation and the bubble chamber medium's intrinsic properties, like density and viscosity, and externally controlled properties, like temperature and pressure. Though not used in this way during its heyday in particle physics, it is possible to robustly tune the latter parameters so that bubble formation can only be initiated by nuclear recoils (see Fig. 10.7). A wide range of liquids can be used — hydrogen, propane, Xe, Ne, CF_3Br, CH_3I, CCl_2F_2, and C_4F_{10} are among those demonstrated. The COUPP (Chicagoland Observatory for Underground Particle Physics) Collaboration has demonstrated the operation of chambers using CF_3Br, CF_3I, and Xe (Sonnenschein, 1996). They are focusing on the use of CF_3I because of the high A of iodine and the high spin of fluorine, providing sensitivity to both spin-independent and proton spin-dependent interactions of WIMPs (see the chapter in this volume by Hooper). They have demonstrated a remarkable rejection of electron recoils, with an upper limit on the misidentification probability of less than 10^{-10} (Ramberg, 2009). The response to nuclear recoils and alpha particles appears to be approximately as expected according to Seitz's theory. The chamber consists of a transparent quartz bell jar with a bellows at the top connecting it to a pressure control system. The active liquid fills most of the chamber with a buffer liquid on top. The chamber temperature environment is well-controlled while a piston provides compression to the appropriate pressure. The chamber stays in the compressed, metastable state until a bubble forms, at which point the liquid "boils". The piston then recompresses the liquid. Video cameras view the chamber at 50 Hz, providing images of bubble formation that localize the event position and time. A key operational issue is the smoothness of the chamber walls: wall imperfections can spontaneously nucleate bubbles. Historically, chambers used in particle physics recompressed so frequently, synchronized with a particle beam, that wall nucleation was not a problem. Wall nucleation has been overcome by appropriate vendor choice and by wetting the wall surface with ultra-pure water to obtain a smoother surface (Ramberg, 2009).

While their electron-recoil rejection is magnificent, bubble chambers are as sensitive to alpha particles as to nuclear recoils. The purity of the chamber liquid in U and Th and the absorption of radon gas are of great importance, as the alpha events due to isotopes in this chain trigger the chamber. The purity of the vessel is also important, not because alpha and nuclear recoil events from the wall would be a background to a WIMP search — reconstruction of the event position easily eliminates them — but because, at too high a rate, they limit the experiment's lifetime by forcing the recompression cycle too frequently. Based on techniques for purifying liquid scintillator and water for solar neutrino experiments, it is expected that the bulk event rate can be reduced sufficiently. Careful sourcing of the vessel material can reduce wall event rates. By retuning the chamber temperature and pressure, one can render the chamber insensitive to nuclear recoils and thereby measure the alpha rate to perform statistical background subtraction. Finally, the COUPP Collaboration has recently confirmed (Dahl, 2010) the earlier demonstration by the PICASSO (Project in Canada to Search for Super-Symmetric Objects) Collaboration of discrimination between alphas and nuclear recoils via the acoustic signature of bubble formation (Aubin *et al.*, 2008). Alpha particle events tend to form two or more proto-bubbles, the first from the nuclear recoil at the site of the parent and the remainder from the stopping of the alpha. These two proto-bubbles are too close to be distinguished optically, but they present a differing acoustic signature than a single-bubble event, primarily with an increased event acoustic "energy".

The COUPP Collaboration has operated a 2-kg and a 4-kg chamber in three runs at 300 mwe in the MINOS near-detector hall at Fermi National Accelerator Laboratory (FNAL). The first run was limited by radon contamination. The second run had radon levels improved by an order of magnitude, though wall events continued to dominate due to the 2-kg vessel's intrinsic background. A third run used a 4-kg target in a synthetic quartz vessel to demonstrate a reduction in the rate of wall events to acceptable levels and to confirm acoustic bubble discrimination. A preliminary analysis yielded sensitivities, respectively, of $3 \times 10^{-42}\,\text{cm}^2$ and $4 \times 10^{-38}\,\text{cm}^2$ for the spin-independent and proton spin-dependent WIMP cross sections at roughly 50–60 GeV c^{-2} without background subtraction (Dahl, 2010). By late 2010, they are moving the 4-kg chamber to a deeper site at SNOLAB (6000 mwe) to escape the cosmogenic neutron background. They are also commissioning larger 20-kg and 60-kg chambers at shallow sites with the expectation of eventually moving them to SNOLAB.

The PICASSO Collaboration has pursued a concept using C_4F_{10} droplets suspended in a polymerized gel, the same principle as radiation dosimeters. Piezoelectric sensors measure the acoustic signal. The medium is globally stable — when an event occurs, a bubble forms at the event location, but its growth is stopped by the polymer matrix. Instead of recompressing after each event, the experiment runs for some amount of time (e.g. 40 hours) and then undergoes a recompression cycle (e.g. 15 hours). The experiment has a total mass of about 2 kg (mostly active

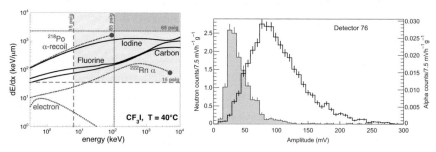

Fig. 10.7 Left: Stopping power versus energy for different particles in CF$_3$I at 40° C, with differing pressures indicated. It is possible to operate in a regime in which electron recoils cannot nucleate bubbles. It is, however, impossible to find a regime that is insensitive to alpha particles from ^{218}Po and ^{222}Rn without also raising the thresholds for WIMP-induced nuclear recoils. **Figure credit:** (Behnke et al., 2008). Right: Demonstration of acoustic discrimination between nuclear recoils and alpha particles by PICASSO. **Figure credit:** (Aubin et al., 2008).

freon) segmented into 32 separate chambers operating at SNOLAB. They were the first group to demonstrate separation of nuclear recoils from alpha particles using an acoustic signature (see Fig. 10.7) (Aubin et al., 2008). They suffer from an alpha particle background due to contamination of the bulk medium, but used alpha background subtraction to obtain sensitivities of approximately 10^{-37} cm^2 for the spin-dependent proton–WIMP cross section and expect another order of magnitude from the above setup.

As for the future, COUPP is making plans to scale up single vessels to as much a 500 kg of target material, making ton-scale targets and cross sections down as 10^{-47} or 10^{-48} cm^2 accessible. PICASSO is developing 25 kg single modules, which would also enable ton-scale target masses. In both cases, further reduction and rejection of the alpha background is necessary before this increase in mass is worthwhile.

10.6.2.2 Sub-Kelvin detectors

Detectors operating below 1 K, also known as "low-temperature" or "cryogenic" detectors, discriminate nuclear recoils from electron recoils by measuring quenching on an event-by-event basis. They measure a phonon or thermal signal, which gives the total deposited energy independent of the type of recoil, as well as a quenched signal such as ionization or scintillation. The ratio of the two signals measures the quenching of an event and thus separates nuclear and electron recoils. Measuring the phonon signal requires sub-Kelvin operation in order to overcome thermal noise.

We first describe the fundamental sub-Kelvin measurement techniques, followed by specific discussion of experiments employing these techniques. An excellent review (Enss, 2005) is available that covers this material and other applications of low-temperature detectors. The proceedings of the Low Temperature Detectors Workshops are also useful (e.g. Cabrera et al., 2009).

The most basic kind of low-temperature detector employs a dielectric absorber coupled to a thermal bath via a weak link. A thermistor monitors the temperature of the absorber. The energy deposited by a particle interaction causes a calorimetric temperature change by increasing the population of thermal phonons. The fundamental sensitivity is

$$\sigma_E^2 = \xi^2 k T [T C(T) + \beta E] \tag{10.11}$$

where C is the heat capacity of the detector, T is the temperature of operation, k is Boltzmann's constant, and ξ is a factor of order unity that depends on the details of the weak link. The first term is imposed by statistical fluctuations on the energy in the absorber due to exchange with the thermal bath (see, e.g. Kittel and Kroemer, 1980 and references therein). The second term is due to statistical fluctuations in the number of phonons excited by the absorbed radiation. The factor β is dimensionless and $\mathcal{O}(1)$ and is also precisely calculable from the nature of the thermal link. The ratio of the second term to the first term is equal to the fractional absorber temperature change due to an energy deposition. Thus, the second term becomes appreciable when this fractional temperature change is appreciable, at which point nonlinear effects also come into play. The above formula typically acquires an additional (quadratic) term due to deviations from an ideal calorimetric model that cause position and/or energy dependence of the signal shape.

The rise time of response is limited by the internal thermal conductivity of the absorber. The decay time constant, describing the time required for the absorbed energy to flow out to the bath, is $\tau = C/G$, where G is the thermal conductance of the weak link. The above formula immediately suggests the use of crystalline dielectric absorbers and low temperatures because of the linear factor of T and because C for crystalline dielectrics drops as T^3 for T well below the material's Debye temperature (Θ_D, typically hundreds of K). Specifically, the Debye model indicates that a crystal consisting of N atoms has

$$C = \frac{12\,\pi^4}{5} N k \left(\frac{T}{\Theta_D}\right)^3 \implies \sigma_E = \sqrt{\frac{12}{5}} \xi \pi^2 k T \sqrt{N} \left(\frac{T}{\Theta_D}\right)^{3/2} \tag{10.12}$$

which gives $\sigma_E = 5.2\,\xi\,\mathrm{eV}$ for 1 kg of germanium operated at $T = 10\,\mathrm{mK}$. In practice, a number of factors degrade the above result by about an order of magnitude (thermistor heat capacity and power dissipation, readout noise, etc.), but the predicted energy resolution for such a large mass remains attractive.

Neutron-transmutation-doped (NTD) germanium and implanted silicon semiconductors are used for thermistors. Conduction is via phonon-assisted hopping between impurity sites, yielding an exponentially decreasing resistance as a function of temperature, $R(T)$, with negative slope, dR/dT. Attachment to the absorber is usually by eutectic bonding or epoxy or by direct implantation into the absorber. Another type of temperature sensor is the superconducting phase-transition thermometers (SPT) or transition-edge sensor (TES). A SPT or TES is a superconducting film operated at its superconducting transition temperature, T_c, where its resistance is a strong function of temperature with positive dR/dT. This can provide strong electrothermal negative feedback, which improves linearity, speeds up response, and mitigates variations in T_c among multiple TESs on the same absorber. Such sensors are made from a variety of materials, including tungsten, titanium, and various bilayers include aluminum/titanium and molybdenum/copper. Nb_xSi_{1-x} is another thermistor material that ranges between the semiconducting and superconducting regimes as a function of the stoichiometry (defined by x). SPTs/TESs and Nb_xSi_{1-x} thermistors are frequently deposited directly onto the absorber by sputtering or evaporation.

The readout method depends on the type of thermometer used. Doped semiconductors typically have high impedances and are well matched to low-noise JFET-based readout while SPTs/TESs are low-impedance devices requiring SQUID00 amplifiers.

The very good energy resolution achievable with thermal phonons has the problem of degrading as $\sqrt{M}$ where M is the detector mass. This motivates the use of athermal phonons. There are three steps in the development of the phonon signal. The recoiling particle deposits energy along its track, with the majority going directly into phonons. (A minority of the energy goes directly into scintillation and ionization. Energy deposited in ionization is recovered when the carriers recombine.) The recoil and band-gap energy scales (keV and higher and eV, respectively) are much larger than phonon energies (meV), so the full energy spectrum of phonons is populated with phase space favoring the most energetic phonons. However, these initial energetic phonons do not propagate because of isotopic scattering (scattering due to variations in lattice ion atomic mass, rate $\propto \nu^4$ where ν is the phonon frequency) and anharmonic scattering (scattering wherein a single phonon splits into two phonons, rate $\propto \nu^5$). Anharmonic scattering downshifts the phonon spectrum, which increases the phonon mean free path, so that eventually phonons can propagate the characteristic dimension of the detector. These phonons travel quasi-ballistically, preserve information about the position of the parent interaction, and are not affected by an increase in detector mass (modulo the concomitant larger distance to the surface where they can be sensed). Anharmonic decay continues until a thermal distribution is reached (μeV at mK temperatures), which is exhibited as a thermal increase in the temperature of the detector. If one can detect the athermal phonons at the crystal surface, keeping the density of such sensors fixed as detector surface area increases with mass, with the crystal pure enough that the athermal phonons can propagate to the surface prior to thermalization, then an increase in detector mass need not degrade energy resolution and can in fact improve position reconstruction.

Superconducting films are frequently used to sense the athermal phonons. These phonons break superconducting Cooper pairs and yield quasiparticles, which are electron-like excitations that can diffuse through the film and that recombine after the quasiparticle lifetime. These quasiparticles can become trapped in a TES or SPT attached to the film, providing a means to sense the energy. Such thin films have diffusion lengths of order 100 μm to 1 mm. Thus, the superconducting film must be segmented on this length scale and have a quasiparticle sensor for each segment. The sensors may, however, be connected in series or parallel in large groups to reduce the readout channel count. As with thermal detectors, the TES/SPT is usually read out using a SQUID amplifier.

While ionization and scintillation detectors usually operate at much higher temperatures, ionization and scintillation can be measured at low temperature and can be combined with a "sub-Kelvin" technique to discriminate nuclear recoils from background interactions producing electron recoils, which is critical for WIMP searches. With ionization, such techniques are based on Lindhard theory (Lindhard et al., 1963), which predicts substantially reduced ionization yield for nuclear recoils relative to electron recoils. For scintillation, application of Birks's law (Johnson, 2010) yields a similar prediction. (These are examples of quenching.)

Specifically, consider the example of measuring thermal phonons and ionization. All the deposited energy eventually appears in the thermal phonon channel, regardless of recoil type (modulo a negligible loss to permanent crystal defect creation). Thus, the ionization yield — the number of charge pairs detected per unit of detected recoil energy in phonons — discriminates nuclear recoils from electron recoils. Similar discrimination is observed with athermal phonons and ionization and with phonons and scintillation.

In semiconducting materials of sufficient purity — germanium and silicon — electron-hole pairs created by recoiling particles can be drifted to surface electrodes by applying an electric field, similar to how this is done at 77 K in high-purity germanium photon spectrometers (Sec. 10.6.1.1). There are three important differences, however, that motivate the use of low fields — of order 1 Vcm^{-1} — instead of the hundreds to thousands of Vcm^{-1} used in 77 K detectors. First, high fields are required at 77 K to deplete the active volume of thermally excited, mobile carriers. At low temperature and in crystals of purity high enough to drift ionization with negligible trapping, the population of thermally excited carriers is exponentially suppressed due to the low ambient thermal energy. Effectively, temperature enforces depletion rather than electric field. Second, high fields in 77 K operation prevent trapping of drifting carriers on ionized impurities and crystalline defects and/or overcome space charge effects. At low temperatures, ionized impurities and space charge can be neutralized (using free charge created by photons from LEDs or radioactive sources) and remain in this state for minutes to hours. Such neutralization increases the trapping length exponentially and enables low-field drift. Third, a high field in a sub-Kelvin detector would result in a massive phonon signal from the drifting carriers, fully correlated with the ionization signal and thereby eliminating nuclear-recoil discrimination, so this must be avoided. Readout of the charge signal is typically done with a conventional JFET-based trans-impedance amplifier.

A number of materials that scintillate on their own (i.e. without doping) continue to do so at low temperatures, including BaF, BGO, CaWO$_4$, ZnWO$_3$, PbWO$_4$, and other tungstates and molybdates. In and of itself, there is little advantage to a low-temperature scintillation measurement because detecting the scintillation is nontrivial, the quanta are large, and the detection efficiency is usually poor. Such techniques are pursued only in order to obtain nuclear-recoil discrimination. Conventional photodetectors do not operate at such low temperatures, so one typically detects the scintillation photons in an adjacent low-temperature detector, that is thermally disconnected from, but resides in, an optically reflective cavity.

Two collaborations, CDMS (Cryogenic Dark Matter Search) and EDELWEISS (Expérience pour DEtecter Les WIMPs En SIte Souterrain) employ simultaneous measurement of phonons and ionization to discriminate nuclear recoils from electron recoils (see Fig. 10.8). Both experiments employ germanium crystal targets of mass 0.25 kg to 0.8 kg to make use of the excellent ionization collection. CDMS employs athermal phonon sensors composed of superconducting aluminum films to absorb phonons and convert them to quasiparticles and tungsten TESs abutting the aluminum to detect the quasiparticles via the thermally induced change in sensor resistance. These sensors enable reconstruction of position to roughly 1 mm precision in xy in addition to sensing total energy. The rise time of the phonon pulse is sensitive to the event depth, with events near the surface having faster phonon rise times because down-conversion to quasi-ballistic phonons is enhanced there. The CDMS detectors must be operated roughly a factor of 2 below the tungsten 80–100 mK transition temperature. EDELWEISS employs NTD Ge thermistors to sense the calorimetric temperature rise of the crystal. They operate at 20–30 mK due to the thermal nature of the detector. CDMS typically operates with a 7–10 keV analysis threshold due to poorer rejection at lower energies, which occurs because the ionization yield band of electron recoils merges with that of nuclear recoils at lower energies due to ionization noise. EDELWEISS typically uses a 20 keV analysis threshold for similar reasons.

Both CDMS and EDELWEISS detectors suffer from poor ionization collection for events near the detector surface. This arises because, when an event occurs, some of the electrons or holes may diffuse into the incorrect ionization electrode before the modest electric field (a few Vcm^{-1}) drives them in the correct direction. Thus, while the electron-recoil discrimination in the detector bulk is excellent, with misidentification probability of less than 10^{-9} and further gains possible, the misidentification probability for surface events is much poorer. Both experiments use amorphous silicon

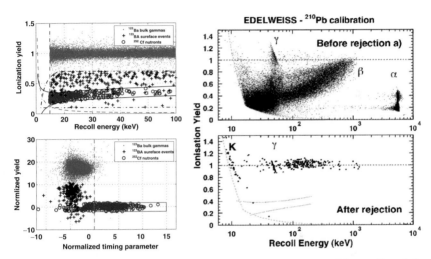

Fig. 10.8 Left top: CDMS ionization-yield-based discrimination. The small, red dots are events from exposure to a continuum photon source. The black +'s are also from exposure to this source, but using a cut that selects surface events. The blue circles are events from exposure to a neutron source to produce nuclear recoils. There is excellent separation in ionization yield ($\equiv$ ratio of electron-equivalent ionization energy to recoil energy) of nuclear recoils from bulk electron recoils, but the suppressed ionization yield for surface electron recoils is clear. Left bottom: CDMS phonon-timing-based discrimination. The vertical axis is the same as for the top figure. The horizontal axis is an energy- and position-corrected parameter that describes the timing of the rising edge of the phonon signal in microseconds. The discrimination between nuclear recoils and both surface and bulk electron recoils in timing is clear. Right: Demonstration of surface-event rejection using complex electrode structure. The axes are the same as for the upper left plot, though note that the horizontal axis is now logarithmic. The top plot shows data from exposure of an EDELWEISS detector to a ^{210}Pb source, which emits beta particles up to 1 MeV, a 46 keV X-ray, and alpha particles near 5 MeV. The bottom plot shows the events that remain after a cut to select bulk events by requiring consistent ionization measurements on the top and bottom electrodes. "K" indicates the Ga K-shell X-ray that ensues from the electron capture decay of cosmogenically created ^{68}Ge in the detector bulk. "γ" indicates the continuum radioactive photon background. **Figure credit:** (CDMS II Collaboration, 2010), left plots; (EDELWEISS Collaboration, 2009), right plots.

and/or germanium layers between the bulk material and the metal electrode to mitigate this problem — the amorphous materials have larger bandgaps than the bulk Ge and thus block carrier diffusion into the electrode — but the misidentification probability is still 0.1 to 0.2. CDMS has employed the rise time of the phonon pulse to reject these events with overall misidentification probability of approximately 10^{-3} (see Fig. 10.8).

Recently, both collaborations have developed more complex electrode configurations, involving alternating ground and positive voltage electrodes on one surface and ground and negative voltage electrodes on the other, to obtain better rejection of surface events. This electrode configuration produces a net drift field in the bulk. Near the surface, though, the field is both high (due to the proximity of different bias electrodes) and transverse (because alternating voltages are present on the same surface). For surface events, this configuration both decreases the lost ionization and ensures that the signal appears only on the electrodes of one side while bulk events have signal on electrodes on both sides. Both experiments have demonstrated substantially improved surface-event discrimination with this design, improving misidentification probabilities to 10^{-4} or better (see Fig. 10.8).

Using 4 kg of Ge detectors with a standard electric field configuration operated for a net 320 kg d exposure in runs at the Soudan Underground Laboratory (2000 mwe), the CDMS Collaboration obtained a 90% CL upper limit on $\sigma_{\delta n}$ of 3.8×10^{-44} cm^2 at 70 GeV c^{-2}. Using 4 kg of its new detectors with the above complex electrode configuration, EDELWEISS reached 4.4×10^{-44} cm^2 at 70 GeV c^{-2} during operation at the Laboratoire Souterrain de Modane (4800 mwe) in a 384 kg d exposure (Armengaud et al., 2011).

Both CDMS and EDELWEISS are developing plans for 100-kg and ton-scale experiments. SuperCDMS SNOLAB is being proposed with 100-kg target mass at SNOLAB. The EURECA (European Rare Event Underground Calorimeter Array) Collaboration is also proposing a 100-kg scale experiment. Longer term plans for ton-scale experiments with a reach of 10^{-47} cm^2 are afoot under the name GEODM (Germanium Observatory for Dark Matter) in the US and as part of EURECA in Europe.

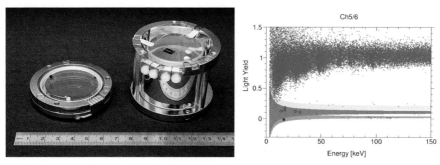

Fig. 10.9 Left: CRESST phonons and scintillation WIMP detector. The left and right assemblies hold the 2-g silicon photon absorber and the 300 g scintillating $CaWO_4$ crystal, respectively. The two assemblies join together to form a light-tight cavity with a reflective liner. Right: Scintillation yield versus recoil energy for a CRESST WIMP-search data set showing the expected locus for electromagnetic interactions (dots near yield of unity), alpha particles (yellow shaded band, top), and O (pink shaded, middle) and W (brown shaded, bottom) recoils. The ^{210}Pb recoils in the W recoil band are evident. The five large dots in the O band belong to the unexplained population of 32 events. **Figure credit:** Seidel (2010).

The CRESST (Cryogenic Rare Event Search with Superconducting Thermometers) Collaboration has developed detectors employing phonons and scintillation in $CaWO_4$. A tungsten superconducting phase-transition thermometer senses the temperature rise of the target scintillating $CaWO_4$ crystal while a similar thermometer senses the temperature rise of an adjacent silicon photon absorber crystal. The two crystals are housed in a light-tight reflective enclosure to obtain maximum light-collection efficiency (see Fig. 10.9). Due to the limited scintillation production and light-collection efficiency, the scintillation-yield band is wider than the ionization-yield band in detectors employing phonons and ionization, but there is no class of surface events with suppressed scintillation yield. The scintillation yields of the oxygen, calcium, and tungsten recoils differ, with tungsten recoils consistent with no scintillation at recoil energies below 100 keV (see Fig. 10.9). The experiment is housed at the Gran Sasso National Laboratory (LNGS) at 3000 mwe. In a 48 kg d exposure, they observed three nuclear recoil candidates, setting a limit of 5×10^{-43} cm^2 on the spin-independent WIMP–nucleon cross section (Angloher *et al.*, 2009). In a more recent 400 kg d exposure using 10 detectors (one of which is $ZnWO_4$), reported on in (Seidel, 2010), they observed a variety of background events. They saw 32 oxygen recoils, a fraction of which may be due to alpha decays, cosmogenic neutrons, or misidentified photon interactions, but the expected number of such events total only 8.7. There is also a large number of events in the expected tungsten recoil band due to ^{206}Pb nuclei generated in alpha decays in the crystal support clamps where the alpha escapes and the ^{206}Pb nucleus enters the crystal. Improvement in this alpha background will provide substantially better WIMP sensitivity. Nevertheless, these event rates are low enough to rule out an inelastic dark matter scenario for explaining the DAMA annual modulation signal.

10.6.2.3 *Liquid noble detectors*

Liquefied noble gases are easily purified underground, can be scaled to large masses in a straightforward manner, and scintillate brightly from ionizing radiation. The scintillating species are excited diatomic molecules (excimers), which are efficiently produced in both singlet and triplet spin states. The decay of a noble gas molecule to two free atoms results in photon emission; the energies of these photons are insufficient to excite the ground-state atoms in the bulk liquid, allowing them to travel to the edges of the liquid and be detected. The heavier noble liquids, such as liquid xenon (LXe) and liquid argon (LAr) also exhibit large ionization yields and electron drift speeds, on the order of $2\,\mathrm{mm}\,\mu\mathrm{m}^{-1}$. The noble liquids also allow efficient discrimination against electron recoil backgrounds, both through measurement of the ionization/scintillation ratio and through measurement of the fraction of scintillation light that is emitted by short-lived (singlet) molecules relative to the total (singlet + triplet). While noble liquids must be operated at cryogenic temperatures, the cost and effort associated with this aspect of the experiments is relatively minor, as at the relevant operation temperatures (165, 87, and 27 K for LXe, LAr, and LNe, respectively), cooling powers of hundreds of watts are easily attained. The basic properties of liquefied noble gases are shown in Table 10.2.

This proliferation of experiments using noble liquids has arisen primarily because noble liquid detectors may be easily scaled to large masses. There are two aspects to this advantage. First, large masses imply that large detector exposures may be attained in a short time. Second, self-shielding may be used as a method of background rejection. Self-shielding has been used to great effect by recent and ongoing neutrino experiments, such as Super-Kamiokande (Fukuda et al., 1999), SNO (SNO Collaboration, 2002), Borexino (Borexino Collaboration, 2008), and KamLAND (Eguchi et al., 2003). Gamma rays emitted by photodetectors and construction materials at the edges of the detector are unlikely to penetrate the outer shell of detection material to reach the inner fiducial volume. If the detection material is low in radioactivity, there is an especially low background rate in the fiducial volume. For neutrino detectors looking for signals in the few hundred keV to MeV energy range, the shielding volume must absorb the gamma rays before they reach the fidudial volume. But in a WIMP detector, with signal in the region of tens of keV, the dominant gamma ray background is due to MeV-scale gamma rays that penetrate the shielding, scatter and deposit a small amount of energy, and then escape the detector without scattering a second time. They must therefore penetrate the outer shielding not just once but twice, making the self-shielding especially effective at low energies. As X-ray photoabsorption cross sections are large, X-rays with energies comparable to those deposited by WIMP–nucleon scattering are extremely unlikely to penetrate very far into the detector. The effectiveness of self-shielding at these low energies was emphasized in the early development of the XMASS (Suzuki, 2000) and CLEAN (McKinsey and Coakley, 2005) programs. For a given event in the noble liquid, the nuclear recoil energy can be determined based on the scintillation signal $S1$. However, it is much more convenient to calibrate the

Table 10.2 Basic properties of liquefied noble gases.

	Liquid density ($g\,cm^{-3}$)	Boiling Point at 1 bar (K)	Electron mobility ($cm^2\,V^{-1}s^{-1}$)	Scintillation wavelength (nm)	Scintillation yield (photons per MeV)	Long-lived radioactive isotopes	Triplet molecule lifetime (μs)
LHe	0.145	4.2	low	80	19 000	none	13 000 000
LNe	1.2	27.1	low	78	30 000	none	15
LAr	1.4	87.3	400	128	40 000	^{39}Ar, ^{42}Ar	1.5
LKr	2.4	120	1200	150	25 000	^{81}Kr, ^{85}Kr	0.09
LXe	3.0	165	2200	178	46 000	^{136}Xe	0.03

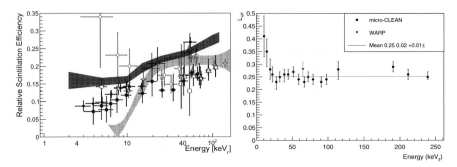

Fig. 10.10 Left: Scintillation efficiency for nuclear recoils relative to that of 122 keV gamma rays in LXe at zero field, comparing (Manzur, 2010) (●) to previous measurements from Arneodo et al. (2000) (△), Akimov et al. (2002) (□), Aprile et al. (2005) (⋆), Chepel et al. (2006) (◇) and Aprile et al. (2009) (○). Figure from (Manzur, 2010). Right: Scintillation efficiency as a function of energy from 10 to 250 keV. The weighted mean (red line) is generated from the data above 20 keVr and indicated the mean $\mathcal{L}_{\mathrm{eff}} = 0.25$. The value measured by WARP is 0.28 at 65 keVr (Brunetti et al., 2005). Figure from (Gastler, 2010).

detector using electron recoil events. The tradition in the field is to base the energy calibration on 122-keV electron recoils from a ^{57}Co γ source. The relative scintillation efficiency, $\mathcal{L}_{\mathrm{eff}}$, defined as the ratio between the electron equivalent energy (E_{ee}) and the true nuclear recoil energy (E_{nr}), becomes necessary for determining the nuclear energy scale and, therefore, the WIMP detection sensitivity. E_{ee} is inferred from the scintillation signal yield resulting from monoenergetic electron recoils. $\mathcal{L}_{\mathrm{eff}}$ has no units and is defined at zero electric field in LXe relative to γ-rays of 122 keV. If an electric field is applied to the LXe, the scintillation yields for both electron and nuclear recoils are suppressed by additional factors S_e and S_n, respectively. The quantity S_e for electron recoils of 122 keV from a ^{57}Co source has been measured very accurately. S_n has been measured for nuclear recoils of 56 keV, with electric fields up to a few kV cm^{-1} in LXe, but measurements at lower energies are more uncertain.

The ionization yields from electron recoils in LXe and LAr and their dependencies on applied electric field have been studied for several decades. However, the analogous ionization yields for low-energy nuclear recoils (the signal for WIMP dark matter searches) have only recently been investigated. Early expectations were for nuclear recoils to produce very little charge, much like alpha events. If in fact the ionization yield were close to zero, this would have led to significant difficulties, as the two-phase detectors rely on the ionization/scintillation ratio for discrimination and for position determination, and any events in zero-field regions at the periphery of the detector would yield events that could be located only with difficulty.

Fortunately, it is now clear that low-energy nuclear recoils in LXe and LAr produce sizable ionization yields, though not as large as for electron recoils. For low-energy events, with energies below the Bragg peak, the specific energy loss dE/dx decreases with energy, and the ionization density also decreases. This allows charge to be more easily extracted from the event before recombination. Furthermore, there is evidence for additional charge extraction in LXe when the electron thermalization distance becomes larger than the track length (Shutt et al., 2007). Recent measurements of the LXe nuclear recoil ionization yield (in electrons keVnr^{-1}) are summarized in Fig. 10.11. Similar behavior is seen in LAr: the WArP Collaboration (Benetti et al., 2005) reports a LAr charge-to-light ratio that can be approximated as $((S2)/(S1)[E_{\text{keV}}] = (a + b/E_{\text{keV}})[1 - \exp(E_{\text{keV}}/10)]$ where E_{keV} is the (S1) (ion) recoil energy in keV, $a = 2.1$ and $b = 670$ keV.)

Because the charge yield ionization stays large even at very low energies, and because charge signal can be amplified through proportional scintillation, charge-only detection, without $S2/S1$ discrimination, may allow more sensitive searches for light WIMPs, where energy threshold is critical. This approach was explored in the XENON10 experiment (Sorensen et al., 2010; Angle et al., 2011).

The ionization yield from nuclear recoils in LNe and LHe have not yet been measured. However, charge drift in these liquids is much slower than in LAr and LXe, as it resides in electron bubble states instead of a conduction band, so in practice charge is more difficult to detect from individual events in the lighter liquefied noble gases.

Because the ionization yield of nuclear recoils is less than that of electron recoils, the ratio of charge-to-light can be used to discriminate nuclear recoils from electron recoils. So far, the charge signal has been amplified through proportional scintillation, and so the discrimination parameter used is $S2/S1$, typically plotted as $\log(S2/S1)$ versus $S1$. A given

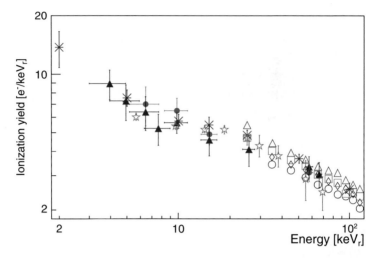

Fig. 10.11 Ionization yield as a function of recoil energy. The unit keVr given in this figure is equivalent to keVnr as defined earlier. Shown are the measured values in Manzur et al. (2010) at 1.00 kV/cm (▲) and 4.00 kV/cm (●), along with previously measured values at 0.10 kV/cm (○), 0.27 kV/cm (□), 2.00 kV/cm (△) and 2.30 kV/cm (◇) from Aprile et al. (2006). Error bars are omitted for clarity. Also shown are the ionization yields calculated by comparing the XENON10 nuclear recoil data and Monte Carlo simulations (Sorensen et al., 2009) for single elastic recoils at 0.73 kV/cm using two different methods (∗ and ⋆). Figure from Manzur et al. (2010)

event type population over a range of energies then produces a band in such a plot, with an example from the XENON10 experiment shown in Fig. 10.12. In Xe two-phase experiments, discrimination efficiency of 99% to 99.99% is typical.

Pulse shape discrimination (PSD) is possible because liquefied noble gases have two distinct mechanisms for the emission of scintillation light due to the radiative decay of singlet and triplet molecules. These two scintillation channels are populated differently for electron recoils than for nuclear recoils, allowing these two types of events to be distinguished on an event-by-event basis. This discrimination method was tested for use in an upcoming neutron EDM experiment (McKinsey et al., 2003), and used in the ZEPLIN-I single-phase LXe experiment (Alner et al., 2005). It was pointed out by McKinsey and Coakley (2005) that the long triplet time constant in LNe should enable superior PSD. Following this observation, Boulay and Hime recognized that the same advantage

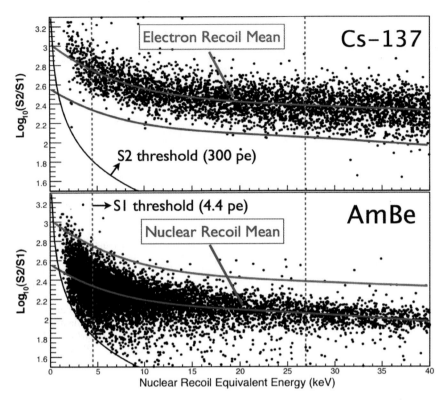

Fig. 10.12 $\log_{10}(S2/S1)$ as a function of energy for electron recoils (top) and nuclear recoils (bottom) from calibration data in the XENON10 experiment (Angle et al., 2008). The colored lines are the mean $\log_{10}(S2/S1)$ values of the electron-recoil (upper) and nuclear-recoil (lower) bands. The region between the two vertical dashed lines is the energy window (4.5–26.9 keVnr equivalent energy) chosen for the XENON10 WIMP search. An *S2* software threshold of 300 photoelectrons was also imposed (black lines). Figure from (Angle et al., 2008).

could apply to LAr, potentially allowing PSD at the very low levels of electronic-recoil contamination needed to perform a competitive dark matter experiment (Boulay and Hime, 2006). The long triplet lifetime in LNe and LAr allows the triplet and singlet components to be distinguished, even in a large detector where photon scattering can smear out the time structure of the original scintillation pulse. The most recent results on PSD and LAr are shown in Fig. 10.13. LXe, with a much shorter triplet lifetime, has less effective PSD for low energy events.

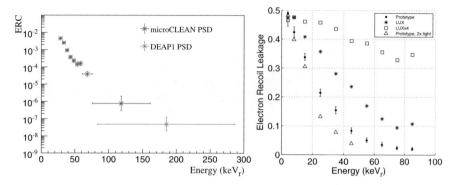

Fig. 10.13 Left: Measured LAr discrimination efficiencies from the MicroCLEAN (Lippincott et al., 2008) and DEAP-I detectors (Boulay et al., 2009). Electron recoil contamination (ERC) is defined as the expected fraction of electron recoils that are misidentified as nuclear-recoils given a 50% nuclear recoil acceptance fraction. Right: Gaussian electron-recoil leakage as a function of recoil energy for a 5 pe/keVee prototype LXe detector and projections for the prototype with two times the primary light collection, LUX, and LUXx4 (a detector with four times the dimensions of LUX) at a drift field of $0.06\,\text{kV}\,\text{cm}^{-1}$. From (Kwong et al., 2010).

Pulse shape discrimination in LXe has been most recently studied by Kwong et al. (2010). Their experimental results for electron recoil contamination, and predictions for future detectors, are shown in Fig. 10.13. In LXe, the authors conclude that use of PSD would clearly not be productive in a large detector searching for low-energy elastic WIMP recoils, so the most effective discrimination method in LXe is based on the $S2/S1$ ratio. However, PSD might be useful in inelastic recoil searches, which predict a peak at higher energy.

10.6.2.4 Two-phase detectors

Two-phase experiments use a liquefied noble gas (LXe or LAr) to search for WIMPs. In this style of detector, both prompt scintillation light ($S1$) and ionization are detected from each event in the liquid. Typically, the ionization signal is detected through proportional light ($S2$) produced when the electrons are extracted into the gas phase. With an electron drift velocity of about 2 mm per microsecond, the time between $S1$ and $S2$ gives the event depth to within an mm. Since electron diffusion is small, the proportional scintillation pulse is produced in a small spot with the same x,y coordinates as the interaction site, allowing two-dimensional localization with an accuracy of order 1 cm. With the more precise z information from the drift time measurement, the three-dimensional event localization provides additional background discrimination via fiducial volume cuts. The ratio of prompt scintillation light to ionization, can be used to discriminate between nuclear recoils and electronic recoils, allowing the reduction of radioactive backgrounds as described above in Sec. 10.6.2.3.

Two-phase xenon detectors

In the XENON10 experiment (Angle et al., 2008), operated in Gran Sasso National Laboratory in Italy, a 15 kg LXe mass was viewed by two PMT arrays, one array immersed at the bottom of the LXe volume and one array suspended above the LXe, in the gas phase. Comparison of the magnitudes of the prompt scintillation and proportional scintillation signals provided effective discrimination at roughly the 99% to 99.9% level between 10 and 50 keVr. XENON10 yielded a highly sensitive search for WIMPs, at the level of $8 \times 10^{-44} \text{cm}^2$ at $100 \,\text{GeV}\, \text{C}^{-2}$.

The ZEPLIN II experiment (Alner et al., 2007) used similar technology to XENON10. The most significant differences were that ZEPLIN II has a larger active mass (35 kg) and less light collection, having no PMT array on the bottom of the detector. ZEPLIN II was operated in the UK Boulby mine, collected over 1 ton-day of WIMP search data, and had highly competitive dark matter sensitivity.

The ZEPLIN III experiment is currently in operation in the Boulby mine, with an active mass of 12 kg and a drift field of $3900 \,\text{Vcm}^{-1}$ (significantly higher than the $700 \,\text{Vcm}^{-1}$ in XENON10 and $1000 \,\text{Vcm}^{-1}$ in ZEPLIN II). An enhanced discrimination efficiency approximately 99.99% is seen in ZEPLIN III (Lebedenko et al., 2009), perhaps due to the high drift field. Initial operation has resulted in WIMP sensitivity of $9 \times 10^{-44} \,\text{cm}^2$. A science run with low radioactivity PMTs is currently underway.

XENON100 is also currently in operation, with a 30 cm diameter, 30 cm deep time projector volume (TPC mass of 64 kg) and a surrounding LXe

Fig. 10.14 Left: Schematic of the LUX experiment. Middle: Photograph of the LUX internals, currently being assembled at the Sanford Laboratory in South Dakata. Right: R8778 photomultiplier mounted in the PMT mounting block, with attached PTFE reflector.

active veto. Recent dark matter results can be found in (Aprile *et al.*, 2011), with a sensitivity of 4×10^{-44} at 100 GeV.

The LUX dark matter experiment is a new 300 kg active mass two-phase LXe detector that will be used to extend current dark matter sensitivity two orders of magnitude beyond the current best limit to a WIMP–nucleon cross section of approximately 7×10^{-46} cm^2. The basic design of LUX is similar to that of XENON10 and ZEPLIN-II. Significant advances in the LUX design include (a) use of a water tank 8 m in diameter to shield against gammas and neutrons from the cavern instead of lead and polyethylene (b) a cathode high voltage feedthrough located outside the shielding, connected with a Xe-filled umbilical to the detector and (c) substantial use of titanium as a construction material. A schematic of the LUX detector is shown in Fig. 10.14, as well as photographs of the LUX internals.

Two-phase argon detectors

The two-phase argon detectors are similar in design to the two-phase xenon detectors, with both prompt ($S1$) and proportional ($S2$) scintillation signals detected. Advantages of argon relative to xenon include less challenging purification, more powerful electron recoil/nuclear recoil discrimination, and lower target material cost. Disadvantages include internal background from ^{39}Ar lower spin-independent WIMP cross section, and reduced sensitivity for spin-independent WIMP interactions due to the expected A^2 dependence of the cross section (because of the lower A^2).

The atmosphere contains about 1% argon, and argon is produced plentifully and cheaply by air liquefaction plants. However, atmospheric argon contains the unstable isotope ^{39}Ar, with a ^{39}Ar/Ar ratio of 8×10^{-16} (Acosta-Kane et al., 2008). The ^{39}Ar decay rate in atmospheric argon is about $1.0\,\text{day}^{-1}\,\text{kg}^{-1}$, which corresponds to 192 events $\text{keVee}^{-1}\,\text{kg}^{-1}\,\text{day}^{-1}$ at low energies relevant to dark matter searches. Hence, a liquid argon detector with 100 kg fiducial mass has about 1.4×10^8 beta decay events per year in a 20 keVee region of interest. In order to attain background-free operation, these beta decay events must be rejected through a combination of pulse-shape discrimination and ionization/scintillation ratio cuts.

Because pulse-shape discrimination dominates the rejection of ^{39}Ar events and the acceptance of a PSD cut improves exponentially with the number of detected scintillation photons, the analysis threshold at which ^{39}Ar events may be sufficiently rejected depends exponentially on the scintillation signal yield per unit energy (typically expressed in photoelectrons keVee^{-1}). This yield may be converted to photoelectrons keVnr^{-1} using the $\mathcal{L}_\text{eff}$ measurements discussed above, with necessary inclusion of any additional quenching of the nuclear recoil scintillation yield due to the drift field. The discrimination efficiency is further weakened at low energies by the convergence of the nuclear recoil and electron recoil prompt-to-late light ratios (Lippincott et al., 2008). The discrimination efficiency required to eliminate the ^{39}Ar background then necessitates raising the analysis threshold significantly relative to the trigger threshold.

In addition to needing to reject this large ^{39}Ar background, the high rate can cause event pile-up and thus places a limit on the detector mass in one module. Some representative numbers: a 300 kg detector, with 40 cm drift length, drift field of $1000\,\text{V}\,\text{cm}^{-1}$, and resulting electron drift velocity of $2\,\text{mm}\,\mu\text{s}^{-1}$, will have an ^{39}Ar event rate of 300 Hz and a typical electron drift time of 100 ms, giving a pile-up fraction of approximately 30%.

Despite this intrinsic background, several projects have begun using two-phase argon detectors. These projects benefit significantly from the several decades of development of liquid argon detectors for the ICARUS program (Arneodo et al., 2003).

The WARP (WIMP Argon Program) detectors are located in the Gran Sasso National Laboratory in Italy and use photomultipliers mounted in the gas phase to detect both S1 and S2 signals. An initial experiment was carried out, with 2.3 liters of liquid argon and an accumulated exposure of about 100 kg d (Benetti et al., 2008). The prompt scintillation signal strength was measured to be 2.5 photoelectrons/keVee. With the experience derived from this prototype detector, a 140-kg detector has been built and is currently being commissioned.

The ArDM program (Haranczyk et al., 2010) differs from WARP in that the charge signal is detected not with photomultipliers via proportional scintillation, but with LEMs (Large Electron Multipliers), which are similar to THGEMs (Thick Gas Electron Multipliers). In these devices, electrons in the gas phase are pulled through holes drilled in an insulating material, resulting in avalanches and substantial electron gain. The LEM-based charge readout is expected to allow high spatial granularity, which may aid in background rejection. As in WARP, the *S1* signal is detected with photomultipliers mounted on the bottom of the detector. Based on initial measurements, an *S1* signal yield of 1.0 photoelectrons keVee^{-1} is projected. The cathode voltage is supplied with a Greinacher (Cockcroft–Walton) circuit, which is immersed in the liquid argon and has been tested up to 400 kV, enabling a drift field of 3 kV cm^{-1}. The ArDM 1 ton prototype instrument is being tested at CERN, and plans for its underground operation are in preparation, to be located in either Canfranc (Spain) or SUNLAB (Poland).

The DarkSide program uses the same basic approach described above, but with the addition of depleted argon, QUPID photosensors, and an organic scintillator veto. It has been demonstrated that certain sources of underground argon have levels of ^{39}Ar that are lower than is found in atmospheric argon, with a factor of ~ 25 depletion demonstrated to-date (Acosta-Kane et al., 2008). This will reduce the necessary rejection power in proportion, as well as the pile-up rate. For the example described above, a 100-kg fiducial mass detector filled with argon reduced in ^{39}Ar by a factor

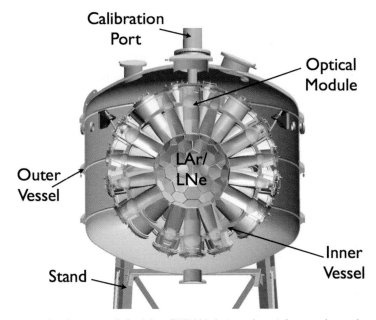

Fig. 10.15 A schematic of the Mini-CLEAN design. A stainless steel vessel contains 92 photomultipliers viewing an active volume of 500 kg of LAr or LNe through acrylic light guides. A wavelength shifting film is coated on the inside surface of acrylic plates, which fit together to form a 92-sided expanded dodecahedron pattern.

of 25 would only have to reject 5.6×10^6 events in one year instead of 1.4×10^8. The QUPID photosensors rely on photoelectron amplification in an avalanche photodiode, giving better defined photoelectron gain. QUPIDs have been shown to have low radioactive background, at the level of $< 1\,\text{mBq}\,\text{kg}^{-1}$ of U. A 50-kg DarkSide detector is in design and is planned to be deployed within the Counting Test Facility in Gran Sasso, which can be used as a neutron veto when filled with boron-doped organic scintillator.

10.6.2.5 *Single-phase argon detectors*

The Mini-CLEAN detector (McKinsey, 2007) will consist of a spherical vessel filled with purified LNe or LAr at a temperature of 27 K or 87 K, respectively. The center of the vessel will be viewed by 92 photomultipliers (PMTs), each 20 cm in diameter and immersed in the cryogen. In the center of the spherical vessel will be mounted a soccer-ball-shaped array of acrylic plates. Tetraphenyl butadiene (TPB), a wavelength shifting fluor, will be evaporated onto the inward-pointing surface of each acrylic plate. Light from each wavelength shifter plate will be transported to the nearest PMT via an acrylic light guide. Each light guide will map the hexagonal or pentagonal wavelength shifter plates onto the hemispherical PMT. The wavelength shifter plates will enclose a central liquid mass of about 500 kg. A conceptual sketch of Mini-CLEAN is shown below in Fig. 10.15. MiniCLEAN is currently under construction at SNOLAB, with operation expected in 2012.

Ionizing radiation events within the wavelength shifter plate array will cause scintillation in the extreme ultraviolet (80 nm in liquid neon or 128 nm in liquid argon). The ultraviolet scintillation light will be absorbed by the wavelength shifter and re-emitted at a wavelength of 440 nm. The photon-to-photon conversion efficiency is close to unity for these wavelengths (McKinsey et al., 1997). The blue light will then be detected by the PMTs. For electron-like events, a signal is projected of more than 5 photoelectrons keVee^{-1} in LAr and more than 2.5 photoelectrons/keV in LNe. The "scintillation-only" event detection in MiniCLEAN will minimize event pile-up since event timing is set by the triplet molecule decay lifetime and not by electron drift speed as in two-phase detectors.

The high light yield enabled by the spherical geometry will maximize pulse-shape discrimination efficiency. Based on previous studies of pulse-shape discrimination in the MicroCLEAN and DEAP-I experiments (see Fig. 10.13), an analysis threshold of about 50 keVnr is needed in order to sufficiently reject the ^{39}Ar background. This gives an expected dark matter sensitivity of 2×10^{-45} cm^2 at 100 GeV c^{-2}, in 300 kg d of low-background operation. Background from radon daughters plated on the TPB will be removed using positon resolution-based PMT hit pattern-based algorithms. Neutron backgrounds will be reduced and rejected through shielding by the acrylic light guides, fiducialization, and tagging of neutron capture.

After initial operation with LAr, MiniCLEAN will be operated with LNe to test the A^2 dependence of any possible WIMP signal and to test liquid neon operation for the future CLEAN experiment (McKinsey and Doyle, 2000; McKinsey and Coakley, 2005). CLEAN will contain 40 tons

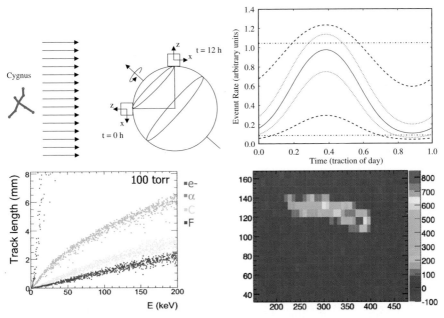

Fig. 10.16 Top left: Illustration of motion of the Earth relative to WIMP "wind" and its orientation relative to the lab frame. Figure from (Ahlen et al., 2010). Top right: Rate as a function of time of the day for seven recoil angles fixed with respect to the Earth's equatorial plane. The solid, dotted, dashed, and dash-dotted lines correspond to 0, ±18, ±54, and ±90 degrees relative to this plane, with negative angles being above the equatorial plane and positive angles below. The ±90 directions are coaligned with the Earth's rotation axis. The calculation uses a 100 GeV c^{-2} WIMP mass and CS_2 target gas. Figure from (Vergados and Faessler, 2007). Bottom left: Stimulated range of particles of various types in CF_4. There are dramatic differences between electrons, alpha particles, and nuclei. Figure from (Sciolla, 2008). Bottom right: CCD image of a 550 keV neutron-induced nuclear recoil in DMTPC. 100 pixels corresponds to 6 mm. The head-tail effect due to variation in dE/dx is apparent. Figure from (Dujmic et al., 2008).

of liquid argon. As with MiniCLEAN, the LNe can be replaced with LNe for A^2 spectral tests. The LNe mass will also be sufficient to measure the pp-solar neutrino flux.

The single-phase DEAP-3600 experiment will also be built at SNOLAB. DEAP-3600 will have a 1 ton fiducial mass and is expected to be sensitive to WIMPs at the level of 1×10^{-46} cm^2. Notable differences from MiniCLEAN include liquid argon-only operation, liquid argon containment within an acrylic vessel, and *in situ* mechanical radon daughter removal from the vessel's inner surface.

10.6.3 *Direction-sensitive detectors*

Direction sensitivity is the ultimate tool for WIMP direct detection because it provides a way to harness the large, unique signal provided by diurnal modulation as well as to study the local three-dimensional velocity structure of the galactic halo from WIMPs if discovered. The potential of directional sensitivity is illustrated in Fig. 10.16, which shows the expected diurnal dependence of the rate of nuclear recoils in CS_2 caused by interactions of a $100\,\text{GeV}\,c^{-2}$ WIMP as a function of the recoil direction relative to the direction of our motion through the galaxy (towards Cygnus). The salient distinction from annual modulation lies in the size of the modulation size of the modulation: while the annual modulation signal is only a few percent of the mean signal, diurnal modulation can be much larger due to the comparable size of the solar system's velocity and the WIMPs' rms velocity. In the lab frame, this feature is manifested as an extreme directionality of the WIMP "wind."

To make use of diurnal modulation, the experiment must measure the direction of the recoiling particle. To date, all directional techniques employ gas targets at a pressure of tens of mbar. At this pressure, the track lengths of recoiling nuclei in gas are of order mm in length, depending on the energy and the nuclear species. In contrast, recoiling electrons of comparable energy have tracks many cm long. This difference is illustrated in Fig. 10.16. As these recoiling particles travel through the gas, they ionize atoms and leave an ionization track. The track is usually imaged via the time-projection chamber technique, wherein one applies an electric field to drift the track to an avalanche region where it can be recorded by detection of the avalanche charge or by imaging of the scintillation light produced in the avalanche.[4] The xy projection of the track is constructed from the detected charge while the z projection is obtained from timing information. The energy loss dE/dx is dependent on the particle speed

---

[4]We note that this avalanche process is different from the electroluminescence used in two-phase liquefied noble gases. Electroluminescence is a linear process providing stable gain while avalanche gain is an exponential process subject to large gain uncertainty and fluctuations.

and thus the recovered ionization shows a dependence on position along the track, making it possible to distinguish the start and end of the track (so-called "head-tail" discrimination).

Diffusion is a challenge when attempting to maintain the shape of nuclear-recoil tracks as they drift to the avalanche region. For a typical target gas like CF_4, this limits the drift region to a height of 20–25 cm. To circumvent this limit, the DRIFT Collaboration has chosen to use CS_2 because of its high affinity for electrons. Free electrons attach to CS_2 and are drifted with the ion. The heavy ion is much less sensitive to diffusion, making drift volumes of 1-m scale possible. However, the large ion mass results in very long drift times, ms instead of the μs typical for electrons, and thus a readout with good low-frequency performance is required.

Another important challenge, and in fact the component that distinguishes the various efforts, is the means used to image the track in the avalanche region at the end of the drift volume. DRIFT uses a multi-wire proportional chamber, a set of three parallel wire grids held at large voltage differences. The large voltage gradient near the anode wires causes a charge avalanche. This avalanche charge at the anode and the associated image charge at the cathode are sensed. The image of the track in the plane parallel to the grid is reconstructed from the charge detected, while the profile in the perpendicular (z) direction is reconstructed from the signal timing. The NEWAGE and MIMAC groups use micro-patterned gaseous detectors to detect and amplify the ionization track, using electronic readout as DRIFT does. DMTPC uses a CCD to image the scintillation light emitted during the avalanche. This is a simple and inexpensive readout technique, but it does not recover the z profile of the track and thus loses important directional information because CCDs are too slow to sense the microsecond-scale time structure in which the z profile of the track is encoded. DMTPC is adding readout of the avalanche charge to recover this information.

All four groups have demonstrated track reconstruction at 100 keV recoil energy. The DRIFT, DMTPC, and MIMAC groups have additionally demonstrated head-tail separation at this recoil energy. The DRIFT and DMTPC groups have done both with neutrons creating nuclear recoils, while MIMAC has used low-energy photons to create electron recoil tracks of comparable length to expected nuclear-recoils. An example of the reconstruction of an energetic nuclear recoil track is shown in Fig. 10.16.

Acknowledgments

The authors gratefully acknowledge the assistance of Ms. Daphne Klemme in the preparation of this manuscript.

References

Aalseth, C. E. *et al.* (2011). *Phys. Rev. Lett.* **106** (2011) 131301.
Acosta-Kane, D. *et al.* (2008). *Nucl. Instr. Meth. A* **587**, p. 46.
Ahmed, Z. *et al.* (2011). *Phys. Rev. Lett.* **106** p. 131302.
Ahlen, S. *et al.* (1987). *Phys. Lett. B* **195**, p. 603.
Ahlen, S. *et al.* (2010). *Int. J. Mod. Phys. A* **25**, p. 1.
Akerib, D. S. *et al.* (2010). *Phys. Rev. D* **82**, p. 122004.
Akimov, D. *et al.* (2002). *Phys. Lett. B* **524**.
Alner, G. J. *et al.* (2005). *Astropart. Phys.* **23**, p. 444.
Alner, G. J. *et al.* (2007). *Astropart. Phys.* **28**, p. 287.
Amare, J. *et al.* (2006). *Journal of Physics: Conference Series* **39**, pp. 123–125.
Angle, J. *et al.* (2008). *Phys. Rev. Lett.* **100**, p. 021303.
Angle, J. *et al.* (2011). arXiv:1104.3088
Angloher, G. *et al.* (2009). *Astrop. Phys.* **31**, p. 270.
Arkani-Hamed, N. *et al.* (2009). *Phys. Rev. D* **79**, p. 015014.
Armengaud, E. (2010). http://indico.in2p3.fr/contributionDisplay.py?contribId=77&sessionId=26&confId=1565, presentation at *Identification of Dark Matter*, Montpellier, July.
Arneodo, F. *et al.*, (2000). *Nucl. Inst. Meth. A* **449**, p. 147.
Arneodo, F. *et al.* (2003). *Nucl. Inst. Meth. A* **508**, p. 287.
Aprile, E. *et al.* (2005). *Phys. Rev. D* **72**.
Aprile, E. *et al.* (2006). *Phys. Rev. Lett.* **97**.
Aprile, E. *et al.* (2009). *Phys. Rev. C* **79**, p. 045807.
Aprile, E. *et al.* (2010). *Phys. Rev. Lett.* **105**, p. 131302.
Aprile, E. *et al.* (2011). arXiv:1104.2549.
Aubin, F. *et al.* (2008). *New J. Phys.* **10**, p. 103017/1.
Baudis, L. *et al.* (1999). *Phys. Rev. D* **59**, p. 022001/1.
Baudis, L. *et al.* (2001). *Phys. Rev. D* **63**, p. 022001/1.
Behnke, E. *et al.* (2008). *Science* **319**, p. 933.
Benetti, P. *et al.* (2008). (WARP Collaboration), *Astopart. Phys.* **28**, p. 495.
Bernabei, R. *et al.* (2003). *Riv. Nuovo Cimento* **26**, p. 1.
Bernabi, R. *et al.* (2010). *Eur. Phys. J. C* **67**, p. 39.
Borexino collaboration (2008). *Phys. Rev. Lett.* **101**, p. 091302.
Boulay, M. G. and Hime, A. (2006). *Astroparticle Physics* **25**, p. 179.

Boulay, M. G. et al. (2009). arXiv:0904.2930.
Brunetti, R. et al. (2005). *New Ast. Rev.* **49**, p. 265.
Cabrera, B., Miller, A. and Young, B. (eds.) (2009). *AIP Conference Proceedings: The Thirteenth International Workshop on Low Temperature Detectors — LTD13*, AIP, Melville, NY, Vol. 1185.
Caldwell, D. O. et al. (1988). *Phys. Rev. Lett.* **61**, p. 510.
CDMS II Collaboration. (2010). *Science* **327**, p. 1619.
Chang, S., Weiner, N. and Yavin, I. (2010). *Phys. Rev. D* **82**, p. 125011.
Chepel, V. et al. (2006). *Astropart. Phys.* **26**, p. 58.
Collar, J. I. and McKinsey, D. N. (2010a). astro-ph.CO/1005.0838.
Collar, J. I. and McKinsey, D. N. (2010b). astro-ph.CO/1005.3723.
Da Silva, A. (1996). Ph.D. thesis, University of British Columbia.
Dahl, E. (2010). http://www.physics.ucla.edu/hep/dm10/talks/dahl.pdf, presentation at *Ninth UCLA Symposium on Sources and Detection of Dark Matter and Dark Energy in the Universe*, Marina del Rey.
Drukier, A. and Stodolsky, L. (1984). *Phys. Rev. D* **30**, p. 2295.
Dujmic, D. et al. (2008). *Astrop. Phys.* **30**, p. 58.
EDELWEISS Collaboration. (2009). *Phys. Lett. B* **681**, p. 305.
Eguchi, K. et al. (2003). *Phys. Rev. Lett.* **90**, p. 021802.
Enss, C. (ed.) (2005). *Cryogenic Particle Detection*, Springer, Berlin.
Formaggio, J. A. and Martoff, C. J. (2004). *Annu. Rev. Nucl. Part. Sci.* **54**, p. 361.
Fukuda, Y. et al., (1999). *Phys. Rev. Lett.* **82**, p. 1810.
Gastler, D. et al. (2010). arXiv:1004.0373.
Glaser, D. A. (1953). *Phys. Rev.* **91**, p. 762.
Goodman, M. W. and Witten, E. (1985). *Phys. Rev. D* **31**, p. 3059.
Haranczyk, M. et al. (2010). arXiv:1006.5335v1.
Heusser, G. (1995). *Annu. Rev. Nucl. Part. Sci.* **45**, p. 543.
Hooper, D. et al. (2010). arXiv:1007.1005.
Johnson, K. F. (2010). *J. Phys. G* **037**, p. 075021/10.
Kim, H. J. et al. (2008). *IEEE Transactions on Nuclear Science* **55**, p. 3.
Kittel, C. and Kroemer, H. (1980). *Thermal Physics*, W. H. Freeman, New York.
Kwong et al. (2010). *Nucl. Inst. Meth.* **612**(2), 328–333.
Kudryavtsev, V. A., Robinson, M. and Spooner, N. J. C. (2009). arXiv:0912.2983.
Lebedenko, V. et al. (2008). arXiv: 0812.1150.
Lebedenko, V. N. et al. (2009). *Phys. Rev. D* **80**, p. 052010.
Lee, H. S. et al. (2007). *Phys. Rev. Lett.* **99**, p. 091301.
Lewin, J. D. and Smith, P. F. (1996). *Astrop. Phys.* **6**, p. 87.
Lindhard, J. et al. (1963). *Mat. Fys. Medd. K. Dan. Vidensk. Selsk.* **33**, p. 10.
Lippincott, W. et al. (2008). *Phys. Rev. C* **78**, p. 035801.
Manzur, A. et al. (2010). *Phys. Rev. C* **81**, p. 025808.
McKinsey, D. N. et al. (1997). *Nucl. Inst. Meth. B* **132**, p. 351.
McKinsey, D. N. and Doyle, J. M. (2000). *J. Low Temp. Phys.* **118**, p. 153.
McKinsey, D. N. et al. (2003). *Phys. Rev. A* **67**, p. 062716.
McKinsey, D. N. and Coakley, K. (2005). *Astroparticle Physics* **22**, p. 355.
McKinsey, D. N. (2007). *Nuclear Physics B (Proc. Suppl.)* **173**, pp. 152–155.
Mei, D. and Hime, A. (2006). *Phys. Rev. D* **73**, p. 053004/1.

Morales, A. et al. (2000). *Phys. Lett. B* **489**, p. 268.
Ni, K. et al. (2007). *Nucl. Inst. Meth. A* **582**, p. 569.
Ramberg, E. (2009). http://snolab2009.snolab.ca/snolab-workshop-09/SNOLAB 2009_Ramberg.pdf, presentation at *Eighth SNOLAB Workshop on Science and Experiments for SNOLAB*, Sudbury.
Sciolla, G. (2008). http://agenda.albanova.se/getFile.py/access?contribId=363&sessionId=252&resId=250&materialId=slides&confId=355, presentation at *Identification of Dark Matter*, Stockholm, August.
Seidel, W. (2010). http://indico.in2p3.fr/contributionDisplay.py?contribId=195&sessionId=9&confId=1565, presentation at *Identification of Dark Matter*, Montpellier, July.
Seitz, F. (1958). *Phys. Fl.* **1**, p. 2.
Shutt, T. et al. (2007). *Nucl. Inst. Meth. A* **579**, pp. 451–453.
Smith, D. and Weiner, N. (2001). *Phys. Rev. D* **64**, p. 043502.
Smith, M. C. et al. (2007). *Mon. Not. Roy. Astron. Soc.* **379**, p. 755.
SNO collaboration (2002). *Phys. Rev. Lett.* **89**, No. 1, p. 011301.
Solovov, V. N. et al. (2004). *Nucl. Inst. Meth. A* **516**, p. 462.
Sonnenschein, A. (1996). http://cryodet.lngs.infn.it/march2006/agenda/ Sonnenschein.pdf, presentation at *Cryogenic Liquid Detectors for Future Particle Physics*, Gran Sasso.
Sorensen, P. et al., *Nucl. Inst. Meth. A* (2009). **601**, p. 339.
Sorensen, P. et al. (2010). arXiv:1011.6439.
Suzuki, Y. (2000). arXiv:hep-ph/0008296.
Vergados, J. D. and Faessler, A. (2007). *Phys. Rev. D* **75**, p. 055007/1.

CHAPTER 11

THE ACCELERATING UNIVERSE

DRAGAN HUTERER
Department of Physics
University of Michigan
Ann Arbor, MI 48109, USA

In this article we review the discovery of the accelerating Universe using Type Ia supernovae. We then outline ways in which dark energy — component that causes the acceleration — is phenomenologically described. We finally describe principal cosmological techniques to measure large-scale properties of dark energy. This chapter therefore complements articles by Caldwell and Linder (2010) in this book who describe theoretical understanding (or the lack thereof) of the cause for the accelerating Universe.

11.1 Introduction and History: Evidence for the Missing Component

Inflationary theory (Guth, 1981) explains how tiny quantum-mechanical fluctuations in the early Universe could grow to become structures we see on the sky today. One of the factors that motivated inflation is that it predicts that the total energy density relative to the critical value is unity, $\Omega \equiv \rho/\rho_{\rm crit} = 1$. This inflationary prediction convinced many theorists that the Universe is precisely flat.

Around the same time that inflation was proposed, a variety of dynamical probes of the large-scale structure in the Universe were starting to indicate that the *matter* energy density is much lower than the value needed to make the Universe flat. Perhaps the most specific case was made by the measurements of the clustering of galaxies, which are sensitive to the parameter combination $\Gamma \equiv \Omega_M h$, where Ω_M is the energy density in matter relative to critical, and h is the Hubble constant in units of 100 km/s/Mpc. The measured value at the time was $\Gamma \simeq 0.25$ (with rather large errors). One way to preserve a flat universe was to postulate that the Hubble constant itself was much lower than the measurements indicated ($h \sim 0.7$), so that $\Omega_M = 1$ but $h \sim 0.3$ (Bartlett *et al.*, 1995). Another possibility was the presence of Einstein's cosmological constant (see the Caldwell article in this book), which was suggested as far back as 1984 as the possible missing ingredient that could alleviate tension between data and matter-only theoretical predictions (Peebles, 1984; Turner *et al.*, 1984) by making the universe older, and allowing flatness with a low value of the matter density.

11.2 Type Ia Supernovae and Cosmology

The revolutionary discovery of the accelerating Universe took place in the late 1990s, but to understand it and its implications, we have to step back a few decades.

Type Ia supernovae

Type Ia supernovae (SN Ia) are explosions seen to distant corners of the Universe, and are thought to be cases where a rotating carbon–oxygen white dwarf accretes matter from a companion star, approaches the Chandrasekhar limit, starts thermonuclear burning, and then explodes. The Ia nomenclature refers to spectra of SN Ia, which have no hydrogen, but show a prominent Silicon (Si II) line at 6150Å.

SN Ia had been studied extensively by Fritz Zwicky who also gave them their name (Baade and Zwicky, 1934), and by Walter Baade, who noted that SN Ia have very uniform luminosities (Baade, 1938). Light from type Ia supernovae brightens and fades over a period of about a month; at its peak flux, an SN Ia can be a sizeable fraction of the luminosity of the entire galaxy in which it resides.

Standard candles

It is very difficult to measure *distances* in astronomy. It is relatively easy to measure the angular location of an object; we can also get excellent measurement of the object's redshift z from its spectrum, by observing the shift of known spectral lines due to expansion of the Universe ($1 + z = \lambda_{\text{observed}}/\lambda_{\text{emitted}}$). But the distance measurements traditionally involve empirical — and uncertain — methods: parallax, period–luminosity relation of Cepheids, main-sequence fitting, surface brightness fluctuations, etc. Typically, astronomers construct an unwieldy "distance ladder" to measure distance to a galaxy: they use one of these relations (say, parallaxes — apparent shifts due to Earth's motion around the Sun) to calibrate distances to nearby objects (e.g. variable stars Cepheids), then go from those objects to more distant ones using another relation that works better in that distance regime. In this process the systematic errors add up, making the distance ladder flimsy.

"Standard candles" are hypothetical objects that have a nearly fixed luminosity (that is, fixed intrinsic power that they radiate). Having standard candles would be useful since then we could infer distances to objects just by using the flux–luminosity inverse square law

$$f = \frac{L}{4\pi d_L^2}, \tag{11.1}$$

where d_L is the luminosity distance which can be predicted given the object's redshift and contents of the Universe (i.e. energy densities of matter and radiation relative to the critical density which makes the Universe spatially flat, as well as other components such as radiation). In fact, we do not even need to know the luminosity of the standard candle to be able to infer *relative* distances to objects.

In astronomy, flux is often expressed in terms of apparent magnitude — a logarithmic measure of flux, and luminosity is related to the absolute magnitude of the object. So, in astronomical units, Eq. (11.1) reads

$$m - M = 5 \log_{10}\left(\frac{d_L}{10\,\text{pc}}\right), \tag{11.2}$$

where the quantity on the left-hand side is also known as the *distance modulus*. For an object that is 10 parsecs away, the distance modulus is zero. For a standard candle, the absolute magnitude M (or, equivalently, luminosity L) is known to be approximately the same for each object. Therefore, measurements of the apparent magnitude to each object provide information about the luminosity distance, and thus the makeup of the Universe.

Finding SN

The fact that SN Ia can potentially be used as a standard candle has been realized long ago, at least as far back as the 1970s (Kowal, 1968; Colgate, 1979). However, a major problem is to find a method to schedule telescopes to discover SN before they happen. If we point a telescope at a galaxy and wait for the SN to go off, we will wait several hundred years. There had been a program in the 1980s to find supernovae (Norgaard-Nielsen *et al.*, 1989) but, partly due to inadequate technology and equipment available at the time, it discovered only one SN, and after the peak of the light-curve.

The first major breakthrough came in the 1990s when two teams of SN researchers Supernova Cosmology Project (SCP; led by Saul Perlmutter and organized in the late 1980s) and High-z Supernova Search Team (Highz; organized in the mid 1990s and led, at the time, by Brian Schmidt) developed an efficient approach to use world's most powerful telescopes working in concert to discover and follow up high-redshift SN, and thus complement the existing efforts at lower redshift led by the Calán/Tololo collaboration (Hamuy *et al.*, 1996). These teams had been able to essentially guarantee that they would find batches of SN in each run. [For popular reviews of these exciting developments, see Kirshner (2002) and Perlmutter and Schmidt (2003).]

The second breakthrough came in 1993 by Mark Phillips, an astronomer working in Chile (Phillips, 1993). He noticed that the SN luminosity — or absolute magnitude — is correlated with the decay time of SN light curve. Phillips considered the quantity Δm_{15}, the attenuation of the flux of SN between the light maximum and 15 days past the maximum. He found that

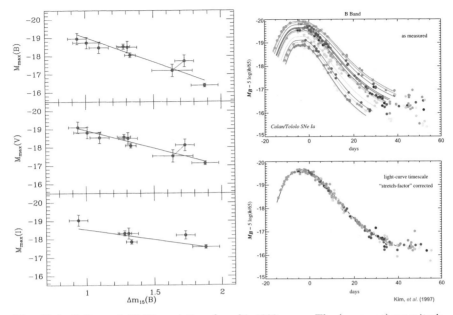

Fig. 11.1 Left panel: Phillips relation, from his 1993 paper. The (apparent) magnitude of Type Ia supernovae is correlated with Δm_{15}, the decay of the light curve 15 days after the maximum. Right panel: Light curves of a sample of SN Ia before correction for stretch (essentially, the Phillips relation; top), and after (bottom). **Figure credit:** Kim (2008).

Δm_{15} is strongly correlated with the intrinsic brightness of SN; see the left panel of Fig. 11.1. The "Phillips relation" roughly goes as

Broader is brighter.

In other words, supernovae with broader light-curves have a larger intrinsic luminosity. One way to quantify this relation is to use a "stretch" factor which is a (calibration) parameter that measures width of a light curve (Perlmutter et al., 1999); see the right panel of Fig. 11.1. By applying the correction based upon the Phillips relation, astronomers found that the intrinsic dispersion of SN, which is of order ~ 0.5 magnitudes, can be brought down to $\delta m \sim 0.2$ magnitudes once we correct each SN luminosity using its stretch factor. Note that the final dispersion in magnitudes corresponds to the error in distance of $\delta d_L/d_L = (\ln(10)/5)\,\delta m \simeq 0.5\,\delta m \sim 0.1$. The Phillips relation was the second key ingredient that enabled SN Ia to achieve precision needed to probe contents of the Universe accurately.

The third key invention was the development of techniques to correct SN magnitudes for dimming by dust, or "extinction", out of multi-color observation of SN light (Riess et al., 1996a,b). Such corrections are an important part of SN cosmology to this day (Jha et al., 2007; Guy et al., 2007; Conley et al., 2008).

Finally, the fourth and perhaps most important ingredient for the discovery of dark energy was development and application of charge-coupled devices (CCDs) in observational astronomy. Both teams of SN hunters used the CCDs, which had originally been installed at telescopes at Kitt Peak and Cerro Tololo (Kirshner, 2009).

Some of the early results came out in the period of 1995–1997; however these results were based on a handful of high-redshift SN and had large errors [e.g. Perlmutter et al. (1997); Garnavich et al. (1998); Perlmutter et al. (1998)].

The discovery of dark energy

The definitive results, based on ~ 50 SN by either team that combined the nearby sample previously observed by the Calán/Tololo collaboration and the newly acquired and crucial sample of high-redshift SN, came out soon thereafter (Riess et al., 1998; Perlmutter et al., 1999). The results of the two teams agreed, and indicated that more distant SN are dimmer than would be expected in a matter-only universe; see Fig. 11.2. In other words, the Universe's expansion rate is speeding up, contrary to expectation from the matter-dominated universe with *any* amount of matter and regardless of curvature.

Phrased yet differently, the data indicate that the Universe is accelerating — that is presence of a new component with strongly negative pressure. This can easily be seen from the *acceleration equation*, which is one of Einstein's equations applied to the case of the homogeneous universe

$$\frac{\ddot{a}}{a} = -\frac{4\pi G}{3}(\rho + 3p) = -\frac{4\pi G}{3}(\rho_M + \rho_{\rm DE} + 3p_{\rm DE}), \qquad (11.3)$$

where ρ and p are the energy density and pressure of components in the universe, assuming they are matter and a new component which we call dark energy (radiation is negligible relative to matter at redshifts much less

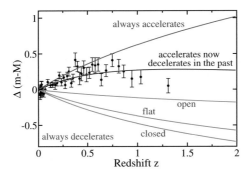

Fig. 11.2 Evidence for transition from deceleration in the past to acceleration today. The blue line shows a model that fits the data, where acceleration happens at late epochs in the history of the Universe (i.e. starting a few billion years ago, and billions of years after the Big Bang). For illustration, we also show three representative matter-only models in green, with open, closed and flat geometry. Finally, the red curve shows a model that always exhibits acceleration, and it too does not fit the SN data which show a characteristic "turnover" in the magnitude versus redshift plot. The plot uses binned data from the Union2 compilation (Amanullah et al., 2010) containing 557 SN.

than $\sim 10^3$, and the pressure of matter is always negligible). If the Universe is accelerating, then $\ddot{a} > 0$, and the only way it can be is when the *pressure of the new component is strongly negative*. Phrased in terms of the equation of state, $w \equiv p_{\rm DE}/\rho_{\rm DE} < -1/3$ regardless of the density of matter ρ_M.

The discovery of the accelerating Universe with supernovae was a watershed event in modern cosmology, and the aforementioned two discovery papers are among the most cited physics papers of all time. This component that makes the Universe accelerate was soon named "dark energy" by the theoretical cosmologist Michael Turner (Huterer and Turner, 1999).

The SN data are illustrated in Fig. 11.2, where upwards of 500 SN measurements from the Union2 compilation (Amanullah et al., 2010) have been binned in redshift. The blue line shows a model that fits the data, where acceleration happens at late epochs in the history of the Universe (i.e. starting a few billion years ago, and billions of years after the Big Bang). For illustration, we also show three representative matter-only models in green, with open, closed and flat geometry, neither of which fit the data well. Finally, the red curve shows a model that always exhibits acceleration, and it too does not fit the SN data which show a characteristic "turnover" in the magnitude versus redshift plot.

Observable and inferred quantities with SN Ia

The luminosity distance d_L is related to the cosmological parameters via

$$d_L = (1+z)\frac{H_0^{-1}}{\sqrt{\Omega_K}} \sinh\left[\sqrt{\Omega_K}\int_0^z \frac{dz'}{\sqrt{\Omega_M(1+z)^3 + \Omega_{\rm DE}(1+z)^{3(1+w)} + \Omega_R(1+z)^4 + \Omega_K(1+z)^2}}\right] \quad (11.4)$$

where the key term in this expression featuring $\sinh(x)$ for $\Omega_K > 0$ (open universe) effectively turns into $\sin(x)$ (closed universe; $\Omega_K < 0$) or just x (flat universe; $\Omega_K = 0$). Here Ω_M, Ω_R, and $\Omega_{\rm DE}$ are the energy densities of matter (visible plus dark), radiation (mainly cosmic microwave background (CMB) photons), and dark energy relative to critical density, and $\Omega_K = 1 - \Omega_M - \Omega_R - \Omega_{\rm DE}$.

Now Eq. (11.2) can be rewritten as

$$m \equiv 5\log_{10}(H_0 d_L) + \mathcal{M}, \quad (11.5)$$

where the "script-M" factor is defined as

$$\mathcal{M} \equiv M - 5\log_{10}\left(\frac{H_0}{\text{Mpc}^{-1}}\right) + 25. \quad (11.6)$$

Note that $\mathcal{M}$ is a dummy parameter that captures *two* uncertain quantities: the absolute magnitude (i.e. intrinsic luminosity) of a supernova, M, and the Hubble constant H_0. We typically do not know $\mathcal{M}$, and we need to marginalize (i.e. integrate) over all values of this parameter in the cosmological analysis.

The situation is now clear: astronomers measure m, which is inferred, for example, from the flux at the peak of the light curve. Then they measure the redshift of SN host galaxy. With the sufficient number of SN measurements, they can marginalize over the parameter $\mathcal{M}$ and be left with, effectively, measurements of luminosity distance versus redshift. A plot of either $m(z)$ or $d_L(z)$ is called the Hubble diagram.

These results have been greatly strengthened since, with many hundreds of SN Ia currently indicating same results, but with smaller errors, compared to the original 1998–1999 papers (Knop *et al.*, 2003, Riess *et al.*, 2004; Astier *et al.*, 2006; Riess *et al.*, 2007; Wood-Vasey *et al.*, 2007; Kessler *et al.*, 2009; Hicken *et al.*, 2009; Amanullah *et al.*, 2010). Meanwhile, other cosmological probes have come in with results confirming the SN results (see the right panel of Fig. 11.4).

Systematic errors

Systematic errors that can creep up in SN observations, and stand in the way of making SN Ia a more precise tool of cosmology. Here we list a few prominent sources of error, and ways in which they are controlled:

- Extinction: Is it possible that SN appear dimmer simply because of extinction by dust particles scattered between us and distant SN? Fortunately there are ways to stringently control (and correct for) extinction, by observing SN in different wavelength bands. But also, if extinction were to be responsible for the appearance of dimming, then we would expect more distant SN to appear uniformly dimmer. Moreover, a "turnover" in the SN Hubble diagram has been clearly observed [e.g. Riess *et al.* (2004)] indicating that the Universe is matter-dominated at high z. The turnover cannot easily be explained by extinction.

- Evolution: Is it possible that SN evolve, so that we are seeing a different population at higher redshift that is intrinsically dimmer (violating the assumption of a standard candle)? SN Ia do not own a "cosmic clock"; rather, they respond to their local environment, in addition to being ruled by the physics of accretion/explosion. So, by observing various signatures, in particular in SN spectra, researchers can identify local environmental conditions, and even go so far to compare only like-to-like SN (resulting, potentially, in several Hubble diagrams, one for each subspecies). First such comparisons have been made recently.

- Typing: Is it possible that non-Ia SN have crept into the samples used for dark energy analysis? This question is rather easy to answer, as SN Ia possess characteristic spectral lines which uniquely identify these SN. Accurate typing, however, becomes more challenging for SN surveys which cannot afford to take the spectra of all SN; upcoming and future imaging surveys such as the Dark Energy Survey (DES) or Large

Synoptic Survey Telescope (LSST) are examples. For those surveys, one will have to apply sophisticated tests based on photometric information alone to establish whether or not a given supernova is Type Ia.

- K-corrections: As SN Ia are observed at larger and larger redshifts, their light is shifted to longer wavelengths. Since astronomical observations are normally made in fixed band passes on Earth, corrections need to be applied to account for the differences caused by the spectrum shifting within these band passes, and error in these corrections needs to be tightly controlled.
- Gravitational lensing: Distant SN are gravitationally lensed by matter along the line of sight, making them magnified or demagnified, and thus appearing brighter or dimmer. The lensing effect goes roughly as z^2 and is non-negligible only for high-z SN; $z \gtrsim 1.2$. The *mean* magnification is zero (owing to a theorem that the total light is conserved), but the distribution is skewed, meaning that most SN get demagnified but occasional ones get strongly magnified. The way to protect against biases due to gravitational lensing is to seek "safety in numbers" (Holz and Linder, 2005): simply put, if we collect enough SN at any given redshift (in practice, ~ 50 SN per $\Delta z = 0.1$), the effects of gravitational lensing will average down to near zero; see Fig. 11.3.

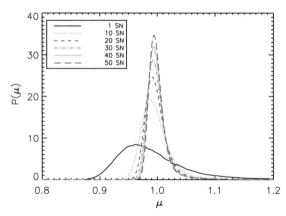

Fig. 11.3 Magnification distribution for lensing of a supernova at $z = 1.5$ in the usual ΛCDM cosmology (black curve). Other curves show how the distribution both narrows and becomes more Gaussian as we average over more SN. **Figure credit:** Holz and Linder (2005).

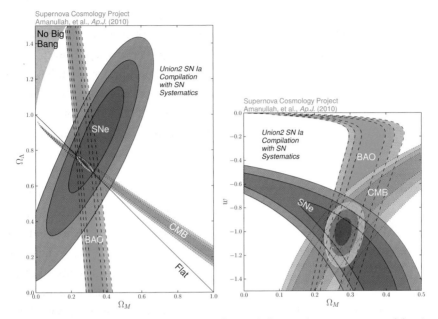

Fig. 11.4 Left panel: Constraints upon Ω_M and Ω_Λ in the consensus model using baryon acoustic oscillation (BAO), CMB, and SN measurements. Right panel: Constraints upon Ω_M and constant w in the fiducial dark energy model using the same data sets. **Figure credit:** Amanullah *et al.* (2010).

At the present time the SN systematic errors are well controlled, and are comparable to the statistical errors. The factor that gives undisputed credence to the result that the universe is accelerating, however, is confirmation with the galaxy clustering (the baryon acoustic oscillations, to be described later in this article), and the CMB constraints; see Fig. 11.4. In fact, even if one completely drops the SN constraints from the analysis, the combination of the galaxy clustering and the CMB firmly points to the existence of dark energy!

11.3 Parametrizations of Dark Energy

Introduction

The absence of a consensus model for cosmic acceleration presents a challenge in trying to connect theory with observations. For dark energy, the equation-of-state parameter w provides a useful phenomenological description; because it is the ratio of pressure to energy density, it is also closely connected to the underlying physics. On the practical side, determining a free function is more difficult than measuring parameters. We now review a variety of formalisms that have been used to describe and constrain dark energy.

First, let us recall some basics. From continuity equation, $\dot{\rho} + 3H(p + \rho) = 0$, we can calculate the dark energy density as a function of redshift for an arbitrary equation of state $w(z)$

$$\frac{\rho_{\rm DE}(z)}{\rho_{\rm DE,0}} = \exp\left(3\int_0^z (1 + w(z'))d\ln(1 + z')\right). \tag{11.7}$$

Parametrizations

The simplest parameterization of dark energy is

$$w = \text{const.} \tag{11.8}$$

This form fully describes vacuum energy ($w = -1$) or topological defects ($w = -N/3$ with N an integer dimension of the defect — 0 for monopoles, 1 for strings, 2 for domain walls). Together with $\Omega_{\rm DE}$ and $\Omega_{\rm M}$, w provides a three-parameter description of the dark-energy sector (two parameters if flatness is assumed). However, it does not describe scalar field or modified gravity models which generically have a time-varying w.

A number of two-parameter descriptions of w have been explored in the literature, e.g. $w(z) = w_0 + w'z$ (Cooray and Huterer, 1999). For low redshift they are all essentially equivalent, but for large z, some lead to unrealistic behavior, e.g. $w \ll -1$ or $\gg 1$. The parametrization (Linder, 2003)

$$w(a) = w_0 + w_a(1-a) = w_0 + w_a \frac{z}{1+z}, \tag{11.9}$$

where $a = 1/(1+z)$ is the scale factor, avoids this problem, fits many scalar fields and some modified gravity behavior, and leads to the most commonly used description of dark energy, namely $(\Omega_{\rm DE}, \Omega_{\rm M}, w_0, w_a)$. The energy density is then

$$\frac{\rho_{\rm DE}(a)}{\rho_{\rm DE,0}} = a^{-3(1+w_0+w_a)}e^{-3(1-a)w_a}. \tag{11.10}$$

More general expressions have been proposed. However one problem with introducing more parameters is that additional parameters make the equation of state very difficult to measure, while the parametrizations are still *ad hoc* and not well motivated from either theory or measurement's point of view.

Finally, it is useful to mention one simple way to elucidate redshift where the measurement accuracy of the equation of state, for a given survey, is highest. Two-parameter descriptions of $w(z)$ that are linear in the parameters entail the existence of a "pivot" redshift z_p at which the measurements of the two parameters are uncorrelated and the error in $w_p \equiv w(z_p)$ reaches a minimum; see the left panel of Fig. 11.5. Writing the equation of state in Eq. (11.9) in the form

$$w(a) = w_p + (a_p - a)w_a \qquad (11.11)$$

it is easy to translate constraints from the (w_0, w_a) to (w_p, w_a) parametrization, as well as determine a_p (or z_p), for any particular data set. This is useful, as measurements of the equation of state at the pivot point might provide most useful information in ruling out models (e.g. ruling out $w = -1$).

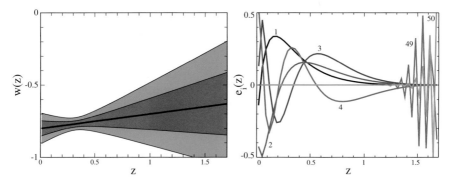

Fig. 11.5 Left panel: Example of forecast constraints on $w(z)$, assuming $w(z) = w_0 + w'z$. The "pivot" redshift, $z_p \simeq 0.3$, is where $w(z)$ is best determined. **Figure credit:** Huterer and Turner (2001). Right panel: The four best-determined (labeled 1–4) and two worst-determined (labeled 49, 50) principal components of $w(z)$ for a future SN Ia survey such as SNAP, with several thousand SN in the redshift range $z = 0$ to $z = 1.7$. **Figure credit:** Huterer and Starkman (2003).

Direct reconstruction

Another approach is to directly invert the redshift–distance relation $r(z)$ measured from SN data to obtain the redshift dependence of $w(z)$ in terms of the first and second derivatives of the comoving distance (Huterer and Turner, 1999; Nakamura and Chiba, 1999; Starobinsky, 1998),

$$1 + w(z) = \frac{1+z}{3} \frac{3H_0^2 \Omega_M (1+z)^2 + 2(d^2 r/dz^2)/(dr/dz)^3}{H_0^2 \Omega_M (1+z)^3 - (dr/dz)^{-2}}. \qquad (11.12)$$

Assuming that dark energy is due to a single rolling scalar field, the scalar potential $V(\phi)$ can also be reconstructed. Others have suggested reconstructing the dark energy density (Wang and Mukherjee, 2004; Wang and Tegmark, 2005)

$$\rho_{\rm DE}(z) = \frac{3}{8\pi G} \left[\frac{1}{(dr/dz)^2} - \Omega_M H_0^2 (1+z)^3 \right]. \qquad (11.13)$$

Direct reconstruction is the only approach that is truly model-independent. However, it comes at a price — taking derivatives of noisy data. In practice, one must fit the distance data with a smooth function, and the fitting process introduces systematic biases. While a variety of methods have been pursued (Huterer and Turner, 2001; Weller and Albrecht, 2002), it appears that direct reconstruction is too challenging and not robust even with SN Ia data of excellent quality (however see Holsclaw et al. (2010)). And while the reconstruction of $\rho_{\rm DE}(z)$ is easier since it involves only first derivatives of distance, $w(z)$ is more useful as a quantity since it contains more information about the nature of dark energy than $\rho_{\rm DE}(z)$. [For a review of dark energy reconstruction methods, see Sahni and Starobinsky (2006).]

Principal components

The cosmological function that we are trying to determine — $w(z)$, $\rho_{\rm DE}(z)$, or $H(z)$ — can be expanded in terms of principal components, a set of functions that are uncorrelated and orthogonal by construction (Huterer and Starkman, 2003). In this approach, the data determine which components are measured best.

For example, suppose we parametrize $w(z)$ in terms of piecewise constant values w_i ($i = 1, \ldots, N$), each defined over a small redshift range (z_i, $z_i + \Delta z$). In the limit of small Δz this recovers the shape of an arbitrary dark energy history (in practice, $N \gtrsim 20$ is sufficient), but the estimates of the w_i from a given dark energy probe will be very noisy. Principal Component Analysis extracts from those noisy estimates the best-measured features of $w(z)$. We find the eigenvectors $e_i(z)$ of the inverse covariance matrix for the parameters w_i and their corresponding eigenvalues λ_i. The equation of state parameter is then expressed as

$$1 + w(z) = \sum_{i=1}^{N} \alpha_i\, e_i(z), \tag{11.14}$$

where the $e_i(z)$ are the principal components. The coefficients α_i, which can be computed via the orthonormality condition

$$\alpha_i = \int (1 + w(z)) e_i(z) dz, \tag{11.15}$$

are each determined with an accuracy $1/\sqrt{\lambda_i}$. Several of these components are shown for a future SN survey in the right panel of Fig. 11.5, while measurements of the first six PCs of the equation of state from the current (and predictions for future) data are shown in Fig. 11.6.

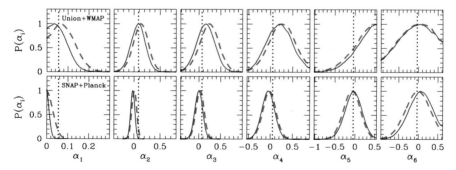

Fig. 11.6 Marginalized 1D posterior distributions for the first 6 PCs of flat (solid blue curves) and nonflat (dashed red curves) quintessence models. Top row: *current* Union+WMAP data; note that all PCs are consistent with $\alpha_i = 0$ (that is, $w(z) = -1$) except perhaps the 5th one. Bottom row: forecasts for *future* SNAP+Planck *assuming* a realization of the data with $\alpha_i = 0$. Vertical dotted lines show the predictions of an example quintessence model. **Figure credit:** Mortonson *et al.* (2010).

There are multiple advantages of using the PCs of dark energy (of either the equation of state $w(z)$, or of $\rho_{\rm DE}(z)$ or $H(z)$):

- The method is as close to "model-independent" as one can realistically get;
- Data tells us what we measure and how well; there are no arbitrary parametrizations imposed;
- One can use this approach to design a survey that is most sensitive to the dark energy equation of state parameter in some specific redshift interval or to study how many independent parameters are measured well by a combination of cosmological probes (i.e. how many PCs have $\sigma(\alpha_i)$ or $\sigma(\alpha_i)/\alpha_i$ less than some threshold value (de Putter and Linder, 2008)).

There are a variety of useful extensions of this method, including uncorrelated measurements of the equation of state parameters in redshift intervals (Huterer and Cooray, 2005).

Figures of Merit

We finally discuss the so-called figures of merit (FoMs) for dark energy experiments. A FoM is a number, or collection of numbers, that serves as simple and quantifiable metrics by which to evaluate the accuracy of constraints on dark energy parameters from current and proposed experiments. For example, marginalized accuracy in the (constant) equation of state, w, could serve as a figure of merit — since a large FoM is "good", we could simply define FoM $= 1/\sigma_w$, or $1/\sigma_w^m$ where m is some positive power.

The most commonly discussed FoM is that proposed by the Dark Energy Task Force (Albrecht et al. (2006), though this proposal goes back to Huterer and Turner (2001)), which is essentially inverse area in the w_0–w_a plane. For uncorrelated w_0 and w_a this would be $\propto 1/(\sigma_{w_0} \times \sigma_{w_a})$; because the two are typically correlated, the FoM can be defined as

$$\text{FoM}^{(w_0-w_a)} \equiv (\det \mathbf{C})^{-1/2} \approx \frac{6.17\pi}{A_{95}}, \qquad (11.16)$$

where $\mathbf{C}$ is the 2×2 covariance matrix in (w_0, w_a) after marginalizing over all other parameters, and A_{95} is the area of the 95.4% CL region in the

w_0–w_a plane. Note that the constant of proportionality is not important, since typically we compare the FoM from different surveys, and the constant disappears when we take the ratio.

While the standard "DETF FoM" defined in Eq. (11.16) keeps some information about the dynamics of DE (that is, the time variation of $w(z)$), several other general FoMs have been proposed. For example, Mortonson et al. (2010) proposed taking the FoM to be inversely proportional to the volume of the n-dimensional ellipsoid in the space of principal component parameters

$$\mathrm{FoM}_n^{(\mathrm{PC})} \equiv \left(\frac{\det \mathbf{C}_n}{\det \mathbf{C}_n^{(\mathrm{prior})}} \right)^{-1/2}, \tag{11.17}$$

where the prior covariance matrix is again unimportant since it would cancel in the comparison of ratios of the FoMs. Figure 11.10, near the end of this chapter, illustrates this FoM for current and future surveys.

11.4 Other Probes of Dark Energy

In addition to Type Ia supernovae, there are several other important probes of dark energy. These probes operate using very different physics, and have very different systematic errors.

The principal probes, in addition to SN Ia, are: baryon acoustic oscillations, weak gravitational lensing, and galaxy cluster abundance. We will now discuss each of those in turn. Additionally, there are secondary probes of dark energy — ones that might be useful for DE, but are currently not as well developed as the primary probes. We will discuss these briefly as well.

Baryon acoustic oscillations (BAO)

BAO refers to the signature of acoustic oscillations which are imprinted into the present-day correlations of galaxies by baryonic physics at the epoch of recombination [for a popular review, see Eisenstein (2005)]. Measurements of the length scale characteristic of these oscillations, roughly $100\,h^{-1}\mathrm{Mpc}$ comoving, enable inferring the angular diameter distance out to galaxies probed in a survey, and thus a robust way to measure the energy contents of the Universe.

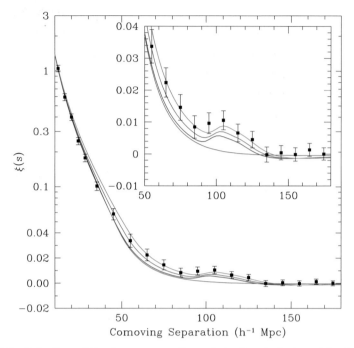

Fig. 11.7 Detection of the baryon acoustic peak in the clustering of luminous red galaxies in the SDSS (Eisenstein *et al.*, 2005). Shown is the two-point galaxy correlation function in redshift space; inset shows an expanded view with a linear vertical axis. Curves correspond to ΛCDM predictions for $\Omega_M h^2 = 0.12$ (green), 0.13 (red), and 0.14 (blue). Magenta curve shows a ΛCDM model without BAO.

Note that the power spectrum of density perturbations in dark matter, $P(k)$, is mainly sensitive to the density in matter (relative to critical), Ω_M. If we assume a flat universe (either motivated by the inflationary "prior", or by recent data), then $\Omega_{\rm DE} = 1 - \Omega_M$ and measurements of the broad-band shape of the power spectrum can get the dark energy density, but not the equation of state w.

However, the small (∼10%) *oscillations* in the power spectrum provide much more information about DE. The BAO data determine the ratio of the sound horizon at last-scattering to the quantity $D_V(z) \equiv [z\, r^2(z)/H(z)]^{1/3}$ at the measured redshift; given that the sound horizon is independently determined rather accurately, the BAO approximately provides measurement of distance to the redshift where the galaxies reside. For example, Percival *et al.* (2010) analyze combined data from two-degree Field

Galaxy Redshift Survey and the Sloan Digital Sky Survey which measure the clustering at mean redshifts $z = 0.2$ and $z = 0.35$, respectively.

Key to successful application of baryon acoustic oscillations are redshift measurements of galaxies in the sample. We need the galaxy redshifts in order to know where to "put them" in three dimensions, and thus to reconstruct the precise length scale at which the slight excess of clustering occurs. Another systematic that needs to be understood is the bias of galaxies in the sample (whose clustering we measure) to the underlying dark matter (whose clustering we can predict); if the bias has scale-dependent features on scales of $\sim 100\,\text{Mpc}$, then the systematic errors creep in. Future surveys that plan to utilize this method typically propose measuring redshifts of millions of galaxies, and the goal is to go deep ($z \sim 1$, and beyond) and have wide angular coverage as well.

Let us finally say a few words about the measured quantity, the power spectrum. In the dimensionless form, it is given by

$$\Delta^2(k) \equiv \frac{k^3 P(k)}{2\pi^2} = A\,\frac{4}{25}\frac{1}{\Omega_M^2}\left(\frac{k}{k_{\text{piv}}}\right)^{n-1}\left(\frac{k}{H_0}\right)^4 D(z)^2\,T^2(k)\,T_{\text{nl}}(k)\,, \tag{11.18}$$

where A is the normalization of the power spectrum (for the concordance cosmology, $A \simeq 2.4 \times 10^{-9}$), k_{piv} is the "pivot" around which we compute the spectral index n ($k_{\text{piv}} = 0.002\,\text{Mpc}^{-1}$ is often used); $D(z)$ is the linear growth of perturbations normalized to unity today; $T(k)$ is the transfer function that describes evolution of fluctuations inside the horizon and across the matter–radiation transition epoch and which encodes the BAOs; T_{nl} is a prescription for the *nonlinear* power spectrum which is relevant at small scales (e.g. $k \gtrsim 0.2\,h\,\text{Mpc}^{-1}$ today). Notice that $\Delta^2 \propto k^{n+3}$, and thus $P(k) \propto k^n$, with $n \simeq 1$, was predicted by Harrison, Zeldovich and Peebles in the late 1960s; this was a decade before inflation was proposed, and about three decades before measurements confirmed that $n \simeq 1$!

Weak gravitational lensing

The gravitational bending of light by structures in the Universe distorts or shears images of distant galaxies; see the left panel of Fig. 11.8. This distortion allows the distribution of dark matter and its evolution with time

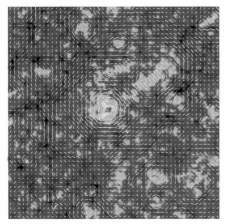 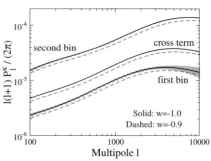

Fig. 11.8 Left panel: Cosmic shear field (white ticks) superimposed on the projected mass distribution from a cosmological N-body simulation: overdense regions are bright, underdense regions are dark. Note how the shear field is correlated with the foreground mass distribution. Figure courtesy of T. Hamana. Right panel: Cosmic shear angular power spectrum and statistical errors expected for LSST for $w = -1$ and -0.9. For illustration, results are shown for source galaxies in two broad redshift bins, $z_s = 0-1$ (first bin) and $z_s = 1-3$ (second bin); the cross-power spectrum between the two bins (cross term) is shown without the statistical errors.

to be measured, thereby probing the influence of dark energy on the growth of structure (for a detailed review, see e.g. Bartelmann and Schneider (2001); for brief reviews, see Hoekstra and Jain (2008) and Huterer (2010)).

Gravitational lensing produces distortions of images of background galaxies. These distortions can be described as mapping between the source plane (S) and image plane (I)

$$\delta x_i^S = A_{ij} \delta x_j^I, \tag{11.19}$$

where $\delta \mathbf{x}$ are the displacement vectors in the two planes and A is the distortion matrix

$$A = \begin{pmatrix} 1 - \kappa - \gamma_1 & -\gamma_2 \\ -\gamma_2 & 1 - \kappa + \gamma_1 \end{pmatrix}. \tag{11.20}$$

The deformation is described by the convergence κ and complex shear (γ_1, γ_2); the total shear is defined as $|\gamma| = \sqrt{\gamma_1^2 + \gamma_2^2}$. We are interested in the weak lensing limit, where $\kappa, |\gamma| \ll 1$. Magnification can be expressed in terms of κ and $\gamma_{1,2}$ as

$$\mu = \frac{1}{|1-\kappa|^2 - |\gamma|^2} \approx 1 + 2\kappa + O(\kappa^2, \gamma^2), \quad (11.21)$$

where the second approximate relation holds in the weak lensing limit.

We can theoretically predict convergence and shear, given a sample of sources with known redshift distribution and cosmological parameter values. The convergence in any particular direction on the sky $\hat{\mathbf{n}}$ is given by the integral along the line-of-sight

$$\kappa(\hat{\mathbf{n}}, \chi) = \int_0^\chi W(\chi') \delta(\chi') \, d\chi', \quad (11.22)$$

where δ is the perturbation in matter energy density and $W(\chi)$ is the geometric weight function describing the lensing efficiency of foreground galaxies. The most efficient lenses lie about half-way between us and the source galaxies whose shapes we measure.

The statistical signal due to gravitational lensing by large-scale structure is termed "cosmic shear". The cosmic shear field at a point in the sky is estimated by locally averaging the shapes of large numbers of distant galaxies. The primary statistical measure of the cosmic shear is the shear angular power spectrum measured as a function of source galaxy redshift z_s. (Additional information is obtained by measuring the correlations between shears at different redshifts or with foreground lensing galaxies.)

The convergence can be transformed into multipole space $\kappa_{lm} = \int d\hat{\mathbf{n}} \, \kappa(\hat{\mathbf{n}}, \chi) \, Y_{lm}^*(\hat{\mathbf{n}})$, and the power spectrum is defined as the two-point correlation function (of convergence, in this case) $\langle \kappa_{\ell m} \kappa_{\ell' m'} \rangle = \delta_{\ell \ell'} \, \delta_{m m'} \, P_\ell^\kappa$. The angular power spectrum is

$$P_\ell^\gamma(z_s) \simeq P_\ell^\kappa(z_s) = \int_0^{z_s} \frac{dz}{H(z) d_A^2(z)} W(z)^2 P\left(k = \frac{\ell}{d_A(z)}; z\right), \quad (11.23)$$

where ℓ denotes the angular multipole, $d_A(z) = (1+z)^{-2} d_L(z)$ is the angular diameter distance, the weight function $W(z)$ is the efficiency for lensing a population of source galaxies and is determined by the distance distributions of the source and lens galaxies, and $P(k,z)$ is the usual power spectrum of density perturbations. Notice the integral along the line of sight: essentially, weak lensing projects the density fluctuations between us and the galaxies whose shear we measure.

The dark-energy sensitivity of the shear angular power spectrum comes from two factors:

- *geometry* — the Hubble parameter, the angular-diameter distance, and the weight function $W(z)$; and
- *growth of structure* — through the redshift evolution of the power spectrum $P(k)$ [or more precisely, from the function $D(z)$ in Eq. (11.18)].

The *three*-point correlation function of cosmic shear is also sensitive to dark energy, and provides important complementary information about dark energy [e.g. Takada and Jain (2004)].

The statistical uncertainty in measuring the shear power spectrum on large scales is

$$\Delta P_\ell^\gamma = \sqrt{\frac{2}{(2\ell+1)f_{\text{sky}}}} \left[P_\ell^\gamma + \frac{\sigma^2(\gamma_i)}{n_{\text{eff}}} \right], \quad (11.24)$$

where f_{sky} is the fraction of sky area covered by the survey ($f_{\text{sky}} = 0.5$ for half-sky, etc.), $\sigma^2(\gamma_i)$ is the variance in a single component of the (two-component) shear (this number is ~ 0.2 for typical measurements), and n_{eff} is the effective number density per steradian of galaxies with well-measured shapes. The first term in brackets dominates on large scales, and comes from sample variance (also known as *cosmic variance*) due to the fact that only a finite number of samples of structures are available in our Universe. The second term dominates on small scales, and represents the shot-noise from the variance in galaxy ellipticities ("shape noise") combined with a finite number of galaxies, hence the inverse proportionality to n_{eff}.

The principal systematic errors in weak lensing measurements come from the limitations in measuring galaxy shapes accurately. There are also systematic uncertainties due to limited knowledge of the redshifts of source galaxies: because taking spectroscopic redshifts of most source galaxies will be impossible (they number in many millions), one has to rely on approximate photometric redshift techniques, where one gets redshift information from multiple-wavelength (i.e. multi-color) observations.

The right panel of Fig. 11.8 shows the dependence on the dark energy of the shear power spectrum and an indication of the statistical errors expected for a survey such as LSST, assuming a survey area of 15 000 square degrees and effective source galaxy density of $n_{\text{eff}} = 30$ galaxies per square arcmin, and divided into two radial slices. Current surveys cover a more modest ~100 square degrees, with a comparable or slightly lower galaxy density. Note that the proportionality of errors to $f_{\text{sky}}^{-1/2}$ means that large sky coverage is at a premium.

Clusters of galaxies

Galaxy clusters are the largest virialized objects in the Universe. Therefore, not only can they be observed, but also their number density can be *predicted* quite reliably, both analytically and from numerical simulations. Comparing these predictions to measurements from the large-area cluster surveys that extend to high redshift ($z \gtrsim 1$) can provide precise constraints on the cosmic expansion history.

The absolute number of clusters in a survey of solid angle Ω_{survey} centered at redshift z and in the shell of thickness Δz is given by

$$N(z, \Delta z) = \Omega_{\text{survey}} \int_{z-\Delta z/2}^{z+\Delta z/2} n(z, M_{\min}(z)) \frac{dV(z)}{d\Omega\, dz} dz, \quad (11.25)$$

where $M_{\min}$ is the minimal mass of clusters in the survey (usually of order $10^{14} M_\odot$). Note that knowledge of the minimal mass is extremely important, since the mass function $n(z, M_{\min}(z))$ is exponentially decreasing with M, so that most of the contribution comes from a small range of masses just above $M_{\min}$. The mass function is key to theoretical predictions, and it is usually obtained from a combination of analytic and numerical results; the original mass function used in cosmology is the 36-year old Press–Schechter mass function (Press and Schechter, 1974), and the more recent work provides fitting functions to simulation results that are accurate to several percent (Tinker et al., 2008). Furthermore, the volume element can easily be related to comoving distance $r(z)$ and the expansion rate $H(z)$ via $dV(z)/(d\Omega\, dz) = r^2(z)/H(z)$, and it is known exactly for a given cosmological model.

The sensitivity of cluster counts to dark energy arises — as in the case of weak lensing — from two factors:

- *geometry* — the term $dV(z)/(d\Omega\,dz)$ in Eq. (11.25) is the comoving volume element;
- *growth of structure* — the mass function $n(z, M_{\min}(z))$ depends on the evolution of density perturbations.

The mass function's near-exponential dependence upon the power spectrum is at the root of the power of clusters to probe dark energy. More specifically, the mass function explicitly depends on the *amplitude of mass fluctuations* smoothed on some scale R

$$\sigma^2(R, z) = \int_0^\infty \Delta^2(k, z) \left(\frac{3j_1(kR)}{kR}\right)^2 d\ln k, \qquad (11.26)$$

where $\Delta^2(k, z)$ is the dimensionless power spectrum defined in Eq. (11.18), while R is traditionally taken to be $\sim 8\,h^{-1}$Mpc at $z=0$ and roughly corresponds to the typical size of a galaxy cluster. The term in angular parentheses is the Fourier transform of the top-hat window that averages out the perturbations over regions of radius R.

Systematic errors in cluster counts mainly concern uncertainty in how to convert from an observable quantity (X-ray light, gravitational lensing signal, etc.) to the mass of a cluster. Current best estimates of mass are at the level of several tens of percent per cluster, and there is ongoing effort to find observable quantities, or combinations there of, that are tightly correlated with mass.

The left panel of Fig. 11.9 shows the sensitivity to the dark energy equation of state parameter of the expected cluster counts for the South Pole Telescope and the Dark Energy Survey. At low to intermediate redshift, $z < 0.6$, the differences are dominated by the volume element; at higher redshift, the counts are most sensitive to the growth rate of perturbations. The right panel shows measurements of the mass function using recent X-ray observations of clusters.

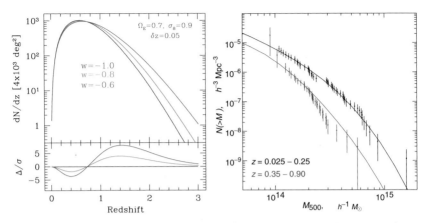

Fig. 11.9 Left panel: Predicted cluster counts for a survey covering 4 000 square degrees that is sensitive to halos more massive than $2\times10^{14} M_\odot$, for three flat cosmological models with fixed $\Omega_M = 0.3$ and $\sigma_8 = 0.9$. Lower panel shows fractional differences between the models in terms of the estimated Poisson errors. **Figure credit:** Mohr (2005) Right panel: Measured mass function, $n(z, M_{\min}(z))$, in our notation, from the 400 square degree survey of ROSAT clusters followed up by Chandra. **Figure credit:** Vikhlinin et al. (2009).

Table 11.1 Comparison of dark energy probes, adopted from Frieman et al. (2008). CDM refers to Cold Dark Matter paradigm, FoM is the Figure-of-Merit for dark energy surveys defined in the Dark Energy Task Force (DETF) report, while SZ refers to Sunyaev-Zeldovich effect.

Method	Strengths	Weaknesses	Systematics
WL	growth + geometry, Large FoM	CDM assumptions	Shear systematics, Photo-z
SN	pure geometry, mature	complex physics	evolution, dust extinction
BAO	pure geometry, low systematics	coarse-grained information	bias, nonlinearity, redshift distortions
CL	growth + geometry, X-ray + SZ + optical	CDM assumptions	mass-observable, selection function

Summary of principal probes

Figure 11.4 adopted from Amanullah et al. (2010) summarizes constraints in the Ω_M–Ω_Λ and Ω_M–w planes (the latter assuming a flat Universe) from CMB, BAO and SN Ia. In Table 11.1 we list the principal strengths and weaknesses of the four principal probes of DE. Control of systematic errors — observational, instrumental and theoretical — is crucial for these probes to realize their intrinsic power in constraining dark energy.

Role of the CMB

While the CMB provides precise cosmological constraints, by itself it has little power to probe dark energy. The reason is simple: the CMB provides a single snapshot of the Universe at a time when dark energy contributed a tiny part of the total energy density (a part in 10^9 if dark energy is the vacuum energy, or when $w = -1$). Nevertheless, the CMB plays a critical supporting role by determining other cosmological parameters, such as the spatial curvature and matter density, to high precision, thereby considerably strengthening the power of the methods discussed above. Essentially, what we get from the CMB is a *single* measurement of the angular diameter distance to recombination, $d_A(z \approx 1000)$; therefore it provides a single very accurate measurement of the parameters: Ω_M, $\Omega_{\rm DE}$ (if we do not assume a flat universe), and w (or $w(z)$ if we do not assume that the equation of state is constant). So, while the CMB alone suffers from degeneracy between the DE parameters, it is indispensable in breaking parameter degeneracies present in other cosmological probes; see Frieman *et al.* (2003) for more details. Data from the Planck CMB mission, launched in 2009, will therefore strongly complement those from dark energy surveys.

Secondary probes

There are a number of secondary probes of dark energy; here we review some of them.

- The Integrated Sachs–Wolfe (ISW) effect provided a confirmation of cosmic acceleration. ISW impacts the large-angle structure of the CMB anisotropy, but low-ℓ multipoles are subject to large cosmic variance, limiting their power of this probe. Nevertheless, ISW is of interest because it is able to reveal the imprint of large-scale dark-energy perturbations (Hu and Scranton, 2004).
- Gravitational radiation from inspiraling binary neutron stars or black holes can, if detected in the future, serve as "standard sirens" to measure absolute distances (Holz and Hughes, 2005). If their redshifts can be determined, then they could be used to probe dark energy through the Hubble diagram (Dalal *et al.*, 2006).

- Long-duration gamma-ray bursts have been proposed as standardizable candles (Schaefer, 2003), but their utility as cosmological distance indicators that could be competitive with or complementary to SN Ia has yet to be established.
- The optical depth for strong gravitational lensing (multiple imaging) of QSOs or radio sources has been proposed and used to provide independent evidence for dark energy, though these measurements depend on modeling the density profiles of lens galaxies.
- The redshift drift effect [also known as the Sandage–Loeb effect (Sandage, 1962; Loeb, 1998)] is the redshift change of an object measured using extremely high resolution spectroscopy over a period of 10 years or more and may some day be useful in constraining the expansion history at higher redshift $2 \lesssim z \lesssim 5$ (Corasaniti et al., 2007).
- Polarization measurements from distant galaxy clusters, which probe the quadrupole of the CMB radiation at the epoch when the cluster light was emitted and therefore the gravitational potential at that epoch, provide in principle a sensitive probe of the growth function and hence dark energy (Cooray et al., 2004).
- The relative ages of galaxies at different redshifts, if they can be determined reliably, provide a measurement of dz/dt and, from

$$t(z) = \int_0^{t(z)} dt' = \int_z^\infty \frac{dz'}{(1+z')H(z')}, \qquad (11.27)$$

measure the expansion history directly (Jimenez and Loeb, 2002).

11.5 The Accelerating Universe: Summary

There are the five important things to know about dark energy:

(1) Dark energy has negative pressure. It can be described with its energy density relative to critical today $\Omega_{\rm DE}$, and equation of state $w \equiv p_{\rm DE}/\rho_{\rm DE}$; the cosmological constant (or vacuum energy) has $w = -1$ precisely and at all times. More general explanations for dark energy may have constant or time-dependent equation of state. Assuming constant w, current constraints roughly give $w \approx -1 \pm 0.1$. Measuring

the equation of state (and its time dependence) may help understand the nature of dark energy, and is a key goal of modern cosmology.

(2) The accelerating Universe quenches gravitational collapse of large structures and suppresses the growth of density perturbation: whenever dark energy dominates, structures do not grow, essentially because the expansion is too rapid.

(3) Dark energy comes to dominate the density of the Universe only recently, at $z \lesssim 1$. At earlier epochs, dark energy density is small relative to matter density.

(4) Dark energy is spatially smooth. It affects both the geometry (that is, distances in the Universe) and the growth of structure (that is, clustering and abundance of galaxies and clusters of galaxies).

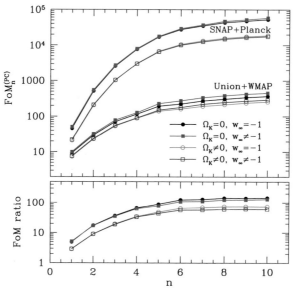

Fig. 11.10 Current and future figures of merit (FoM) for dark energy surveys, based on the principal-component (PC) based FoM from Eq. (11.17). Top panel: PC figures of merit FoM$_n^{(PC)}$ with forecasted uncertainties for a combination of planned and ongoing space telescopes SNAP(SN)+Planck and with measured uncertainties for already completed surveys Union+WMAP. Bottom panel: Ratios of FoM$_n^{(PC)}$ forecasts to current values. In both panels, point types indicate different quintessence model classes: flat (solid points) or non-flat (open points), either with (squares) or without (circles) early dark energy. **Figure credit:** Mortonson *et al.* (2010).

(5) Dark energy can be probed using a variety of cosmological probes that measure geometry (i.e. the expansion history of the Universe) and the growth of structure. Control of systematic errors in these cosmological probes is key to their success in measuring the properties of dark energy.

Acknowledgments

I thank Bob Kirshner, Eric Linder and Adam Riess for many useful comments on an earlier version of this manuscript, and to Josh Frieman and Michael Turner for collaboration on an earlier review (Frieman et al., 2008) that helped me organize thoughts about dark energy. I am supported by DOE OJI grant under contract DE-FG02-95ER40899, NSF under contract AST-0807564, and NASA under contract NNX09AC89G.

References

Albrecht, A. et al. (2006). Report of the dark energy task force, arXiv:astro-ph/0609591 .
Amanullah, R. et al. (2010). Spectra and light curves of six Type Ia supernovae at $0.511 < z < 1.12$ and the Union2 Compilation, *Astrophys. J.* **716**, pp. 712–738.
Astier, P. et al. (2006). The supernova legacy survey: Measurement of ω_m, ω_λ and w from the first year data set, *Astron. Astroph.* **447**, pp. 31–48.
Baade, W. (1938). The absolute photographic magnitude of supernovae. *Contributions from the Mount Wilson Observatory/Carnegie Institution of Washington* **600**, pp. 1–20.
Baade, W. and Zwicky, F. (1934). On super-novae, *Proceedings of the National Academy of Science* **20**, pp. 254–259.
Bartelmann, M. and Schneider, P. (2001). Weak gravitational lensing, *Phys. Rept.* **340**, pp. 291–472.
Bartlett, J. G. et al. (1995). The case for a Hubble constant of $30\,\mathrm{km s^{-1} Mpc^{-1}}$, *Science* **267**, pp. 980–983.
Colgate, S. A. (1979). Supernovae as a standard candle for cosmology, *ApJ* **232**, pp. 404–408.
Conley, A. et al. (2008). SiFTO: An empirical method for fitting SN Ia light curves, *ApJ* **681**, pp. 482–498.
Cooray, A., Huterer, D. and Baumann, D. (2004). Growth rate of large scale structure as a powerful probe of dark energy, *Phys. Rev. D* **69**, p. 027301.
Cooray, A. R. and Huterer, D. (1999). Gravitational lensing as a probe of quintessence, *ApJ* **513**, pp. L95–L98.
Corasaniti, P.-S., Huterer, D. and Melchiorri, A. (2007). Exploring the dark energy redshift desert with the sandage-loeb test, *Phys. Rev. D* **75**, p. 062001.

Dalal, N. et al. (2006). Short GRB and binary black hole standard sirens as a probe of dark energy, *Phys. Rev. D* **74**, p. 063006.

de Putter, R. and Linder, E. V. (2008). To bin or not to bin: Decorrelating the cosmic equation of state, *Astropart. Phys.* **29**, p. 424.

Eisenstein, D. J. (2005). Dark energy and cosmic sound [review article], *New Ast. Rev.* **49**, pp. 360–365.

Eisenstein, D. J. et al. (2005). Detection of the baryon acoustic peak in the large-scale correlation function of SDSS luminous red galaxies, *ApJ* **633**, pp. 560–574.

Frieman, J., Turner, M. and Huterer, D. (2008). Dark energy and the accelerating universe, *Ann. Rev. Astron. Astrophys.* **46**, pp. 385–432.

Frieman, J. A. et al. (2003). Probing dark energy with supernovae: Exploiting complementarity with the cosmic microwave background, *Phys. Rev. D* **67**, p. 083505.

Garnavich, P. M. et al. (1998). Constraints on cosmological models from Hubble Space Telescope observations of high-z supernovae, *Astrophys. J.* **493**, pp. L53–57.

Guth, A. H. (1981). The inflationary universe: A possible solution to the horizon and flatness problems, *Phys. Rev. D* **23**, pp. 347–356.

Guy, J. et al. (2007). SALT2: using distant supernovae to improve the use of type Ia supernovae as distance indicators, *Astron. Astroph.* **466**, pp. 11–21.

Hamuy, M. et al. (1996). BVRI light curves for 29 Type IA supernovae, *AJ* **112**, p. 2408.

Hicken, M. et al. (2009). Improved dark energy constraints from ~100 new CfA supernova Type Ia light curves, *ApJ* **700**, pp. 1097–1140.

Hoekstra, H. and Jain, B. (2008). Weak gravitational lensing and its cosmological applications, *Annual Review of Nuclear and Particle Science* **58**, pp. 99–123.

Holsclaw, T. et al. (2010). Nonparametric reconstruction of the dark energy equation of state, arXiv:1009.5443.

Holz, D. E. and Hughes, S. A. (2005). Using gravitational-wave standard sirens, *ApJ* **629**, pp. 15–22.

Holz, D. E. and Linder, E. V. (2005). Safety in numbers: Gravitational lensing degradation of the luminosity distance–redshift relation, *ApJ* **631**, pp. 678–688.

Hu, W. and Scranton, R. (2004). Measuring dark energy clustering with CMB-galaxy correlations, *Phys. Rev. D* **70**, p. 123002.

Huterer, D. (2010). Weak lensing, dark matter and dark energy, *Gen. Rel. Grav.* **42**, pp. 2177–2195.

Huterer, D. and Cooray, A. (2005). Uncorrelated estimates of dark energy evolution, *Phys. Rev. D* **71**, p. 023506.

Huterer, D. and Starkman, G. (2003). Parameterization of dark-energy properties: A principal-component approach, *Phys. Rev. Lett.* **90**, p. 031301.

Huterer, D. and Turner, M. S. (1999). Revealing quintessence, *Phys. Rev. D* **60**, p. 081301.

Huterer, D. and Turner, M. S. (2001). Probing the dark energy: Methods and strategies, *Phys. Rev. D* **64**, p. 123527.

Jha, S., Riess, A. G. and Kirshner, R. P. (2007). Improved distances to Type Ia Supernovae with multicolor light-curve shapes: MLCS2k2, *ApJ* **659**, pp. 122–148.

Jimenez, R. and Loeb, A. (2002). Constraining cosmological parameters based on relative galaxy ages, *ApJ* **573**, pp. 37–42.

Kessler, R. et al. (2009). First-year Sloan Digital Sky Survey-II (SDSS-II) supernova results: Hubble diagram and cosmological parameters, *Astrophys. J. Suppl.* **185**, pp. 32–84.

Kim, A. (2008). Stretched and non-stretched B-band supernova light curves, *LBNL Report LBNL-56164*.

Kirshner, R. P. (2002). *The Extravagant Universe: Exploding Stars, Dark Energy and the Accelerating Cosmos*, Princeton University Press.

Kirshner, R. P. (2009). Foundations of supernova cosmology, arXiv:0910.0257.

Knop, R. A. et al. (2003). New constraints on Ω_M, Ω_Λ, and w from an independent set of eleven high-redshift supernovae observed with HST, *Astrophys. J.* **598**, p. 102.

Kowal, C. T. (1968). Absolute magnitudes of supernovae. *AJ* **73**, pp. 1021–1024.

Linder, E. V. (2003). Exploring the expansion history of the universe, *Phys. Rev. Lett.* **90**, p. 091301.

Linder, E. V. (2010). Frontiers of dark energy, arXiv:1009.1411.

Loeb, A. (1998). Direct measurement of cosmological parameters from the cosmic deceleration of extragalactic objects, *ApJL* **499**, p. L111.

Mohr, J. J. (2005). Cluster survey studies of the dark energy, in S. C. Wolff and T. R. Lauer (eds.), *Observing Dark Energy, Astronomical Society of the Pacific Conference Series* **339**, p. 140.

Mortonson, M. J., Huterer, D. and Hu, W. (2010). Figures of merit for present and future dark energy probes, *Phys. Rev. D* **82**, p. 063004.

Nakamura, T. and Chiba, T. (1999). Determining the equation of state of the expanding universe: Inverse problem in cosmology, *Mon. Not. Roy. Astron. Soc.* **306**, pp. 696–700.

Norgaard-Nielsen, H. U. et al. (1989). The discovery of a Type IA supernova at a redshift of 0.31, *Nature* **339**, pp. 523–525.

Peebles, P. J. E. (1984). Tests of cosmological models constrained by inflation, *ApJ* **284**, pp. 439–444.

Percival, W. J. et al. (2010). Baryon acoustic oscillations in the Sloan Digital Sky Survey data release 7 galaxy sample, *Mon. Not. Roy. Astron. Soc.* **401**, pp. 2148–2168.

Perlmutter, S. and Schmidt, B. P. (2003). Measuring cosmology with supernovae, in K. Weiler (ed.), *Supernovae and Gamma-Ray Bursters*, pp. 195–217. *Lecture Notes in Physics*, Vol. 598, Berlin, Springer Verlag.

Perlmutter, S. et al. (1997). Measurements of the cosmological parameters omega and lambda from the first seven supernovae at $z \geq 0.35$, *ApJ* **483**, p. 565.

Perlmutter, S. et al. (1998). Discovery of a supernova explosion at half the age of the universe and its cosmological implications, *Nature* **391**, pp. 51–54.

Perlmutter, S. et al. (1999). Measurements of omega and lambda from 42 high-redshift supernovae, *Astrophys. J.* **517**, pp. 565–586.

Phillips, M. M. (1993). The absolute magnitudes of Type Ia supernovae, *ApJL* **413**, pp. L105–L108.

Press, W. H. and Schechter, P. (1974). Formation of galaxies and clusters of galaxies by selfsimilar gravitational condensation, *Astrophys. J.* **187**, pp. 425–438.

Riess, A. G., Press, W. H. and Kirshner, R. P. (1996a). A precise distance indicator: Type IA supernova multicolor light-curve shapes, *ApJ* **473**, p. 88.

Riess, A. G., Press, W. H. and Kirshner, R. P. (1996b). Is the dust obscuring supernovae in distant galaxies the same as dust in the milky way? *ApJ* **473**, p. 588.

Riess, A. G. et al. (1998). Observational evidence from supernovae for an accelerating universe and a cosmological constant, *AJ* **116**, pp. 1009–1038.

Riess, A. G. et al. (2004). Type ia supernova discoveries at $z > 1$ from the hubble space telescope: Evidence for past deceleration and constraints on dark energy evolution, *ApJ* **607**, pp. 665–687.

Riess, A. G. et al. (2007). New Hubble Space Telescope discoveries of Type Ia supernovae at $z \geq 1$: Narrowing constraints on the early behavior of dark energy, *ApJ* **659**, pp. 98–121.

Sahni, V. and Starobinsky, A. (2006). Reconstructing dark energy, arXiv:astro-ph/0610026.

Sandage, A. (1962). The change of redshift and apparent luminosity of galaxies due to the deceleration of selected expanding universes, *ApJ* **136**, pp. 319–333.

Schaefer, B. E. (2003). Gamma-ray burst Hubble diagram to $z = 4.5$, *ApJL* **583**, pp. L67–L70.

Starobinsky, A. A. (1998). How to determine an effective potential for a variable cosmological term, *JETP Lett.* **68**, pp. 757–763.

Takada, M. and Jain, B. (2004). Cosmological parameters from lensing power spectrum and bispectrum tomography, *Mon. Not. Roy. Astron. Soc.* **348**, p. 897.

Tinker, J. L. et al. (2008). Toward a halo mass function for precision cosmology: The limits of universality, *Astrophys. J.* **688**, pp. 709–728.

Turner, M. S., Steigman, G. and Krauss, L. M. (1984). Flatness of the universe — Reconciling theoretical prejudices with observational data, *Phys. Rev. Lett.* **52**, pp. 2090–2093.

Vikhlinin, A. et al. (2009). Chandra cluster cosmology project III: Cosmological parameter constraints, *Astrophys. J.* **692**, pp. 1060–1074.

Wang, Y. and Mukherjee, P. (2004). Model-independent constraints on dark energy density from flux-averaging analysis of Type Ia supernova data, *ApJ* **606**, pp. 654–663.

Wang, Y. and Tegmark, M. (2005). Uncorrelated measurements of the cosmic expansion history and dark energy from supernovae, *Phys. Rev. D* **71**, p. 103513.
Weller, J. and Albrecht, A. (2002). Future supernovae observations as a probe of dark energy, *Phys. Rev. D* **65**, p. 103512.
Wood-Vasey, W. M. *et al.* (2007). Observational constraints on the nature of the dark energy: First cosmological results from the ESSENCE supernova survey, *Astrophys. J.* **666**, pp. 694–715.

CHAPTER 12

FRONTIERS OF DARK ENERGY

ERIC V. LINDER

Berkeley Lab & University of California,
Berkeley, CA 94720, USA
and
Institute for the Early Universe WCU,
Ewha Womans University,
Seoul, Korea

12.1 Introduction to Dark Energy

Emptiness — the vacuum — is a surprisingly rich concept in cosmology. A Universe devoid of all matter and radiation can still have evolution in space and time. In fact there can be very distinct empty Universes, defined through their geometry. One of the great modern quests in science is to understand the hidden constituents of the Universe, neither matter nor radiation, and their intimate relation with the nature of the quantum vacuum and the structure of spacetime itself.

Cosmologists are just beginning to probe the properties of the cosmic vacuum and its role in reversing the attractive pull of gravity to cause an acceleration in the expansion of the cosmos. The cause of this acceleration is given the generic name of dark energy, whether it is due to a true vacuum, a false, temporary vacuum, or a new relation between the vacuum and the force of gravity. Despite the common name, the distinction between these origins is of utmost interest and physicists are actively engaged in finding ways to use cosmological observations to distinguish which is the true, new physics. See Huterer (2011) in this volume for further details on the observational probes, respectively, and Caldwell and Kamionkowski (2009); Durrer and Maartens (2010); Frieman *et al.* (2008); Silvestri and Trodden (2009) for other recent reviews, including a variety of theories.

Here we will discuss how to relate the theoretical ideas to the experimental constraints, how to understand the influences of dark energy on the expansion and structure in the Universe, and what frontiers of new physics are being illuminated by current and near-term data. In Sec. 12.2 we consider the vacuum, quantum fields, and their interaction with material components. The current level of our understanding about the properties of dark energy is reviewed in Sec. 12.3, and we relate this to a few, robust theories for the origin of dark energy. Looking to the frontiers of exploration, Sec. 12.4 anticipates what we may learn from experiments just now underway or being developed.

12.2 The Dynamics of Nothing

Emptiness, in general relativity, merely means that nothing has been put on the stage of space and time. In the framework, however, the structure of space and time and their relation into spacetime, is part of the theory itself. A Universe devoid of matter, radiation, all material contents still has geometry. We will consider here only the highly symmetrical case of a simply-connected Universe (no holes or handles) that is homogeneous (uniform among spatial volumes) and isotropic (uniform among spatial directions). A Universe with only spatial (not spacetime) curvature is called a Milne Universe, or often just an empty Universe. If even spatial curvature vanishes, then this is a Minkowski Universe, a relativistic generalization of Euclidean space.

Suppose we now consider an energy completely uniform everywhere in space. One possibility for this is the energy of the spatial curvature itself, for example in the Milne Universe. In evolving toward a lower energy state, the Universe reduces the curvature energy, proportional to the inverse square radius of curvature, a^{-2}, by expanding. That is, the factor a increases with time (and is often called the expansion factor or scale factor). Since the dynamical time-scale of a self-gravitating system is proportional to the inverse square root of the energy density, then $a \propto t$. We see that there is no acceleration, i.e. $\ddot{a} = 0$, and the expansion continues at the same rate, $\dot{a} = $ constant forever.

Now imagine another uniform energy not associated with spatial curvature: a vacuum energy, a nonzero ground state level of energy. If this were negative, it would reduce the curvature energy and could counteract the expansion, possibly even causing collapse of the Universe. That is, a reduces with time until it reaches zero. Such a uniform energy is called a negative cosmological constant. However suppose the vacuum energy were positive: then it would add to the energy and the expansion, increasing the rate such that we have $\ddot{a} > 0$ — acceleration.

Finally, remove the spatial curvature completely. With just the positive cosmological constant one still has the acceleration (recall the spatial curvature did not contribute to the acceleration positively nor negatively). Nothing is in the Universe but a positive, uniform energy. This is called a de Sitter Universe. Most interesting, though, is what happens when we restore the matter and radiation into the picture. Matter and radiation have the usual gravitational attraction that pulls objects together, fighting against expansion. They act to decelerate the expansion. Depending on the relative contributions then between matter etc. and the vacuum, the final result can be either a decelerating or accelerating Universe. One of the great paradigm shifts in cosmology was the realization and experimental discovery (Perlmutter et al., 1999; Riess et al., 1998) that we live in a Universe that accelerates in its expansion, where gravity is not predominantly attractive.

This is really quite striking a development, opening up whole frontiers of new physics. At its most personal, it reminds us of the "principle of cosmic modesty". Julius Caesar (at least through George Bernard Shaw) defined a barbarian as one who "thinks that the customs of his tribe and island are the laws of nature." After Copernicus we have moved beyond thinking the Earth is the center of the Universe; with the development of astronomy we know that the Milky Way Galaxy is not the center of the Universe; through physical cosmology we know that what we are made of — baryons and leptons — is not typical of the matter in the Universe; and now we even realize that the gravitational attraction we take as commonplace is not the dominant behavior in the Universe. We are decidedly on the doorstep of new physics.

How then do we elucidate the role of the vacuum? A first step is certainly to determine whether we are indeed dealing with a uniform, constant energy filling space. The vacuum is the lowest energy state of a quantum field. One can picture this as a field of harmonic oscillators, imaginary springs at every point in space, and ask whether these springs are identical and frozen, or whether they have some spatial variation and motion. An assemblage of values defined at points in space and time is basically a scalar field, and we seek to know whether dark energy is a true cosmological constant or a dynamical entity, perhaps one whose energy is not in the true ground state but is temporarily lifted above zero and is changing with time.

The scalar field approach is a fruitful one since one can use it as an effective description of the background dynamics of the Universe even if the origin of acceleration is from another cause. That is, one can define an effective energy density and effective pressure (determining how the energy density changes with time), and use that in the equations governing the expansion (although the growth of inhomogeneities can be influenced by other degrees of freedom). This description of the cosmic expansion holds even if there is no physical field at all, such as in the case of a modification of the gravitational theory. (We discuss some of the ways to distinguish between explanations in Sec. 12.4.)

Indeed, it is instructive to review some historical cases where dynamics indicated new physics beyond what was then known. In the 18th century, the motion of the planet Uranus did not accord with predictions of Newton's laws of gravitation applied from the other material contents of the Solar System. Two choices presented themselves: the laws were inadequate, or the knowledge of the material contents was incomplete. Keeping the laws intact and asking what new material content was needed to explain the anomaly led to the discovery of Neptune. In the 19th century, the motion of Mercury disagreed with the laws and material contents known. While some again sought a new planet, Einstein developed extensions to Newtonian gravity — the solution lay in new laws. For dark energy, we do not know whether we need to add new contents — a quantum scalar field, say — or an extension to Einsteinian gravity. However what is certain is that we are in the midst of a revolution in physics. While Einstein's correction to Mercury's orbit led to a minuscule 43'' per century of extra precession, dark energy turns cosmology upside down by changing gravitational attraction into accelerated expansion, dominates the expansion rate, and determines the ultimate fate of the Universe.

We can investigate dark energy's dynamical influence in more mathematical detail through the scalar field language (without assuming a true, physical scalar field). The Lagrangian density for a scalar field is just

$$\mathcal{L} = \frac{1}{2}\phi_{;\mu}\phi^{;\mu} + V(\phi), \qquad (12.1)$$

where ϕ is the value of the field, V is its potential, and $;\mu$ denotes derivatives with respect to the time and space coordinates. Using the Noether construction of the energy–momentum tensor, one can identify the energy density ρ and isotropic pressure p (all other terms vanishing under homogeneity and isotropy) as

$$\rho = \frac{1}{2}\dot{\phi}^2 + \frac{1}{2}(\nabla\phi)^2 + V, \qquad (12.2)$$

$$p = \frac{1}{2}\dot{\phi}^2 + \zeta\,(\nabla\phi)^2 - V. \qquad (12.3)$$

Here $\zeta = -1/6$ (1/2) depending on whether the field is treated as spatially incoherent or coherent. If the spatial gradient terms dominated, then the pressure-to-density ratio would be $-1/3$ (i.e. acting like spatial curvature) or $+1$ (i.e. acting like a stiff fluid or gradient tilt) in the two cases. However, in the vast majority of cases the spatial gradients are small compared to the other terms and are neglected.

It is convenient to discuss the scalar field properties in terms of the equation of state parameter

$$w \equiv \frac{p}{\rho} = \frac{(1/2)\dot{\phi}^2 - V}{(1/2)\dot{\phi}^2 + V}, \qquad (12.4)$$

where the first equality is general and we neglect spatial gradients in the second equality, as in the rest of the article. When the kinetic energy term dominates, then w approaches $+1$; when the potential energy dominates, then $w \to -1$; and when they balance (as in oscillating around the minimum of a quadratic potential) then $w = 0$, like non-relativistic matter. Acceleration occurs when the total equation of state, the weighted sum (by energy density) of the equations of state of each component, is $w_{\rm tot} < -1/3$. The Friedmann equation for the acceleration of the expansion factor is

$$\frac{\ddot{a}}{a} = -\frac{4\pi G}{3}(\rho_{\rm tot} + 3p_{\rm tot}) = -\frac{4\pi G}{3}\sum \rho_w(1+3w), \qquad (12.5)$$

where G is Newton's constant and we set the speed of light equal to unity.

The other equation of motion is either the Friedmann expansion equation

$$H^2 \equiv \frac{\dot{a}^2}{a^2} = \frac{8\pi G}{3}\rho_{\rm tot}, \qquad (12.6)$$

where we can include any curvature energy density in $\rho_{\rm tot}$, or the energy conservation or continuity equation

$$\dot{\rho} = -3H(\rho + p) \quad \text{or} \quad \frac{d\ln\rho}{d\ln a} = -3(1+w). \qquad (12.7)$$

The continuity equation holds separately for each individually conserved component. In particular, for a scalar field we can write the continuity equation as a Klein–Gordon equation

$$\ddot{\phi} + 3H\dot{\phi} + dV/d\phi = 0. \qquad (12.8)$$

To examine the dynamics of the dark energy, one can solve for $w(a)$; it is also often instructive to work in the phase-space of w'-w, where a prime denotes $d/d\ln a$. For example, many models can be categorized as either thawers or freezers (Caldwell and Linder, 2005): their behavior either starts with the field frozen by the Hubble friction of the expanding Universe (so kinetic energy is negligible and $w = -1$), and then at late times the field begins to roll, moving w away from -1, or the field starts off rolling and gradually comes to settle at a minimum of the potential, asymptotically reaching $w = -1$.

Since the dark energy does not always dominate the energy budget and expansion of the Universe, it is also useful to examine the dynamics of the full system of components. One can define variables representing each contribution to the energy density, say, and obtain a coupled system of equations (Copeland, Liddle and Wands, 1998). For example, for a scalar field

$$x' = -3x + \lambda\sqrt{\frac{3}{2}}y^2 + \frac{3}{2}x[2x^2 + (1+w_b)(1-x^2-y^2)], \qquad (12.9)$$

$$y' = -\lambda\sqrt{\frac{3}{2}}xy + \frac{3}{2}y[2x^2 + (1+w_b)(1-x^2-y^2)], \qquad (12.10)$$

where $x = \sqrt{\kappa\dot\phi^2/(2H^2)}$, $y = \sqrt{\kappa V/H^2}$, $\kappa = 8\pi G/3$, and $\lambda = -(1/V)dV/d(\phi\sqrt{3\kappa})$, with w_b being the equation of state of the background, dominating component (e.g. matter, with $w_b = 0$, during the matter-dominated era). To solve these equations one must specify initial conditions and the form of $V(\phi)$, i.e. λ.

The fractional dark energy density $\Omega_w = x^2 + y^2$, so its evolution is bounded within the first quadrant of the unit circle in the x–y plane (taking $\dot\phi > 0$; it is simple enough to generalize the equations), and the dark energy equation of state is $w = (x^2 - y^2)/(x^2 + y^2)$. So the dynamics can be represented in polar coordinates, with the density being the radial coordinate and the equation of state the angular coordinate (twice the angle with respect to the x axis is $2\theta = \cos^{-1} w$). Figure 12.1 illustrates some dynamics in the y–x energy density component (or w–Ω_ϕ) plane.

The term in square brackets in Eqs. (12.9) and (12.10) is simply $1+w_{\rm tot}$. Another way of viewing the dynamics is through the variation of the dark energy equation of state

$$w' = -3(1-w^2) + \lambda(1-w)x\sqrt{2}. \qquad (12.11)$$

One can readily see that $w = -1$ (and hence $x = 0$) is a fixed point, with $w' = 0$. It can either be a stable attractor (in the case of freezing fields) or unstable (in the case of thawing fields). Figure 12.2 illustrates some dynamics in the phase plane w'–w — an alternate view to Fig. 12.1. Considerably more detail about classes of dynamics is given in Caldwell and Linder (2005); Linder (2006). For example, through nonstandard kinetic terms one can get dynamics with $w < -1$, sometimes called phantom fields (Caldwell, Kamionkowski and Weinberg, 2003).

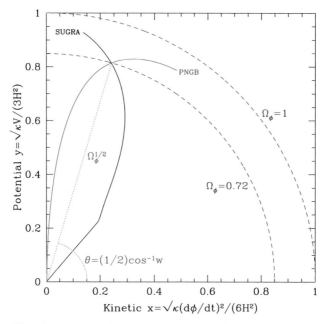

Fig. 12.1 The dynamics in the y–x, or potential–kinetic energy phase-space for example of a thawing (PNGB) and a freezing (SUGRA) field. The dark energy density Ω_ϕ acts as a radial coordinate, while the dark energy equation of state w acts as an angular coordinate. Note the constancy of w (i.e. the angle θ) for SUGRA at early times, when it is on the attractor trajectory. The models have been chosen to have the same values today, $\Omega_{\phi,0} = 0.72$ and $w_0 = -0.839$ (where the curves cross). The curves end in the future at $a = 1.47$.

The dynamical view of dark energy identifies several key properties that would lead to insight into the nature of the physics behind acceleration. Since $w = -1$ is a special state, we can ask whether the dark energy always stays there, i.e. is it a cosmological constant? Does dark energy act like a thawing (roughly $w' > 0$) or freezing (roughly $w' < 0$) field? Is it ever phantom ($w < -1$)? One could also look back at the continuity equation and ask whether each component is separately conserved or whether there is interaction. Keeping overall energy conservation, one could write

$$\frac{d\ln\rho_w}{d\ln a} = -3(1+w) + \frac{\Gamma}{H}, \tag{12.12}$$

$$\frac{d\ln\rho_m}{d\ln a} = -3(1+w_m) - \frac{\Gamma}{H}, \tag{12.13}$$

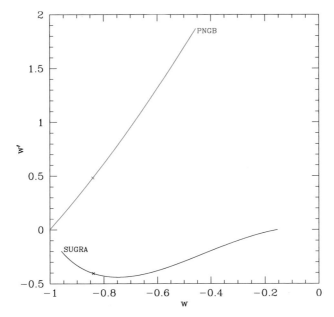

Fig. 12.2 The dynamics in the w'–w phase plane for example a thawing (PNGB) and a freezing (SUGRA) field. The right or left curvature in Fig. 12.1 here translates into $w' > 0$ or < 0. The thawer starts in a frozen state ($w = -1$, $w' = 0$) and evolves away from the cosmological constant behavior, while the freezer starts at some constant w given by an attractor solution and then evolves as its energy density becomes more substantial, eventually approaching the cosmological constant state. The x's mark the present state, and the curves here end in the future at $a = 1.47$.

for the dark energy and (dark) matter components, where Γ represents the interaction. The impact of this is to shift each equation of state, such that $w_{\text{eff}} = w - \Gamma/(3H)$ and $w_{m,\text{eff}} = w_m + \Gamma/(3H)$.

Such interactions act as a fifth force violating the Equivalence Principle if dark energy responds to different components in different ways. Certainly interaction with baryons is highly constrained otherwise we would have found dark energy from particle physics experiments. The shift in equation of state could make dark energy that intrinsically has $w > -1$ look like a phantom field, and vice versa [see Wei (2010) for some current constraints]. Dynamical analysis does allow us to make some general statements: for example, consider a phantom field arising from a negative kinetic term. The dynamical variable $y = \sqrt{\kappa V/H^2}$ has a fixed point when $y'_c = 0$, so the potential obeys $V'/V = 2H'/H \equiv -3(1+w_{\text{tot}})$. However, such negative kinetic term fields roll *up* the potential so V' is positive. Therefore $w_{\rm tot}$ must be less than -1 and the field must remain asymptotically phantom, even in the presence of interactions.

12.3 Knowing Nothing

The existence of dark energy was first discovered through the geometric probe of the distance–redshift relation of Type Ia supernovae (Perlmutter *et al.*, 1999; Riess *et al.*, 1998). Such data have been greatly expanded and refined so that now the analysis of the Union2 compilation of supernova data (Amanullah *et al.*, 2010), together with other probes, establishes that the energy density contribution of dark energy to the total energy density is $\Omega_{de} = 0.719 \pm 0.017$ and the dark energy equation of state, or pressure-to-density ratio, is $w = -1.03 \pm 0.09$ (assumed constant).

Other cosmological probes are now investigating cosmic acceleration, although none by themselves have approached the leverage of supernovae. Experiments underway use Type Ia and Type II supernovae, baryon acoustic oscillations, cosmic microwave background measurements, weak gravitational lensing, and galaxy clusters with the Sunyaev–Zel'dovich effect and X-rays. [See Huterer (2011) for more detailed discussion.]

Observables such as the distance–redshift relation and Hubble parameter–redshift relation, and those that depend on these in a more complex manner, can be used to test specific models of dark energy. For some examples of this, see Rubin *et al.* (2009); Sollerman *et al.* (2009); Mortonson, Hu and Huterer (2009). However it is frequently useful to have a more model-independent method of constraining dark energy properties. We have already seen in the previous section that one can classify many models into the general behaviors of thawers and freezers. There appears diversity within each of these classes, but de Putter and Linder (2008) found a calibration relation between the dark energy equation of state value and its time variation that defines homogeneous families of dark energy physics. Figure 12.3 illustrates both the diversity and the calibration.

This calibration provides a physical basis for a very simple but powerful relation between the equation of state value and time variation in the dark

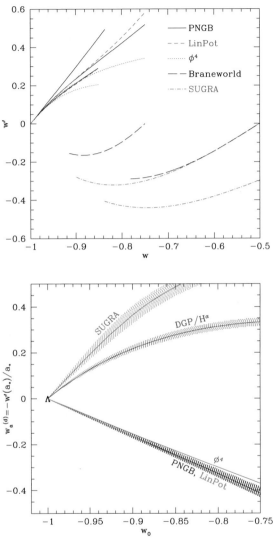

Fig. 12.3 (Top panel) Representative models exhibiting a diversity of dynamics are plotted for various parameter values in the w–w' phase-space, including braneworld/H^α models ($\alpha = 1$ DGP and $\alpha = 0.5$). (Bottom panel) Using the calibrated dark energy parameters w_0 and w_a, dark energy models and families lie in tightly homogeneous regions. Contrast this with the top panel, showing the same models before calibration (note w_a has the opposite sign from w'). We here vary over all parameters in the potentials. Shading shows the effect of scanning over ± 0.03 in Ω_m (we omit the shading for ϕ^4 and linear potential models to minimize confusion; the width would be about half that shown for PNGB). Distinctions between thawing and freezing models, and between freezing models, become highlighted with calibration. **Figure credit:** de Putter and Linder (2008).

energy dynamics phase plane. The resulting parameterization

$$w(a) = w_0 + w_a(1 - a) \tag{12.14}$$

gives a highly accurate match to the observable relations of distance $d(z)$ and Hubble parameter $H(z)$. This form, emphatically not a Taylor expansion, achieves 10^{-3} accuracy on the observables and matches the w_0–w_a parameterization devised to fit the exact solutions for scalar field dynamics (Linder, 2003).

Current data constrains w_0 to within ~ 0.3 and w_a to within ~ 1, which is insufficient to answer any of the questions raised in the previous section, e.g. whether dark energy is a cosmological constant or not, is thawing or freezing, etc. To give a clear picture of our current state of knowledge, Fig. 12.4 displays the constraints from all current data in several different ways.

For example, for w held constant, Amanullah et al. (2010) find that the energy density contribution of dark energy to the total energy density is $\Omega_{de} = 0.719 \pm 0.017$ and the dark energy equation of state, or pressure-to-density ratio, is $w = -1.03 \pm 0.09$ (68% confidence level, including systematic uncertainties). While viewing the constraints on w under the assumption that it is constant (upper left panel) gives an impression of substantial precision, in fact none of the key physical questions have been answered. The upper right panel shows that when we leave open the values of w in different redshift ranges (redshift $z = a^{-1} - 1$), then we have no reasonable constraints on whether w is in fact constant in time. Recall that for a simple scalar field, w is bounded from below by -1, and must be less than $-1/3$ to provide acceleration. So the panoply of current data does not give much evidence for or against the constancy of w.

The bottom left panel demonstrates that we have no constraints at all on dark energy above $z \approx 1.6$, neither knowing its properties nor even whether it exists. In the bottom right panel it is clear that the situation at low redshift (near the present time) is also quite uncertain: does w differ from -1, and if so in which direction?

On the theoretical front, no consensus exists on any clear concept for the origin of dark energy. Any expansion history can be accommodated by a combination of potential and kinetic terms, but it is really not a

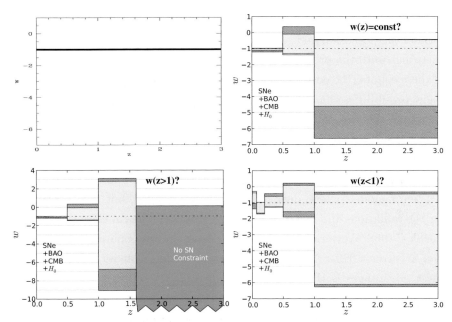

Fig. 12.4 Constraints from the Union2 supernova compilation, WMAP7 CMB, SDSS DR7 baryon acoustic oscillation, and Hubble constant data on the dark energy equation of state $w(z)$, in redshift bins. Top left plot appears to show that data have zeroed in on the cosmological constant value of $w = -1$, but this assumes w is constant. When one allows for the values of w to be different in different redshift bins, our current knowledge of dark energy is seen to be far from sufficient. Top right plot shows that we do not yet have good constraints on whether $w(z)$ is constant. Bottom left plot (note change of scale) shows we have little knowledge of dark energy behavior, or even existence, at $z > 1$. Bottom right plot shows we have little detailed knowledge of dark energy behavior at $z < 1$. Outer (inner) boxes show 68% confidence limits with (without) systematics. The results are consistent with $w = -1$, but also allow considerable variation in $w(z)$.
Figure credit: Amanullah *et al.* (2010).

case of an embarrassment of riches. There are two main problems: any potential that one writes down should receive quantum corrections at high energies and so end up different from the original intent, and the energy scale corresponding to dark energy is much lower (by many tens of orders of magnitude) than scales associated with initial conditions in the early Universe. How do we cue dark energy and cosmic acceleration to appear on the stage of the Universe at the right moment? That is, one generically requires fine-tunings to describe the Universe today starting from high-energy physics.

To surmount these difficulties requires some symmetry to preserve the form of the potential, and some tracking mechanism to keep dark energy in the wings until the proper moment. Simple scalar fields fail on one or both of these counts (the cosmological constant fails on both). However there are a few possibilities that might offer guidance toward a more robust theory.

Some theories, such as the pseudo-Nambu Goldstone boson (PNGB) model (Frieman et al., 1995), impose a symmetry that protects the form. Such theories are known as natural theories. However to achieve acceleration at the right time still requires a restricted range of initial conditions. Attractor models where dark energy is kept off stage, but not too far off, for the radiation- and matter-dominated eras, are a useful class (Ratra and Peebles, 1988; Wetterich, 1988; Zlatev, Wang and Steinhardt, 1999; Liddle and Scherrer, 1999). An intriguing class of models that incorporates both these advantages is the Dirac–Born–Infeld (DBI) action based on higher dimension theories (Alishahiha, Silverstein and Tong, 2004; Martin and Yamaguchi, 2008; Ahn, Kim, and Linder, 2009; Ahn, Kim and Linder, 2010). This employs a geometric constraint to preserve the potential and a relativistic generalization of the usual scalar field dynamics to provide the attractor property. The attraction to $w = -1$ actually occurs in the future, but prevents the dynamics from diverging too far from $w = -1$ at any time. Another class of interest, although not arising directly from high-energy physics, is that of barotropic models. In the barotropic aether model the equation of state naturally rapidly transitions from acting like another matter component to being attracted to $w = -1$, thus "predicting" $w = -1$ for much of the observational redshift range and ameliorating the coincidence of recent acceleration (Linder and Scherrer, 2009); see Fig. 12.5.

Theories along the lines of DBI or barotropic dark energy seem promising guideposts to a natural physical origin for acceleration, at least within the "new component" approach to dark energy. Interestingly, both of them also make predictions for the microphysics of the dark energy distinct from simple scalar fields. Minimally coupled, canonical (standard kinetic term) scalar fields have a sound speed of field perturbations equal to the speed of light, and hence do not cluster except on near horizon scales. Both the DBI and barotropic theories have sound speeds that instead approach

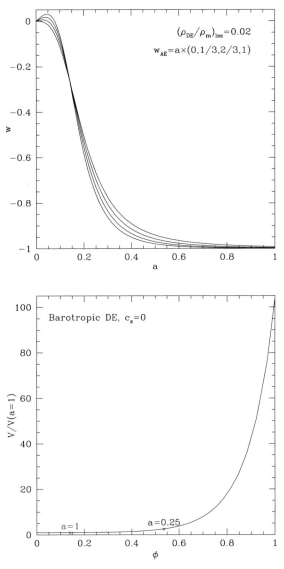

Fig. 12.5 Barotropic models make a rapid transition from $w = 0$ at high redshift ($a \ll 1$) to $w \approx -1$ more recently: the transition from $w = -0.1$ to $w = -0.9$ always takes less than 1.5 e-folds. This is inherent in the barotropic physics and, in distinction to quintessence, gives a prediction that observations of the recent Universe should find $w \approx -1$. (Bottom panel) Effective potential corresponding to a barotropic model with $c_s = 0$. The x's mark where the field is today and at $a = 0.25$, showing that it has reached the flat part of the potential, and so $w \approx -1$ for the last $\sim 90\%$ of the age of the Universe.

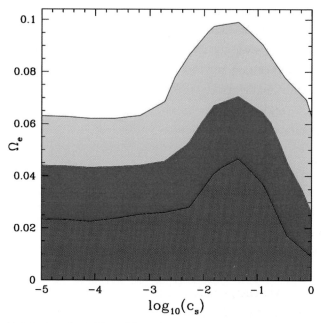

Fig. 12.6 68.3, 95.4, and 99.7% confidence level contours in the early dark energy model with constant sound speed c_s and early dark energy density fraction Ω_e. The constraints are based on current data including CMB, supernovae, galaxy power spectrum and cross-correlation of CMB with matter tracers. **Figure credit:** de Putter, Huterer and Linder (2010).

zero (and hence could cluster) for at least part of their dynamics. We explore this further in the next section, but Fig. 12.6 shows limits due to current data on the sound speed c_s for a barotropic-type model (DBI models have even weaker constraints on c_s).

12.4 The Frontiers of Nothing

From the previous section it may seem like our knowledge of nothingness is close to nothing. But the past dozen years of experimental work and theoretical investigation have ruled out large classes of models, though these are perhaps the simplest ones and of course the ones with the most obvious experimental signatures. The increasing difficulty has led some to pessimism, but the advancing network of diverse observational probes to be carried out over the next dozen years can be a source of hope.

Recall the story of Auguste Comte, who in 1835 declared that "we shall never be able to know the composition of stars". It was only 14 years later that it was discovered that the spectrum of electromagnetic radiation encodes the composition of material. Perhaps within the next 14 years there will be an analogous breakthrough of theoretical and experimental techniques for dark energy. Such progress in scientific discovery has become gratifyingly habitual, as in Richard Feynman's quote:

> Yesterday's sensation is today's calibration and tomorrow's background.

Just as perturbations in the cosmic microwave background (CMB) radiation were undetected (and beginning to be despaired of) in the 1980s, discovered in the 1990s, and are today sometimes regarded as "background noise" relative to the signatures of galaxy cluster physics, so may the homogeneous background of dark energy and the value of $w(a)$ be treated in the future.

What lies beyond w? Even for the expansion history $a(t)$, i.e. the homogeneous dynamics of expansion, there are the questions of whether dark energy makes a contribution at high redshift, and whether in an accelerating form or not. This is called early dark energy and current constraints are at the few percent level (Doran, Robbers and Wetterich, 2007) — by contrast the cosmological constant would contribute a fractional density of 10^{-9} at the CMB last-scattering surface at $z \approx 1090$. Within a few years, CMB data from the Planck satellite should tighten the constraints by a factor 10.

There is the issue of whether dark energy interacts with any other component other than through gravity. This could become apparent through a situation such as cosmological neutrino mass bounds being at variance with laboratory measurements [if dark energy interacts with neutrinos, e.g. Amendola, Baldi and Wetterich (2008); Wetterich (2007)], or through features in the matter density perturbation power spectrum [if dark energy interacts with dark matter, e.g. Bean, Flanagan and Trodden (2008)].

Does dark energy cluster? This could come about either through a low sound speed (although it also requires that w deviate appreciably from −1) or a coupling to other components. Observationally this can be probed through detailed measurements of matter clustering on various length scales, using the next generation of galaxy surveys.

Perhaps the most intriguing possibility is new laws of physics: in the "Neptune versus post-Newton" alternative to end up with extensions to the laws of gravitation beyond Einstein's general relativity rather than a new quantum scalar field. It is not easy to find viable theories of gravity that accord with observations, and most of the ones that do exist are driven toward similarity with general relativity (GR). Again, we seek a model-independent approach that might identify some key features that a fundamental extended theory would need.

The simplest generalization is to take a phenomenological approach of asking what feature of the observations could be shifted by a non-GR theory. As mentioned in Sec. 12.2, any modification of the expansion history is identical to an effective $w(a)$, so we must look further for an observational distinction. General relativity predicts a definite relation between the expansion history of the homogeneous Universe and the growth history of energy density perturbations. Other theories of gravity may deviate from this relation so we can define a gravitational growth index γ that accounts for effects on growth beyond the expansion influence, seeing if it is consistent with the GR prediction.

Parameterization of the growth of linear matter density perturbations $\delta\rho$ can be written as

$$g(a) = e^{\int_0^a (da'/a') \left[\Omega_m(a')^\gamma - 1\right]}, \qquad (12.15)$$

where $g(a) = (\delta\rho/\rho)/a$. This separates out the expansion history (which enters $\Omega_m(a)$) effects on growth from any extra-gravitational influences (entering γ). The gravitational growth index γ is substantially independent of other cosmological parameters and can be determined accurately. This form of representing deviations through γ, a single constant, reproduces the growth behavior to within 0.1% accuracy for a wide variety of models (Linder, 2005; Linder and Cahn, 2007). Note that other changes to the gravitational driving of growth besides the theory of gravity, such as other clustering components or couplings, can also cause γ to deviate from its standard general relativity value of 0.55.

Moreover, gravitational modifications do more than affect growth: they alter the light deflection law in lensing and the relation between the matter density and velocity fields. This can introduce both time and scale-dependent terms. In particular, the two potentials, appearing in the time-time and space-space terms of the metric, may no longer be equal as they are in general relativity, and the Poisson-type equations connecting them to the matter density and velocity fields could change. Among other approaches [e.g. Hu (1998); Hu and Sawicki (2007)], one can define new functions to account for these differences as (Daniel and Linder, 2010)

$$-k^2(\phi + \psi) = 8\pi G_N a^2 \bar{\rho}_m \Delta_m \times \mathcal{G}, \qquad (12.16)$$

$$-k^2 \psi = 4\pi G_N a^2 \bar{\rho}_m \Delta_m \times \mathcal{V}, \qquad (12.17)$$

where ϕ and ψ are the metric potentials, $\bar{\rho}_m \Delta_m$ the gauge-invariant matter density perturbations, k is the wavenumber, and G_N is Newton's constant. In general relativity, the time- and scale-dependent functions $\mathcal{G}$ and $\mathcal{V}$ are identically unity.

Within a given theory of gravitation, the deviations $\mathcal{G}$ and $\mathcal{V}$ will be specified, but if we are searching for general deviations from Einsteinian gravity then we should take model-independent forms for these functions. Allowing their values to float in bins in redshift and in scale (wavenumber k) gives considerable freedom and does not prejudice the search for concordance or contradiction with general relativity. Figure 12.7 shows both the current constraints and those expected from the next generation galaxy redshift surveys.

Considerable current data exist to constrain gravity and cosmology, including the cosmic microwave background (CMB), supernova distances, weak gravitational lensing, galaxy clustering statistics, and cross-correlation between the CMB photon and galaxy number density fields. Nevertheless, although this now constrains the sum of the potentials, and hence $\mathcal{G}$ [see Eq. (12.16)], fairly well, the growth of structure, in terms of $\mathcal{V}$, is still poorly known. This should change with the next generation of large

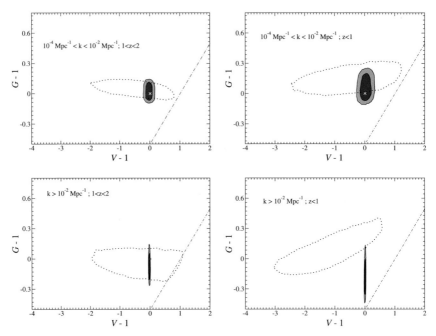

Fig. 12.7 Filled contours show 68% and 95% cl constraints on $\mathcal{V}-1$ and $\mathcal{G}-1$ for the two redshift and two wavenumber bins using mock future BigBOSS galaxy, PlanckCMB, and JDEM supernova data. The dotted contours recreate the 95% cl contours from Figs. 8 of Daniel and Linder (2010) using current data (note the offset from (0,0) may be from systematics within the CFHTLS weak lensing data) to show the expected improvement in constraints. The x's denote the fiducial GR values. **Figure credit:** Daniel and Linder (2010).

volume, three-dimensional galaxy mapping surveys. We see that the eight "beyond GR" gravity parameters ($\mathcal{G}$ and $\mathcal{V}$ each in two redshift and two wavenumber bins) could be determined to within $\sim 10\%$ or better.

Testing gravity on cosmic scales is an area of intense interest at the moment; previous works using current data (Daniel *et al.*, 2010; Bean and Tangmatitham, 2010; Zhao *et al.*, 2010; Reyes *et al.*, 2010) find consistency with GR, although again deviations are certainly allowed. Better data from growth probes could play a key role in tightening constraints or uncovering new physics. A particularly exciting prospect is comparing the density, velocity, and potential field information through combining imaging and spectroscopic surveys (Jain and Zhang, 2008; Reyes *et al.*, 2010; Jain and Khoury, 2010). Extending our probes and understanding into the nonlinear density structure regime is another area of active exploration (Oyaizu, Lima and Hu, 2008; Schmidt, 2009).

12.5 Conclusions

Beyond the atoms and photons that make up our familiar world, and all the particles of the Standard Model of particle physics, the nature of the vacuum and spacetime is a mystery that has come to the forefront of physics. Gravity, the most familiar and omnipresent of all the forces, is not behaving as we expected. More than 70% of the energy density in the Universe is made of nothing — nothing we have experienced before. Conditions are ripe for a true adventure in cosmology.

While current data are consistent with a cosmological constant as a source for dark energy, a cornucopia of other physical origins are in agreement as well. We do not yet know whether the dark energy is uniform, is dynamic, disappears at early times, is of a quantum origin or a gravitational origin. All are valid possibilities, and carry profound implications for the frontiers of physics and the fate of the Universe.

A key question is whether we are dealing with a new physical ingredient or new physical laws — or both. For example, the dark energy may interact with neutrinos through a novel interaction; is dark energy really a completely separate sector of physics or are there new forces and symmetries as intricate as in known particle physics? We are very much at the beginning of our explorations of the frontier physics of dark energy and cosmic acceleration.

The exciting goal of future observations is to explore this wonderland of physics. We have few robust models but some general concepts, and some excellent model-independent parameterizations. For the dynamical aspects of cosmic expansion, next generation measurements of the equation of state and its time variation, w and w', in the calibrated form of w_0 and w_a describe the experimental reach to the subpercentage level of observational accuracy. Comparison of tests of growth and expansion could give key clues to the underlying physics, as can contrasting the density, velocity, and gravitational potential fields of large scale structure. These should be enabled by a diverse network of future observations, delineating the physical properties of dark energy and testing general relativity. At the same time, these measurements deliver information of great value to many other astrophysical explorations as we map the structure, motion, and growth in our Universe.

Settling the frontier will require challenging efforts by both observers and theorists. One must not only measure the expansion history $w(a)$, growth history γ, gravity $\mathcal{G}$ and $\mathcal{V}$, couplings, early dark energy, etc. — but also understand them. Even if we fail to detect deviations from a cosmological constant, we cannot say the revolutionary physics of dark energy is known until we explain it. As two British Astronomers Royal said in the 19th century:

> "I should not have believed it if I had not seen it!" — Sir G.B. Airy

and the reply

> "How different we are! My eyes have too often deceived me. I believe it because I have proved it." — Sir W.R. Hamilton

Acknowledgments

I thank Dragan Huterer for useful comments and the Centro de Ciencias Pedro Pascual in Benasque, Spain for hospitality during the writing of part of this chapter. This work has been supported in part by the Director, Office of Science, Office of High Energy Physics, of the U.S. Department of Energy under Contract No. DE-AC02-05CH11231, and World Class University grant R32-2009-000-10130-0 through the National Research Foundation, Ministry of Education, Science and Technology of Korea.

References

Ahn, C., Kim, Linder, C. and E. V. (2009). *Phys. Rev. D* **80**, p. 123016.
Ahn, C., Kim, C. and Linder, E. V. (2010). *Phys. Lett. B* **648**, p. 181.
Alishahiha, M., Silverstein, E. and Tong, D. (2004). *Phys. Rev. D* **70**, p. 123505.
Amanullah, R. *et al.* (2010). *Astrophys. J.* **716**, p. 712.
Amendola, L., Baldi, M. and Wetterich, C. (2008). *Phys. Rev. D* **78**, p. 023015.
Bean, R., Flanagan, E. E. and Trodden, M. (2008). *Phys. Rev. D* **78**, p. 023009.
Bean, R. and Tangmatitham, M. (2010). *Phys. Rev. D* **81**, p. 083534.

Caldwell, R. R. and Kamionkowski, M. (2009). *Ann. Rev. Nucl. Part. Sci.* **59**, p. 397.
Caldwell, R. R., Kamionkowski, M. and Weinberg, N. N. (2003). *Phys. Rev. Lett.* **91**, p. 071301.
Caldwell, R. R. and Linder, E. V. (2005). *Phys. Rev. Lett.* **95**, 141301 (2005).
Copeland, E. J., Liddle, A. R. and Wands, D. (1998). *Phys. Rev. D* **57**, p. 4686.
Daniel, S. F. *et al.* (2010) *Phys. Rev. D* **81**, p. 123508.
Daniel, S. F. and Linder, E. V. (2010). arXiv:1008.0397.
de Putter, R. and Linder, E. V. (2008). *J. Cosmol. Astropart. Phys.* **0810**, p. 042.
de Putter, R., Huterer, D. and Linder, E. V. (2010). *Phys. Rev. D* **81**, p. 103513.
Doran, M., Robbers, G. and Wetterich, C. (2007). *Phys. Rev. D* **75**, p. 023003.
Durrer, R. and Maartens, R. (2010). in *Dark Energy: Observational and Theoretical Approaches*, ed. Ruiz-Lapuente P., Cambridge University Press, pp. 48–91.
Frieman, J. A. *et al.* (1995). *Phys. Rev. Lett.* **75**, p. 2077.
Frieman, J., Turner, M. and Huterer, D. (2008). *Ann. Rev. Astron. Astrophys.* **46**, p. 385.
Hu, W. (1998). *Astrophys. J.* **506**, p. 485.
Hu, W. and Sawicki, I. (2007). *Phys. Rev. D* **76**, p. 104043.
Huterer, D. (2011). *Adventures in Cosmology*, ed. Goodstein, D., World Scientific.
Jain, B. and Khoury, J. (2010). *Annals of Physics* **325**, p. 1479.
Jain, B. and Zhang, P. (2008). *Phys. Rev. D* **78**, p. 063503.
Liddle, A. R. and Scherrer, R. J. (1999). *Phys. Rev. D* **59**, 023509.
Linder, E. V. (2003). *Phys. Rev. Lett.* **90**, p. 091301.
Linder, E. V. (2005). *Phys. Rev. D* **72**, p. 043529.
Linder, E. V. (2006). *Phys. Rev. D* **73**, p. 063010.
Linder, E. V. and Cahn, R. N. (2007). *Astropart. Phys.* **28**, 481.
Linder, E. V. and Scherrer, R. J. (2009). *Phys. Rev. D* **80**, 023008.
Martin, J. and Yamaguchi, M. (2008). *Phys. Rev. D* **77**, p. 123508.
Mortonson, M. J., Hu, W. and Huterer, D. (2010). *Phys. Rev. D* **81**, p. 063007.
Oyaizu, H., Lima, M. and Hu, W. (2008). *Phys. Rev. D* **78**, p. 123524.
Perlmutter, S. *et al.* (1999). *Astrophys. J.* **517**, p. 565.
Ratra, B. and Peebles, P. J. E. (1988). *Phys. Rev. D* **37**, p. 3406.
Reyes, R. *et al.* (2010). *Nature* **464**, p. 256.
Riess, A. G. *et al.* (1998). *Astron. J.* **116**, p. 1009.
Rubin, D. *et al.* (2009). *Astrophys. J.* **695**, p. 391.
Schmidt, F. (2009). *Phys. Rev. D* **80**, p. 123003.
Silvestri, A. and Trodden, M. (2009). *Rept. Prog. Phys.* **72**, 096901.
Sollerman, J. *et al.* (2009). *Astrophys. J.* **703**, p. 1374.
Wei, H. (2010). *Phys. Lett. B* **691**, p. 173.
Wetterich, C. (1988). *Nucl. Phys. B* **302**, p. 668.
Wetterich, C. (2007). *Phys. Lett. B* **655**, p. 201.
Zhao, G-B. *et al.* (2010). *Phys. Rev. D* **81**, p. 103510.
Zlatev, I., Wang, L. and Steinhardt, P. J. (1999). *Phys. Rev. Lett.* **82**, p. 896.

CHAPTER 13

THE FIRST SUPERMASSIVE BLACK HOLES IN THE UNIVERSE

XIAOHUI FAN
Steward Observatory, University of Arizona
Tucson, AZ 85721, USA

Black holes, one of the most intriguing predictions of Einstein's general relativity, have been a focus of astrophysical research for more than 50 years. Study of astrophysical black holes covers a wide range of mass scales, employing different detection methods. Stellar mass black holes — products of collapse of massive stars at the end of their lifetimes — are detected as invisible but massive members of binary systems that usually emit high-energy photons through gas accretion around the black holes; their masses range from a few to a few tens of solar masses. Intermediate mass black holes are thought to reside in centers of globular clusters or dwarf galaxies, with masses of the order hundreds to tens of thousands solar masses. For the cosmologist, however, the key interests lie in the study of supermassive black holes located in the nuclei of galaxies, with masses ranging from million to billion solar masses. Their mass can be measured by observing gas and stellar motion in the center of nearby galaxies. During the periods when these supermassive black holes are actively accreting matter, the resultant radiation from the accretion process can often outshine the entire galaxy — which we usually refer to as quasars, or active galactic nuclei (AGNs).

The formation of supermassive black holes is now believed to have played a key role in the evolution of galaxies, and in the evolution of the thermal state of the Universe. The most luminous quasars are powered by black holes in the billion solar mass range, and are tens of thousands times brighter than normal galaxies, therefore detectable at great distance, providing powerful light beacons to probe the most distant and oldest epoch of the Universe. One of the key interests in modern cosmology is to understand when and how the first generation supermassive black holes occurred in the Universe, and how they are related to the first generation of galaxies. In this chapter, we first review the observational signatures of supermassive black holes and their connections to galaxy evolution, then discuss the current observations of the most distant quasars and their supermassive black holes, and the questions these observations posed to early cosmic evolution. We will discuss the basic physical processes that produced the earliest supermassive black holes and how the next generation observing facilities will be able to enable us to probe the evolution of the first supermassive black holes in the Universe.

13.1 Supermassive Black Holes and Galaxy Formation

A quasar is originally defined as an astronomical object which appears starlike on ground-based images but possesses many other characteristics, in particular a large redshift, that proves that it is not a star. The name quasar is a contraction of the term quasistellar object (QSO), or quasistellar radio source. Quasars appear starlike because their angular diameters are less than about 1 second of arc, which is the resolution limit of ground-based optical telescopes imposed by atmospheric effects. However, high resolution images using space telescopes and ground-based telescopes equipped with adaptive optics have revealed that quasars reside in the center of active galaxies. They are the most luminous members of a larger family of objects generally referred to as Active Galactic Nuclei (AGNs). People first discovered active galaxies through spectroscopy in the early 1900s by identifying unusually strong emission lines in some galaxy spectra. As a separate class, quasars were discovered as a result of new radio surveys in 1950s. By then, radio interferometry technique allowed more precise position measurements of radio sources and identified most of them as radio galaxies. Sometimes, however, the optical counterparts of radio sources are stellar-like, with blue colors. Early spectroscopy revealed wide, strong emission lines that defied easy classification into transitions of known elements. The first breakthrough came in 1963, when Maarten Schmidt realized that the emission lines he saw in radio source 3C273 were actually the Balmer series emission lines of hydrogen, only redshifted to an unprecedented value of 0.158, implying a tremendous distance and luminosity, if the redshift were cosmological. Many other starlike radio sources were soon identified as similar objects at high redshift, with the maximum redshift exceeding 2 discovered in less than two years. The name quasar came from "quasi-stellar radio source". But people discovered even larger numbers of sources with similar optical properties, but have no detectable radio emission. They are called "quasar-stellar objects", or "QSOs". It is now recognized that quasars and QSOs are essentially the same phenomena except for different levels of observed activity in radio wavelength. These two terms are generally used interchangeably in the literature.

If quasar redshifts were cosmological, that is, they are results of the expansion of the Universe, then the inferred distance is usually billions of light years. The luminosity of quasars based on this cosmological distance, combined with their compact size, suggests an extraordinarily powerful energy source, more than normal stellar population would be able to provide. This puzzle led intense debates in the astronomical community about the nature of their redshift in the 1960s and 1970s. Some argued that the large redshift was kinematic, or due to Doppler effect from a large ejection velocity, while the true distances to quasars were modest. However, later observations show clear evidences that quasar redshifts are indeed cosmological. Their high luminosity is believed to be provided by converting gravitational potential energy to radiation when hot gas falls into a very massive compact object in the galactic center, most likely a massive black hole with masses ranging from a million to a few billion solar masses. This process turns out to be the most efficient astrophysical process of converting rest-mass to radiation. Assuming a radiative efficiency η, then the output luminosity: L = dE/dt = η dM/dt, where dM/dt is the black hole accretion rate. For nuclear reaction that fuses hydrogen into helium, the equivalent $\eta = 0.007$. But through accretion onto a compact object, when material is falling into a very deep potential, η could be as high as 10–30%. In this process, the deep potential results in very high velocity of infalling material. Viscosity in the accretion process, often thought to be related to magnetohydrodynamic (MHD) processes, converts the random motion into heat that is radiated out as high-energy photons. It is easy to show that if this conversion process can proceed all the way to around the last stable orbit of the black hole, than the radiative efficiency is $\eta \sim 0.1$. This rate can be even higher for rotating black holes with smaller horizon sizes. While the detailed physical processes for energy conversion is complicated, this mechanism provides a valid explanation of quasar luminosity powered by black holes. For a typical quasar, it requires a few solar masses of gas a year to be accreted to the black hole to provide its observed luminosity. This explanation for quasar luminosity was first proposed by Lynden-Bell (1969) and is now the accepted theory for quasar emission and supermassive black hole growth.

The basic scales in our problem are the characteristic mass and lifetime of black hole activity in quasars. Black holes cannot accrete gas and grow arbitrarily fast. A fundamental limit for luminosity produced by a given mass in a plasma medium is the Eddington limit: hydrodynamical equilibrium requires gravity to be balanced by pressure, which in the plasma is provided by radiation pressure, or Thomson scattering of photons by electrons. This balance works out to yield the maximum Eddington limiting luminosity proportional to the black hole mass: $L_{edd} = 1.3 \times 10^{38}$ M/M_sun erg/s. For a typical quasar with luminosity of 10^{44} erg/s, this gives the characteristic black hole mass of 10^8 solar masses. Assuming that quasar is accreting at Eddington luminosity, its luminosity is proportional to the black hole mass, $L \sim M$; meanwhile, for accretion process, the luminosity is also proportional to the accretion rate, $L \sim dM/dt$. Thus, $M \sim dM/dt$. Therefore, black holes in this state will grow exponentially: for $\eta = 0.1$, the e-folding time-scale, or Salpeter (1964) time, is about 40 million years. In other words, supermassive black hole in active quasar phase will e-fold its mass every 40 million years — this is generally regarded as the lifetime for quasar growth activity. For comparison, the time-scale for galaxy evolution is much longer, usually hundreds of millions of years.

The spatial density of quasars is about two orders of magnitudes lower than that of normal galaxies. In other words, only 1% of galaxies have an actively accreting black hole at its center at a given time. Does that mean only 1% of galaxies contain supermassive black holes? Or that the duty cycle of black hole accretion is low and every galaxy would go through quasar phase, which only lasts 1% of the galaxy lifetime?

For a long time, quasar evolution and galaxy evolution were treated as separate problems. However, a major discovery in 1990s, largely due to the high spatial resolution of the Hubble Space Telescope, changed the landscape. HST was able to resolve the motion of stars and gas, through the Doppler effect, at very small radii in distant galaxies, allowing measurements of mass distributions in galactic centers. It was found that essentially all massive normal galaxies contain massive dark objects at their centers; for the nearby cases, including the Milky Way galaxy, the density of this central galaxy is so high that the only possible explanation is a supermassive black hole with millions of solar masses. Even more strikingly, as shown in Fig. 13.1, the masses of central black holes are strongly correlated with the luminosity, mass, or central velocity dispersion (depth of potential) of the galaxy. This is usually referred to as the M–σ relation. This relation suggests that the growth of black holes and the growth of galaxies are strongly related. It is surprising because the scale of black hole accretion is the Schwarzschild radius, which for supermassive black holes is less than the size of our solar system, while the scale of galaxies is tens of thousands of light years, many orders of magnitude larger. How can those two scales be coordinated so coherently?

The likely answer to the this question is that black holes in galaxies grow through radiatively efficient accretion in quasar or active galaxy phase; meanwhile, the enormous energy output during this phase will have significant impact to the overall evolution of the galaxy, commonly referred to as the feedback effect. For example, radiative energy from quasars can be coupled to interstellar medium in the galaxy and eventually unbind

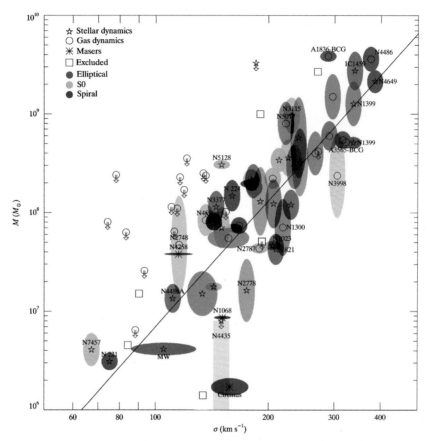

Fig. 13.1 M–σ relation for the masses of black holes and central velocity dispersion of their host galaxies. Different symbol indicates the method of BH mass measurement. The color of the error ellipse indicates the Hubble type of the host galaxy: elliptical (red), S0 (green), and spiral (blue). The line is the best-fit relation to the full sample: $M_{\rm BH} = 10^{8.12}\,\mathcal{M}_\odot\,(\sigma/200\,{\rm km\,s^{-1}})^{4.24}$. Adapted from Gültekin et al. (2009).

interstellar medium, thus stop both star formation in the galaxy and gas supply to feed the black hole. This negative feedback effect is able not only to explain the M–σ relation, but also solve a number of puzzling observations in galaxy evolution that usually indicate massive galaxies have their star formation process stopped too early and too rapidly comparing to theoretical predictions without feedback. Detailed physics of feedback is a matter of intense debate; but the fact that feedback effects involving

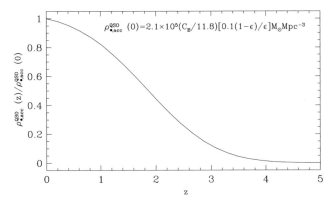

Fig. 13.2 The evolution of total mass density of black hole through accretions in quasar activity as a function of redshift, normalized by the local black hole mass density. The accreted black hole mass density is calculated using quasar luminosity function; the local black hole mass density is calculated using local $M-\sigma$ relation to convert galaxy mass to black hole mass. These two densities agree under reasonable assumptions. Adapted from Yu and Tremaine (2002).

black hole activity played a crucial role in galaxy formation is generally agreed upon.

The connection between supermassive black holes in normal galaxies and quasar activity is further illustrated in Fig. 13.2, which shows the overall accretion history of black holes through quasar activity. It follows Soltan's (1982) argument, that if black holes in galaxies acquired most of their masses in quasar phase, then one can convert the total luminosity density of quasars in the Universe to a total accretion rate, assuming a certain radiative efficiency. This accretion rate, when integrated over the entire history of the Universe, shall reproduce the local mass density for black holes in the center of galaxies, active or not. Yu and Tremaine (2002) showed that this is indeed the case with reasonable assumptions on the evolution of quasar luminosity density and local black hole mass density.

In addition to $M-\sigma$ relation and Soltan's argument, observations also show that the overall evolution of total star formation rate in the Universe and total density of quasars follow similar trends: they have a broad peak at $z = 1$–4 (2–6 billion years after the big bang) and declines towards both lower and higher redshifts. Therefore, there is a coevolution of galaxies and supermassive black holes. In order to understand galaxy evolution, one has to understand black hole evolution as well, and vice versa. It also raises a number of questions regarding the evolution of the early Universe, at the epoch of reionization when the first generation of cosmic objects formed:

- When and how did the first supermassive black holes form in the Universe?
- What kind of cosmic environment did they live in?
- Which appeared first: quasars with massive black holes, or galaxies with intense star formation? Or did they evolve in close locksteps?
- What role did quasar activity play in the overall reionization of the Universe?

13.2 Observations of the Most Distant Quasars

In this section, we first present the observations of the highest redshift, most distant quasars and their environment, then discuss their implications to the black hole/galaxy co-evolution at early epochs. Figure 13.3 presents

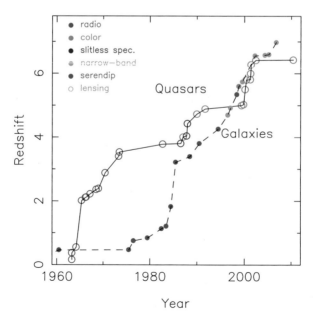

Fig. 13.3 The highest redshift quasars and galaxies known, as a function of discovery years. Different selection techniques are labeled with different symbols.

the record of the highest redshift quasars and galaxies discovered over the last 50 years, labeled by their selection techniques. At the highest redshift, quasar density is so low that the only possible way to look for them is to carry out wide area surveys of sufficient depth to cover large-enough volume. Therefore, in the last 20 years, studies of luminous quasars are dominated by data from large projects using wide-field optical and near-infrared imaging and spectroscopy. The best examples are the 2dF quasar survey (Croom et al., 2004) and the quasar survey in the Sloan Digital Sky Survey (SDSS, Schneider et al., 2010), which discovered 25 000 and 100 000 quasars, respectively. Figure 13.4 shows a composite spectra constructed from a large sample of SDSS quasars, illustrating the evolution of quasar spectra, in particular strong emission lines, as a function of redshift.

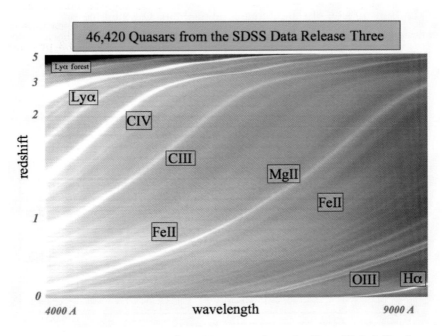

Fig. 13.4 Composite spectra of 46 000 quasars from the Sloan Digital Sky Survey quasar catalog. Strong emission lines from H, C, O, Mg, and Fe match through the wavelength as redshift increases. The highest-redshift quasar known at 2010 is $z = 6.4$, only 900 million years after the big bang. **Figure credit:** Sloan Digital Sky Survey, http://www.sdss.org/includes/sideimages/quasar_stack.html.

Most of the highest redshift quasars are selected using the so-called Lyman break technique. At $z > 3$, the intergalactic intervening absorption systems along the line of sight in quasar spectrum, which have in the rest-frame $\lambda < 1216\,\text{Å}$, are redshifted into optical wavelength, and cause a large continuum break in the quasar spectral energy distribution. High-redshift quasar surveys look for this continuum break using multicolor photometry. Optical techniques can find quasars up to $z \sim 6.5$. As of Fall 2010, there have been about 30 quasars discovered at $z > 6$, 60 at $z > 5.5$, and over 100 at $z > 5$. The highest redshift quasars is at $z = 6.44$, discovered in the Canada–France High-z Quasar Survey project (Willott et al., 2010). Figure 13.5 presents moderate resolution spectroscopy of a sample of 27 quasars at $z \sim 6$ from the Sloan Digital Sky Survey sample (Fan et al., 2006). These spectra illustrated two crucial points regarding cosmic evolution at $z \sim 6$: (a) the spectra show strong redshift evolution of the transmission of the intergalactic medium. On the blue side of Ly-α emission line at $\lambda \sim 8500\,\text{Å}$, almost all flux is absorbed by the atomic,

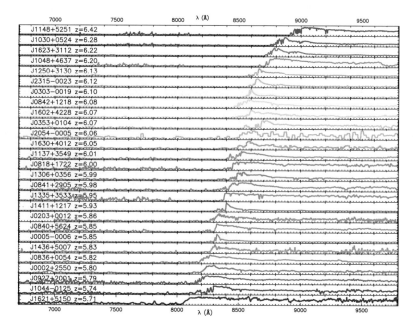

Fig. 13.5 Moderate resolution spectra of 27 quasars at $z = 5.7$–6.4 selected from the SDSS sample. **Figure credit:** Fan et al. (2006).

or neutral hydrogen in the interstellar medium. The absorption troughs deepen for the highest-redshift quasars, producing the so-called Gunn and Peterson (1965) effect. There is a rapid increase of atomic hydrogen density in this era, consistent with the signatures expected at the end of reionization epoch: we are at the threshold of reionization. (b) These spectra also exhibit strong emission lines from heavy elements other than hydrogen, including O, C, Si, Mg, and even Fe. These heavy elements are not results of Big Bang nucleosynthesis, and must have been produced in the star formation process. Even at $z \sim 6$, when the Universe was only about one billion years old, there have already been multiple generations of star formation and rapid chemical enrichment. Observations of these most distant quasars provide crucial constraints to the questions we posed at the end of last section.

Rapid formation of supermassive black holes in the early Universe

The very existence of luminous quasars at $z \sim 6$ poses significant challenges to supermassive black hole formation theory. These quasars have black hole masses estimated to be up to a few billion solar masses. We argued in Sec. 13.1 that the e-folding time for black hole growth is 40 million years. At $z \sim 6$, there are less than 25 e-folding times for the entire history of the Universe, or about 20, if one assumes that the first generation of star formation can start as early as $z \sim 20$. It is usually assumed that quasars form from seed stellar black holes with masses of the order 100 solar masses, and grow through Eddington-limited accretion. If this were the case, it would have taken the entire history of the Universe at $z \sim 6$ to grow the black holes residing in and powering the $z \sim 6$ luminous quasars, with 100% duty cycle and no negative feedback. As we will discuss in Sec. 13.3, this is difficult to accomplish under conventional assumptions. Figure 13.6 presents the overall evolution of quasar density as a function of redshift. If at $z > 6$, quasar growth is indeed limited by the number of e-folding times available, then future detections (or lack) of quasars at $z > 7$ will put the strongest constraints on the epoch when the earliest generation of billion solar mass black holes was formed in the Universe.

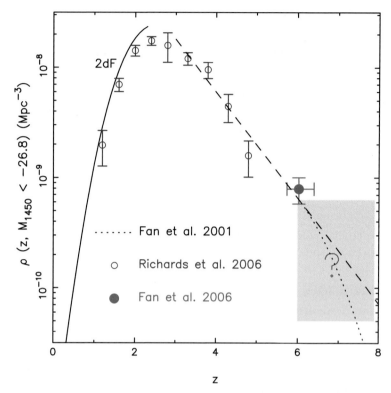

Fig. 13.6 The evolution of the spatial density of the most luminous quasars. Whether there is accelerated evolution at $z > 7$ in quasar densities will indicate whether the formation of the first generation supermassive black holes are results of Eddington-limited accretion from stellar seed black holes. **Figure credit:** Fan *et al.* (2004).

Where are the youngest quasars?

Here we try to answer a different question: does the structure of quasars, probed by their observed spectral energy distribution, evolve with time? Do the highest-redshift quasars differ from their lower-redshift counterparts in their internal properties, and exhibit signatures of youth consistent with the young age of the Universe at high redshift? This has been a long-standing puzzle in quasar research: as shown in Figs. 13.4 and 13.5, rest-frame UV spectra of quasars at $z \sim 6$ are very similar, in fact almost identical, to those at $z \sim 1$–2, with no detectable evolution. Meanwhile, the Universe went through rapid evolution, and the density of quasars increased by orders of magnitude. Relative strengths of emission lines can be used to measure the chemical abundance, or metallicity, of quasar environment: they are similar at low and high redshift, and all significantly exceed the solar value. Quasars must have formed and matured on a very short time-scale. This lack of spectral evolution, combined with the rapid density evolution of quasars, suggests that the feeding and growth of supermassive black holes and the formation of quasar structure happen at very different time-scales, a puzzling result.

Multi-wavelength observations of the highest-redshift quasars are beginning to shed new light on this issue. In general, high-energy photons, such as those emitted in X-ray and UV wavelengths, come from the inner part of the emitting region close to the black hole itself; longer wavelength radiation, such as those emitted in the infrared, are reprocessed emissions on large scale when UV and X-ray photons pass through the gas and dust structure surrounding the central quasar engine. Using the Spitzer Infrared Observatory, Jiang *et al.* (2010) discovered that some of the highest-redshift quasars at $z \sim 6$ lack the infrared emission from the dusty surrounding structure that is thought to be universal among quasars at all epochs. Furthermore, they showed that those lacking dust emission are the quasars with the lowest black hole mass and highest relative accretion rate, therefore the shortest growth time-scale; in other words, likely the youngest quasars. On the other hand, the infrared emission is independent of black hole mass at lower redshift. This strongly suggests that the age of the Universe at $z \sim 6$, or one billion years after the big bang, marks the time-scale for a quasar, powered by a central supermassive black hole, to fully form in the galaxy; they have identified what might be the youngest quasars. Fig. 13.7 is an artist conception of a young supermassive black hole without surrounding dusty structure similar to those found by Jiang *et al.*

Supermassive black holes, or galaxies — which comes first?

To answer this question, one needs measurements of the properties of quasar host galaxies, such as their masses and star formation rates, at the highest possible redshift — an observation that is extremely difficult. At high redshift, quasar host galaxies are very faint. But more importantly, the strong emission from the quasar itself, often more than one thousand times

Fig. 13.7 Artist conception of a primitive supermassive black hole in the center of a young galaxy without dust. Adapted from http://www.jpl.nasa.gov/news/news.cfm?release=2010-088.

that of the host galaxy, needs to be removed accurately from the data in order to reveal the faint galaxy. Even with the HST, it is technically challenging. One way to get around this problem is to carry out observations in much longer wavelength, such as submillimeter and radio wavelength, where the radiation from the quasar itself is very weak. Most of the light at those wavelengths comes from cool dust and gas in the host galaxy. The submillimeter and radio radiation therefore also provides estimates on the star formation rate and on the amount of molecular gas needed to support star formation (and feed supermassive black holes) in the quasar host galaxies.

Observations show that a significant fraction of quasars at $z \sim 6$ are strong submillimeter emitters, with an average star formation rate of a few hundred solar masses per year. They are comparable to the most intense star forming galaxies in the local Universe, and are consistent with being in the process of building up their galactic bulges. Radio observations of the same objects are able to detect CO and ionized Carbon ([CII]) emissions in the interstellar medium of the quasar host galaxies. In particular, CO traces the total mass of molecular gas in the galaxy. Radio CO and [CII]

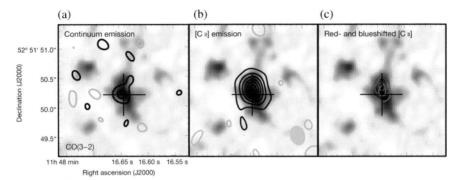

Fig. 13.8 Millimeter observations of dust and gas in quasar SDSS J1148+5251 at $z = 6.42$, 900 million years after the big bang. The left panel shows strong continuum emission from cool dust (~ 50 K) in the quasar host galaxy, suggesting an ongoing star formation rate of ~ 2000 solar mass per year. The right panel shows the emission from ionized Carbon ([CII]), a major coolant in the interstellar medium. The [CII] emission has also been resolved kinematically, and allows measurement of gas motion and dynamical mass of the system. **Figure credit:** F. Walter *et al.* (2009).

line emissions are also used to measure large scale motion of gas in the quasar host galaxy. This can be used as a proxy to their velocity dispersion to measure the mass of a host galaxy. Figure 13.8 presents deep millimeter observations of one of the most luminous quasars at $z = 6.42$. It shows strong dust emission powered by star formation rate exceeding 2000 solar masses per year; it also shows a compact structure, with the size of dust and molecular gas emission in the central 2 kpc (6000 light years), significantly smaller than the typical size of low redshift galaxies. Furthermore, CO and [CII] observations show very narrow emission lines, indicating relatively low velocity dispersion, or low host galaxy mass: the host galaxy mass is estimated to be about an order magnitude lower than local galaxies with similar black hole masses. This result strongly suggests that at high redshift, at least in the luminous quasars that we can observe, black holes accrete more rapidly than the assembly of galaxies. Quasar built-up precedes galaxy built-up in the earliest epoch.

High-redshift quasars and reionization

Observations of the highest-redshift quasars provide tests to the reionization of the Universe. Gunn and Peterson (1965) first proposed using Ly-α resonance absorption in the spectra of distant quasars as a direct probe

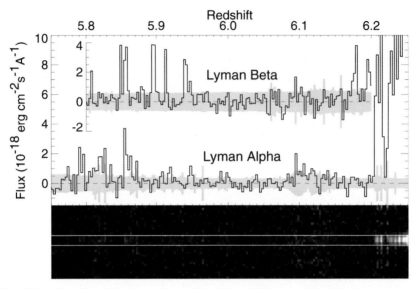

Fig. 13.9 The first discovery of complete Gunn–Peterson absorption in a $z \sim 6.3$ quasar. There is no detectable Ly-α transmission in the spectrum at $z > 6.0$, indicating a rapid increase in the neutral fraction of the intergalactic medium at the end of cosmic reionization. **Figure credit:** Becker *et al.* (2001).

to the atomic, or neutral hydrogen density in the intergalactic medium at high-redshift. For objects beyond reionization, neutral hydrogen in the intergalactic medium creates complete Gunn–Peterson absorption troughs in the quasar spectrum blue ward of Ly-α emission. Observations of the Gunn–Peterson effect directly constrain the evolution of neutral hydrogen fraction and the ionization state of the intergalactic gas. Becker *et al.* (2001) discovered the first complete Gunn–Peterson absorption in the spectrum of a $z \sim 6.3$ quasar (Fig. 13.9). Fan *et al.* (2006) measured the evolution of Gunn–Peterson optical depths along the line of sight of the 19 $z > 5.7$ quasars from the SDSS. They found a rapid transition in the evolution of intergalactic medium optical depth, showing more than an order of magnitude increase in the amount of neutral hydrogen from $z \sim 5.5$ to $z \sim 6.5$. This result is consistent with conditions expected at the end of reionization as suggested by cosmological simulations.

Observations of the highest-redshift quasars also provide constraints on what caused the cosmic reionziation and ended the cosmic dark ages. Due to the rapid decline in quasar density at $z > 5$, as shown in Fig. 13.6, unless quasar luminosity function was extremely steep at the very faint end, they are unlikely to provide enough photons to ionize the Universe. Therefore, the transition from a neutral to ionized Universe was largely accomplished by strong UV radiations from star formation galaxies.

13.3 Growing the First Supermassive Black Holes in the Universe

Observational results presented in Sec. 13.2 provide crucial constraints as well as challenges to how the first supermassive black holes formed and grew in the early Universe. In this section, we look at the theoretical understandings of this problem, from the formation of the supermassive black hole seeds, to the modes of black hole growth and feedback, and the relation between black hole growth and the evolution of their host galaxies.

Seed black holes

Which physical processes could lead to the initial collapse of a seed black hole in the early Universe? Figure 13.10 is the famous flow chart of Rees (1984). There are number of possible routes, from the collapse of a massive star, to dynamical evolution of star cluster, to direct collapse of gas cloud. However, the only direct observational evidence for seed black hole we have in local Universe is the existence of stellar mass black holes as a result of evolution of massive stars in the Galaxy. Production of core collapse supernova, which could result in seed black hole formation, is observed at high redshift, and can be modeled in detail. It also follows from the natural evolution of first stars in the Universe. Therefore, most models of supermassive black holes first consider stellar seed black holes.

Theoretical calculations show that the first "minihalo" with masses of the order one million solar masses will collapse at $z = 20-40$, only a couple of hundred million years after the Big Bang. Cooling from molecular hydrogen in these minihalos will result in further fragmentation of molecular cloud, providing the sites of first stars in the Universe. Simulations have shown that these first stars are likely to be massive, exceeding 100 solar masses, due to the slow contractions from molecular hydrogen cooling.

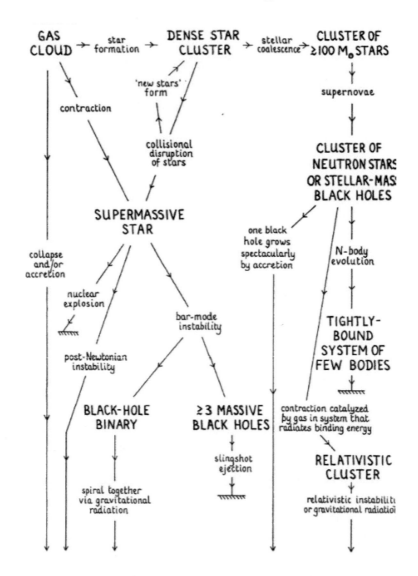

Fig. 13.10 Schematic diagram showing possible routes of the formation of massive black holes. **Figure credit:** Rees (1984).

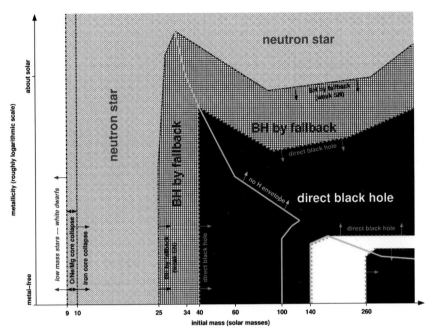

Fig. 13.11 Remnants of stellar evolution as a function of initial mass and metallicity. In low-metallicity environment, direct collapse to black holes with mass ranging from 10 to 200 solar masses is possible. **Figure credit:** Heger *et al.* (2003).

These metal-free, massive stars are often referred to as Population III stars. Some of them will evolve into the first black holes in the Universe. Figure 13.11 presents the final collapse and fate of low metallicity stars as a function of their initial mass. Between 25 and 140 solar masses, they can form black holes directly, with remnant mass in the range of 10 to 40 solar masses. At higher mass, between 140 and 260 solar masses, star death will produce pair instability supernovae; they will be completely disrupted with no remnants left. At even higher mass range, larger than 260 solar masses, stars will collapse into black holes with $M > 150\,M_{\rm sun}$ after about 2 million years of evolution. While the mass distribution of the first stars is still completely theoretical predictions with little observational constraints at this point, popular theory does suggest that seed black holes with masses up to 200 solar masses could be directly produced from the formation of first stars in the first dwarf galaxies.

Black hole growth

The simplest scenario for the growth of supermassive black holes is to start with a seed black hole, and supply it with ample gas that it will accrete at the Eddington rate through radiatively efficient accretion. Soltan's argument, as well as measurements of quasar black hole mass and accretion rate distribution, are showing that this is indeed happens during the quasar phase of black hole growth, when the supermassive black holes acquired most of its mass. However, as we discussed in Sec. 13.2, the existence of very luminous and massive quasars at $z > 6$ is challenging this simple scenario. Detailed calculation of black hole growth has to include additional feedback effects. For example, radiation field from quasars that could interrupt gas supply and therefore slow down accretion. Rapidly spinning black holes, resulted from continuous accretion through accretion disks, would increase the radiative efficiency η, thus increase the Salpeter growth time and reduce the number of e-folding available for growth. When two galaxies merge, their central black holes will merge too. In this process, asymmetric gravitational wave radiation could result in recoil. This so-called rocket effect could remove some merged black holes out of galactic centers, reducing or eliminating gas supplies for further growth.

Overall, the existence of billion solar mass black holes strongly suggests that the growth of black holes in the early Universe is not simply by Eddington accretion from stellar seed black holes. It should be noted that the difficulty is not the last few e-folding which is much better understood, but the early phase of black hole growth, from 100 to 100 000 solar masses. It is theoretically possible to form much more massive seed black holes through either gas dynamical or stellar dynamical processes, as suggested in Rees's flow chart (Fig. 13.10). A number of alternative seed black hole formation and accretion models are being considered. One such process being discussed recently is the growth of "quasi-stars". Such objects could only have existed at very high redshift with zero metallicity. They are thought to have stellar seed black hole in the center, and a massive, radiation-pressure supported envelop that shields the central black hole and allows it to grow rapidly (Begelman et al., 2008). They are not subjected to the Eddington limit for the central black hole and can reach 10^{3-5} solar masses on very short time-scale, after which it could grow to supermassive black holes through conventional accretion process.

Supermassive black hole growth in cosmological context

Assuming massive seed black holes existed in early Universe, they can be incorporated into detailed cosmological simulations that include galaxy merger, star formation and feedback processes. Figure 13.12 shows the results of one of such simulations by Li *et al.* (2007). In the simulation, they followed the hierarchical assembly of massive dark halos in a large cosmological volume to search for the site of possible massive galaxy formation. For those regions that collapsed first, they then followed the evolution of baryons (gas) in the system, and calculated the star formation history of first galaxies. Meanwhile, they assumed massive seed black holes in the first galaxies, which went through both Eddington limited accretion, and merger when two young galaxies merge. They also include feedback effect from the growth of black holes self-consistently. They found that the observational properties of the most luminous quasars discovered at $z > 6$ can be reproduced by the growth of a massive galaxy with a maximum accreting black hole in a dark matter halo of $\sim 10^{13}$ solar masses. The host galaxy went through a series of major mergers before $z \sim 6.5$, and experienced vigorous star formation with a rate up to 10 000 solar masses per year. The quasar phase corresponds to the final breakout of radiation from the accreting black hole through its dense gas and gas environment. This also provides the feedback that blows away most of the cold gas in the host galaxy, therefore stops star formation and black hole growth.

Such simulations will provide testable predictions about the properties of the most distant quasars. They predict the maximum mass as well as mass function of quasar black holes, together with the evolution of M–σ relation, star formation rate, and size and morphology of quasar host galaxies that are all directly compared with observations.

13.4 Future Prospects

Formation and evolution of supermassive black holes in the early Universe is a complex process. It involves stellar, gas and dynamical evolution of

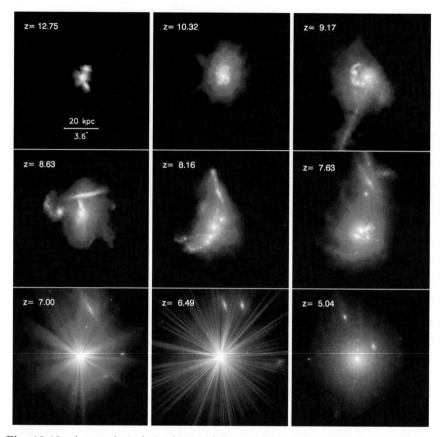

Fig. 13.12 A cosmological simulation of the coevolution of supermassive black holes and their host galaxies. It reproduces observational characteristics of the most luminous quasars at $z \sim 6$. The images show projected stellar density. Quasar host galaxies went through a series of merger, during which the supermassive black holes grew through accretion and merger. Finally, a mature quasar was able to breakout at $z < 7$ which also provided the feedback that stopped galaxy growth. **Figure credit:** Li *et al.* (2007).

the environment where the first generation of star formed, which produced seed black holes. The subsequent growth of these seed black holes into supermassive black holes powering high-redshift quasars is a result of the merger and feedback processes in the evolultion of early massive galaxies. Supermassive black hole formation is a key ingredient of the formation of the earliest generation of cosmic structures, and provides crucial tests to their key epoch of cosmic evolution.

Surveys of the new highest redshift quasars: searching for the first supermassive black holes

Detection and observations at $z \sim 6$ suggest that this is the crucial redshift for quasar evolution. Black hole growth might begin to be limited by the number of e-folding available in early Universe. Quasar spectra begin to show evolutionary trends that suggest we are reaching the earliest, youngest systems. These highlight the need to push to even higher redshift, where constraints to the model of black hole growth are exponentially stronger, and there might be new surprises. Optical surveys are only sensitive to $z < 6.5$. At higher redshift, quasars become rarer, fainter, and redder. This requires wide-angle, infrared sky surveys, which remain to be a major challenge for ground-based astronomy. Surveys such as the Panoramic Survey Telescope & Rapid Response System (Panstarrs), and The Large Synoptic Survey Telescope (LSST) will obtain deep 1 μm image over large fraction of the sky, enabling discovery of quasars up to $z \sim 7.5$. The Visible and Infrared Survey Telescope for Astronomy (VISTA), which is a 4 m wide-field near-IR survey telescope located in Chile, is planning on a number of tiered surveys in near-infrared bands with one of the main goals being the discovery of luminous quasars at $z = 7-9$. Eventually, space-based surveys, such as those planned with ESA's Euclid mission and NASA's WFIRST concept, will reach the depth to uncover faint quasars at $z > 7$.

Because of the tight time constraints in early epoch, the characteristic mass, or luminosity of quasars will rapidly decline with increasing redshift. Searching for the earliest supermassive black holes with $10^6 - 10^7$ solar masses and luminosity significantly lower than a normal quasar will be a major task of the NASA's James Webb Space Telescope (JWST, Fig. 13.13), which is to be launched in 2017. JWST will have both a sensitive near-infrared camera and near-infrared spectrograph on board. It will enable detections and identification of quasars and active galaxies at $z \sim 10-15$, only 200–400 million years after the Big Bang. At this redshift, JWST in principle can detect black holes with masses as low as a few times 10^5 solar masses if they are shining at the Eddington limit, although spectroscopic confirmation is only possible at higher luminosities.

Fig. 13.13 Artist's conceptions of the James Webb Space Telescope (JWST), which will allow detections of the quasars and active galaxies at $z > 10$, and the Atacama Large Millimeter Array (ALMA), which will map gas dynamics and star formation in the host galaxies of the highest redshift quasars. Both will be fully operational in the second half of 2010s. Images are taken from http://www.esa.int/images/4AJWST_SCIDisplay_H.jpg, http://www.nrao.edu/pr/2008/alma.aaas/ALMA-2.jpg

Studies of quasar host galaxies: testing feedback and coevolution

Cosmological simulations, such as those presented in Fig. 13.12, have suggested a very modest evolution or no evolution in the relation between black hole mass and galaxy mass, or the M–σ relation. Observations (e.g. Fig. 13.8), on the other hand, imply a strong evolution of this relation, in the sense that black hole growth in the most massive systems preceded the assembly of its host galaxies. Is this an observational selection effect because current facilities are only sensitive to the most extreme systems at high redshift, or an indication of unknown ingredients in black hole formation and feedback process in galaxy formation? JWST, as well as large ground-based telescope with advanced adaptive optics system, will allow high angular resolution observations in the near-infrared for direct detections of stellar light from quasar host galaxies. Even more sensitive observations of quasar host galaxies will be provided by the Atacama Large Millimeter Array (ALMA), a sub/millimeter interferometer array that consists of 66 radio telescopes located in the high Chilean desert with unprecedented sensitivity and resolution (Fig. 13.13). ALMA will provide detailed maps of star formation and gas dynamics in quasar host galaxies and trace the coevolution of supermassive black holes and galaxies to the earliest epoch.

Gravitational wave experiment: discriminating black hole seed formation mechanism

The most difficult, yet most important task is to characterize the mass distribution and formation mechanism of black hole seeds and their earliest growth. Seed black hole masses are estimated to range between 10 to 10^5 solar masses. This is lower than even what JWST can detect. The best hope to characterize them is through gravitational wave radiation when early black holes merge. The Laser Interferometer Space Antenna, or LISA (Fig. 13.14), a joint ESA–NASA mission, is expected to be sensitive to gravitational waves from binary black holes of 10^3-10^6 solar masses at any reasonable redshift. The detection rate of coalescence events can directly constrain black hole formation process. LISA will allow

Fig. 13.14 LISA will detect gravitational wave radiation generated in the in-spiral during black hole–black hole merger when supermassive black holes were growing in the early Universe. It will be sensitive to black holes at lower than one million solar masses that would be otherwise beyond the limit of electromagnetic probes. **Image credit:** http://lisa.nasa.gov.

measurements of black hole–black hole merger rate as a function of redshift and mass. The merger rate and mass distribution is expected to be very different for different models of black hole seed formation and evolution, allowing discrimination of models with Population III star seed black holes, and much more massive seed black holes such as those produced by quasi-stars.

It has been almost half a century since Maarten Schmidt first identified the nature of the first quasar 3C273, and more than 40 years since Donald Lynden-Bell first postulated that quasars are powered by accretion to supermassive black holes. Since then, tremendous progress has been made in characterizing both the properties of supermassive black holes in galaxies and the evolution of quasars, traced all the way to the first 6 percentile of the cosmic history. In the next decade, new facilities will bring order-of-magnitude improvements in sensitivities and survey speed in infrared and radio wavelengths, and open up astrophysical research using gravitational waves. Advance in numerical techniques will greatly improve detailed simulations of the formation and evolution of early black holes and early galaxies in cosmological context. Together, we will be able to identify the epoch of first supermassive black hole formation, and map the history of the evolution of the first generation quasars and their connection to early galaxy formation. These new observations will test models of black hole and galaxy formation, and likely reveal new physical processes or new phenomena in the early Universe.

References

Becker, B. H. *et al.* (2001). Evidence for reionization at $z \sim 6$: Detection of a gunn-Peterson trough in a $z = 6.28$ quasar, *Astrophysical Journal* **122**, p. 2850.

Begelman, M. C., Rossi, E. M. and Armitage, P. J. (2008). Quasi-stars: accreting black holes inside massive envelopes, *Monthly Notice of the Royal Astronomical Society* **387**, p. 1649.

Croom, S. *et al.* (2004). The 2dF QSO redshift survey — XII. The spectroscopic catalogue and luminosity function, *Monthly Notice of the Royal Astronomical Society* **606**, p. 126.

Fan, X. *et al.* (2004). A survey of $z > 5.7$ quasars in the sloan digital sky survey. III. Discovery of five additional quasars, *Astronomical Journal* **128**, p. 515.

Fan, X. et al. (2006). Constraining the evolution of the ionizing background and the epoch of reionization with $z \sim 6$ quasars. II. A sample of 19 quasars, *Astronomical Journal* **132**, p. 117.

Gunn, J. E. and Peterson, B. A. (1965). On the density of neutral hydrogen in intergalactic space, *Astrophysical Journal* **142**, p. 1633.

Gültekin, K. et al. (2009). The M–σ and M–L relations in galactic bulges, and determinations of their intrinsic scatter, *Astrophysical Journal* **698**, p. 198.

Heger, A. et al. (2003). How massive single stars end their life, *Astrophysical Journal* **591**, p. 288.

Jiang, L. et al. (2010). Dust-free quasars in the early Universe, *Nature* **464**, p. 280.

Li, Y. et al. (2007). Formation of $z \sim 6$ quasars from hierarchical galaxy mergers, *Astrophysical Journal* **665**, p. 187.

Lynden-Bell, D. (1969). Galactic nuclei as collapsed old quasars, *Nature* **223**, p. 690.

Rees, M. J. (1984). Black hole models for active galactic nuclei, *Annual Reviews of Astronomy and Astrophysics* **22**, p. 471.

Salpeter, E. E. (1964). Accretion of interstellar matter by massive objects, *Astrophysical Journal* **140**, p. 796.

Schneider, D. P. et al. (2010). The sloan digital sky survey quasar catalog. V. Seventh data release, *Astronomical Journal* **139**, p. 2360.

Soltan, A. (1982). Masses of quasars, *Monthly Notice of the Royal Astronomical Society* **200**, p. 115.

Walter, F. et al. (2009). A kiloparsec-scale hyper-starburst in a quasar host less than 1gigayear after the Big Bang, *Nature* **457**, p. 699.

Willott, C. J. et al. (2010). Eddington-limited accretion and the black hole mass function at redshift 6, *Astronomical Journal* **140**, p. 546.

Yu, Q. and Tremaine, S. (2002). Observational constraints on growth of massive black holes, *Monthly Notice of the Royal Astronomical Society* **335**, p. 965.

INDEX

21-cm, 42, 44, 61, 69, 75, 76, 78–87
21-cm radiation, 128

acceleration, 321, 322, 326, 346, 348, 355, 357–360, 362, 364, 366–368, 375
active galactic nuclei, 379, 380
Advanced Camera for Surveys, 216
ALMA, 402
alpha particle, 277, 290, 291, 293
Andromeda, 214

Baade, Walter, 227, 322
background rejection, 281, 283
background subtraction, 280–283, 291
barotropic dark energy, 368–370
baryon, 48, 50, 52, 53, 56, 67, 70, 73, 83–85
baryon acoustic oscillation, 337–339, 345
beta decay, 176
    neutrinoless, 177
big bang, 43, 45, 59, 67, 73
black hole, 9, 12, 25–32, 57, 58, 60, 61, 63, 65–67, 379–386, 389–393, 395–404
    first, 132, 139, 143, 154, 156, 159, 166, 170, 171
Boulby mine, 309
bremsstrahlung emission, 94
bubble chamber, 290
bubble discrimination, 292
bulk flows, 103

Canfranc Underground Laboratory, 286
cepheids, 196, 198, 199, 201, 205–209, 229, 230, 232
    distance scale, 208, 209, 216, 217, 231, 232
Chandrasekhar limit, 226
charge-coupled devices, 326
Cherenkov radiation, 175
Chicagoland Observatory for Underground Particle Physics, 290
cold dark matter, 174
cold dark matter model, 112
CoRE, 160, 161, 163, 166, 171
cosmic dawn, 143–146, 148–159, 161, 165, 167, 170
cosmic expansion, 358, 375
cosmic microwave background, 41, 44, 47, 76, 86, 141, 142, 180, 181, 184–186, 188–190, 237, 345, 346
    anisotropies, 232
    fluctuations, 112
cosmic rays, 261
cosmogenic activation, 275
cosmogenic production, 276

dark ages, 7–9, 12, 113
dark energy, 89, 100, 107, 321, 326, 327, 329, 331, 332, 334, 336–338, 340, 342, 344–347, 349, 355, 356, 358–368, 371, 375, 376
dark matter, 8–10, 13–17, 20, 24–26, 29, 32, 44–46, 48, 50, 52–54, 56–58,

62, 71, 72, 81–85, 89–91, 94, 96, 98, 99, 105–107, 241–265
dark matter halo, 112
de Sitter, 194, 195
density fluctuation, 89, 91, 92, 99, 100, 102, 104
DES, 189

early dark energy, 370, 371, 376
Eddington limit, 382, 389, 390, 398, 399, 401
Eddington luminosity, 122, 133
Eddington, Arthur, 203, 204
EDGES, 160, 161, 163, 164, 166, 171
Einstein, Albert, 195
electron recoil, 281, 283, 286, 289, 290, 292, 293, 296–299, 302, 304–311, 318
electron-recoil discrimination, 285, 299
equation of state, 359–364, 367, 368, 375
Ewen, Harold, 146, 148
expansion, 193
experiment
    KATRIN, 177, 189
    MINOS, 176
    NEMO, 178
    Planck, 189
    Sudbury Neutrino Observatory, 175
    Super-Kamiokande, 175
extinction, 326, 329
extra dimension, 251, 252

feedback, 8, 14, 16, 24, 25, 27, 28, 31, 32
feedback effect, 383, 384, 389, 395, 398–400, 402
feedback processes, 120
Field, George, 148
first stars, 113
Freedman, Wendy, 209
freeze-out, 246–248, 250, 253
Friedmann equation, 179, 360
Friedmann, Alexander, 196

fundamental plane, 198, 210, 218–220, 231, 232

galactic globular clusters, 213
galaxy, 5–11, 14–26, 28, 29, 31–35, 201
    cluster, 89–109, 181–183, 185–190, 242–244, 343
galaxy formation
    first, 134
gamma rays, 259, 260
gamma-ray bursts, 61
general relativity, 179, 180, 356, 372, 373, 376
germanium, 284
germanium spectrometer, 284, 285
GMRT, 162, 163, 168, 169
Goodricke, John, 201
Gran Sasso National Laboratory, 301, 309, 312, 313
gravitational growth index, 372
gravity, 355–359, 364, 371–376
Gunn–Peterson effect, 389, 394

H II regions, 222
Harvard College Observatory, 202
head-tail discrimination, 317
helium flash, 211
hot dark matter, 174
Hubble constant, 196, 229, 231, 322, 328
Hubble diagram, 194, 230
Hubble sequence, 17, 18, 25
Hubble Space Telescope, 193, 197, 209, 228, 235
Hubble, Edwin, 193, 195

inelastic dark matter, 274
inelastic dark matter scenario, 302
inflation, 51, 52, 72, 83, 89, 101, 103, 105
initial mass function, 114, 122
interferometer, 163, 167, 171
interferometry, 162, 163
intergalactic gas, 57
intergalactic medium, 46, 57, 59–62, 67, 71–78, 81, 123, 143–145, 150

ionization, 283, 284, 289, 293, 295–302, 304–306, 308, 309, 311, 316, 317
ionosphere, 164, 165, 170

Jeans instability, 117
JWST, 130, 401–403

Kaluza–Klein state, 252, 253
Kamioka mine, 288
Kennicutt, Robert, 210
Kowal, Charles, 227, 228

Laboratoire Souterrain de Modane, 300
Large Hadron Collider, 264, 265
Large Magellanic Cloud, 201, 202, 207
large scale structure, 90, 102–105
Las Campanas Observatory, 200
Leavitt law, 206, 207, 230, 232
Leavitt, Henrietta, 202
Lemaître, Georges, 196
LISA, 403
Local Group, 16, 17
Local Group dwarf galaxies, 213
LOFAR, 162–164, 167–170
Lowell Observatory, 196
Lyman-α, 44, 59, 60, 62, 63, 69, 70, 73, 74, 77, 83, 85–87
Lyman-alpha forest, 187, 189, 190

Majorana, 280
Massive Compact Halo Object survey, 207
MegaZ, 188
Messier 5, 210, 211
Milgrom, 243
Milky Way, 193, 201, 203
modified gravity, 105, 107
modulation, 282
molecular hydrogen, 115
MOND, 243, 244
morphology, 17, 18
Mould, Jeremy, 209
M–σ relation, 383–385, 399, 402
muons, 278
MWA, 140, 146, 162–164, 168, 169

N-body simulation, 181, 186, 242
neutralinos, 249–251, 256–259, 265
neutrino, 244, 247–250, 252, 258, 259, 262, 263
    eigenstates, 176
    mass hierarchy, 176
    oscillations, 175
    solar neutrino problem, 175
neutrino mass, 90, 99, 100
NGC 4258, 229, 232
NGC 6397, 235
noble liquid, 302, 303, 306, 314, 315
nuclear recoil, 277, 278, 281–283, 286, 287, 289–293, 296–299, 301, 303–311, 315–317
nuclear recoil discrimination, 281, 283, 289, 297, 298
number counts, 97, 99–102, 107

Oort, Jan, 146
Optical Gravitational Lensing Experiment, 207

pair-instability supernova, 129
PAPER, 162–164, 169
particle physics
    beyond the Standard Model, 175
Phillips, Mark, 324
phonon, 293–301
population III star, 113, 397, 404
    formation, 115
power spectrum, 180, 183–186
protosellar accretion, 120
pulsation, 199, 205
pulse shape discrimination, 306, 308, 311, 314
Purcell, Edward, 146, 148

quasar, 44, 61–67, 71, 73, 74, 83, 379–383, 385–395, 398–402, 404
quasi-stars, 398, 404

radiation hydrodynamics, 119, 120
radio telescope, 160, 162, 163, 168–170
recoil energy, 283, 297, 299, 301, 306, 308, 317

recoil energy spectrum, 270–272, 279, 282
recombination, 43, 45, 46, 49, 52, 53, 67, 70, 71, 76, 142, 143
recombination timescale, 126
reconstruction, 334
redshift, 7–9, 11, 28, 32–34, 154
reionization, 8, 15, 16, 43, 44, 46, 60–62, 66–72, 75, 76, 78, 79, 81–83, 85, 86, 123, 124, 144, 151, 152, 154, 158–160, 166, 167, 386, 389, 393, 394
relic abundance, 244, 246, 251, 253, 259
rotation curves, 241, 243
R-parity, 249, 252, 254

Salpeter timescale, 382
Sandage, Allan, 210, 228
scalar field, 358–361, 366, 368, 372
Schneider, Donald, 214
scintillation, 283, 285, 288, 289, 293, 295–298, 301–312, 314, 316, 317
scintillator, 291, 313
SDSS, 123, 188, 387, 388, 393, 394
Small Magellanic Cloud, 200, 202
Soltan's argument, 385, 398
Soudan Underground Laboratory, 300
sound speed, 368, 370, 371
South Pole Telescope, 344
spin temperature, 75–80
spin-flip transition, 145–148, 150, 152, 155
Spitzer Space Telescope, 223
stars
    first, 139, 153–155, 157–159, 166, 170, 171
stellar archaeology, 130
strong gravitational lenses
    time delay, 232
structure
    growth, 180, 181, 183, 184
structure formation, 90–92, 96
Sunyaev–Zeldovich effect, 96
supernovae, 49, 58, 65, 82, 85, 179, 364, 367, 370, 373, 374
    Type I, 227
    Type Ia, 198, 210, 224–226, 228–232, 236, 237, 321, 322, 324, 328–330, 334, 337, 345, 347
supersymmetry, 249, 251, 254, 264
Surface Brightness Fluctuation, 198, 210, 215–217, 231, 232

terrestrial interference, 164, 170
thermistor, 294, 295
Tip of the Red Giant Branch, 198, 212, 213, 231
Tonry, John, 214
triple-alpha reaction, 199
Tully–Fisher relation, 198, 210, 221–223, 231, 232

vacuum, 355–358, 375
van de Hulst, Hendrik, 146
Very Long Baseline Array, 208
Virgo, 214, 216
virialization, 112

Warsaw telescope, 200
weak gravitational lensing, 339–342, 344
weak lensing, 184–187, 189, 190
white dwarf, 234
WIMP, 269–273, 275, 276, 279–286, 288–292, 296, 301–305, 307–309, 314–316
    inelastic, 274, 275, 302
    interaction, 274, 275, 277, 282, 311
WIMP–nucleon scattering, 303
WIMP–nucleus scattering events, 269, 270
WMAP, 111, 112, 127, 128, 188
Wouthuysen–Field effect, 77

YangYang underground laboratory, 287

Zwicky, Fritz, 322